D1519820

Springer Series in Optical Sciences Volume 52
Edited by Arthur L. Schawlow

Springer Series in Optical Sciences

Editorial Board: J.M. Enoch D.L. MacAdam A.L. Schawlow K. Shimoda T. Tamir

Volume 42 **Principles of Phase Conjugation**
By B. Ya. Zel'dovich, N. F. Pilipetsky, and V. V. Shkunov

Volume 43 **X-Ray Microscopy**
Editors: G. Schmahl and D. Rudolph

Volume 44 **Introduction to Laser Physics**
By K. Shimoda 2nd Edition

Volume 45 **Scanning Electron Microscopy**
Physics of Image Formation and Microanalysis
By L. Reimer

Volume 46 **Holography and Deformation Analysis**
By W. Schumann, J.-P. Zürcher, and D. Cuche

Volume 47 **Tunable Solid State Lasers**
Editors: P. Hammerling, A. B. Budgor, and A. Pinto

Volume 48 **Integrated Optics**
Editors: H. P. Nolting and R. Ulrich

Volume 49 **Laser Spectroscopy VII**
Editors: T. W. Hänsch and Y. R. Shen

Volume 50 **Laser-Induced Dynamic Gratings**
By H. J. Eichler, P. Günter, and D. W. Pohl

Volume 51 **Tunable Solid State Lasers for Remote Sensing**
Editors: R. L. Byer, E. K. Gustafson, and R. Trebino

Volume 52 **Tunable Solid-State Lasers II**
Editors: A. B. Budgor, L. Esterowitz, and L. G. DeShazer

Volumes 1–41 are listed on the back inside cover

Tunable Solid-State Lasers II

Proceedings of the OSA Topical Meeting,
Rippling River Resort, Zigzag, Oregon,
June 4–6, 1986

Editors:
A.B. Budgor, L. Esterowitz, and
L.G. DeShazer

With 285 Figures

Springer-Verlag Berlin Heidelberg New York
London Paris Tokyo

Dr. Aaron B. Budgor
Electronics Division, Northrop Corp., P.O. Box 5032,
Hawthorne, CA 90250, USA

Dr. Leon Esterowitz
Naval Research Lab., Code 6550, Washington, DC 20375, USA

Dr. Larry G. DeShazer
Spectra Technology, Inc., 2755 Northup Way, Bellevue, WA 98004, USA

Editorial Board

ARTHUR L. SCHAWLOW, Ph. D.
Department of Physics, Stanford University
Stanford, CA 94305, USA

JAY M. ENOCH, Ph. D.
School of Optometry,
University of California
Berkeley, CA 94720, USA

Professor KOICHI SHIMODA
Faculty of Science and Technology,
Keio University, 3-14-1 Hiyoshi, Kohoku-ku
Yokohama 223, Japan

DAVID L. MACADAM, Ph. D.
68 Hammond Street,
Rochester, NY 14615, USA

THEODOR TAMIR, Ph. D.
981 East Lawn Drive,
Teaneck, NJ 07666, USA

ISBN 3-540-17320-X Springer-Verlag Berlin Heidelberg New York Tokyo
ISBN 0-387-17320-X Springer-Verlag New York Heidelberg Berlin Tokyo

This work is subject to copyright. All rights are reserved, whether the whole or part of the material is concerned, specifically those of translation, reprinting, reuse of illustrations, broadcasting, reproduction by photocopying machine or similar means, and storage in data banks. Under § 54 of the German Copyright Law, where copies are made for other than private use, a fee is payable to "Verwertungsgesellschaft Wort", Munich.

© Springer-Verlag Berlin Heidelberg 1986
Printed in Germany

The use of registered names, trademarks, etc. in this publication does not imply, even in the absence of a specific statement, that such names are exempt from the relevant protective laws and regulations and therefore free for general use.

Offset printing: Druckhaus Beltz, 6944 Hemsbach/Bergstr.
Bookbinding: J. Schäffer GmbH & Co. KG., 6718 Grünstadt.
2153/3150-543210

Pstack
SCIMON

Preface

In June 1984 a conference on visible and IR tunable solid-state lasers was held in La Jolla, California. The proceedings were published as the first volume of this series, *Tunable Solid State Lasers**. The emphasis of this meeting focused on discerning unified themes in the generic areas of:

- Laser host/dopant identification and growth procedures,
- Theoretical analysis to elucidate fundamental tunable laser principles,
- Experimental investigations on laser spectroscopy to which theoretical analyses and models can be anchored, and
- Auxiliary technology developments in efficient laser pumping sources (diodes, flashlamps).

Subsequent to the La Jolla conference, two topical meetings were held, co-sponsored by the Lasers and Electro-Optics Society of the Institute of Electrical and Electronics Engineers, and the Optical Society of America (OSA). The contents of *Tunable Solid-State Lasers II* comprise the proceedings of the second of these two, held at Rippling River Resort, Zigzag, Oregon, June 4–6, 1986.

In addition to the four areas of attention in the La Jolla meeting, papers on color-center and new rare-earth lasers, and on nonlinear frequency shifting were also given. In a fashion similar to the previous meetings, the informal atmosphere of the conference and meeting site was conducive to constructive interaction among the attendees. A total of 54 papers were scheduled for presentation, 20 of which were invited and 34 contributed. The program was divided up into sessions on general spectroscopic methods applicable to solid- state tunable lasers, chromium spectroscopy, crystal growth, chromium-doped alexandrite and non-alexandrite lasers, titanium sapphire lasers, color-center lasers, neodymium and other rare-earth lasers, and applications and nonlinear optics. In addition, a panel discussion on chromium lasers was held.

---

*Springer Series in Optical Sciences, Vol. 47

Some of the highlights were the talks on laser action of Cr, Nd, Tm, and Ho-doped garnets (Part VIII); unique properties of Cr as activator (Part II); excited state absorption (Parts I and II); advances in alexandrite lasers, including injection-locking operation using semiconductor diode lasers (Parts V and X); new Cr tunable laser hosts (Part IV); advances in Ti sapphire growth and laser performance, including the first CW operation at room temperature (Part VI); OPO and harmonic generation in the IR using $AgGaSe_2$ (Part X); new color-center lasers, including the demonstration of a high-power (~1 W) source emitting at wavelengths $> 1.6\,\mu m$ (Part VII); diode-pumped Ho laser (Part VIII); laser operation of Cr-sensitized Er and Ho in YSGG (Part VIII); and Nd-doped fiber laser (Part IX).

Of particular interest were the applications of 2- and 3-photon spectroscopy to provide detailed information on excited-state absorption spectra and the dynamical energy-relaxation processes of these states; the growth and lasing of a promising new tunable, relatively long lifetime (~$115\,\mu s$ at room temperature) material – $Cr{:}ScBO_3$ – with fluorescence extending between 720 and 1000 nm; and the generation of CW-tunable oscillation in the vicinity of $0.9\,\mu m$ utilizing a Nd-doped fiber laser.

The conference editors would like to thank the sponsors, the Optical Society of America and the Lasers and Electro-Optics Society of the Institute of Electrical and Electronics Engineers; session chairpersons, R. Buser (NVEOL), W. Krupke (LLNL), W. Sibley (Oklahoma State University), G. Huber (University of Hamburg), A. Mooradian (MIT Lincoln Laboratory), K. Schepler (WPAL), J. Walling (Light Age), L. DeShazer (Spectra Technology), L. Esterowitz (NRL), and A. Budgor (Northrop); the Technical Program Committee; and Panel participants for outstanding efforts in making this meeting a success. Special thanks go to the co-chairs, L. DeShazer and G. Huber, for their administrative efforts, and to C. Heath of Northrop; and to J. Kushnick and R. Morris of the Allied-Signal Corporation who saw to the financial support without which this book could not have been published. Additional support was provided by Lawrence Livermore National Laboratory, Spectra Technology, and Union Carbide Corporation.

Hawthorne,
Washington,
Bellevue, October 1986

A. Budgor
L. Esterowitz
L. DeShazer

Contents

Part VI Titanium Sapphire Lasers

Part VII Color Center Lasers

Part VIII Rare Earth Lasers

Part IX Neodymium Lasers

Part X Applications and Nonlinear Optics

Part I

Spectroscopy

Two-Photon Spectroscopy Using Infrared Photons

D.S. McClure

Department of Chemistry, Princeton University, Princeton, NJ 08544, USA

Spectra of ions substituted at centrosymmetric sites in crystals ought to be the easiest cases to analyze because they have the smallest number of crystal field parameters. This high symmetry should also be of advantage, if one wants to understand the forces resulting from electronic excitations which cause the Jahn-Teller effect and lattice mode excitations. For intraconfigurational transitions d→d, f→f and for other cases in which parity is the same for ground and excited state, such as d→s, the spectra observed are disappointingly complicated, because odd parity-vibronic activity is a necessary accompaniment of one-photon transitions. In the few cases where magnetic dipole transitions can be observed, they are limited to the electronic origins and even parity-vibronic structure is usually obscured by odd parity structure.

Spectral structure in these cases is beautifully revealed by two-photon spectroscopy, since parity is preserved in two-photon transitions. Several recently published examples of ultraviolet spectra using visible dye laser photons illustrate this fact.[1,2,3] Most transition metal spectra occur in the visible or near infrared, however, and it would be necessary to use infrared photons to access their two-photon spectra.

We have begun a study of two-photon spectra in the visible range in collaboration with Paul Rabinowitz and Bruce Perry at Exxon Research Laboratories. At Exxon a high-powered YAG and dye laser combination had been assembled with a specially designed multiple pass Raman shifting cell using high-pressure hydrogen, thus providing tunable infrared after one or more Stokes shifts.[4] Two-photon absorption was detected by way of the fluorescence of the sample excited by the focussed infrared beam. A similar system is now being built in our laboratory at Princeton.

The results are illustrated by our investigation of Cs_2GeF_6:Mn^{4+} a system in which the Mn^{4+} at a Ge site is at the center of an octahedron of F^- ions, which itself is enclosed in a cube of Cs^+ ions. The Mn^{4+} ion has a $A_{2g}(d^3)$ ground state and excited states ^2E_g, $^2T_{1g}$, $^4T_{2g}$ and $^4T_{1g}$, all of which are analogous to the states of Cr^{+3} in strong crystal fields. All of the transitions from the ground state to these excited states were observed in both one- and two-photon spectra.[5]

Figure 1 shows the two types of spectra for the $^4A_{2g}\rightarrow{}^2T_{1g}$ transition. There are almost no coincidences between these spectra as one would find for the Raman and infrared vibrational spectra of a centrosymmetric molecule. The two electronic origins representing the spin orbit splitting of the doublet state are clearly present in the two-photon spectrum, and entirely absent from the one-photon spectrum. Only a_g vibrational additions are seen in the two-photon spectrum, while only odd parity vibrational additions are seen in the one-photon spectrum. The latter are present because of vibronic coupling to odd parity electronic states,

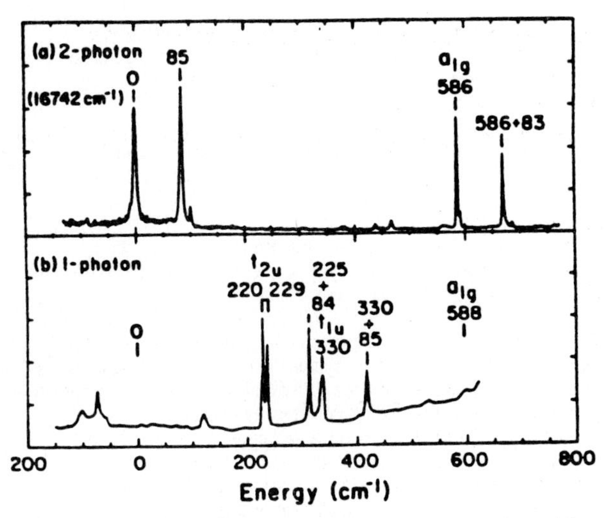

Fig. 1. The $^4A_{2g} \rightarrow {}^2T_{1g}$ transition in $Cs_2GeF_6{:}Mn^{4+}$. a. Two-photon spectrum at 10K, b. One-photon spectrum at 2K. Origin at 16742 cm^{-1}

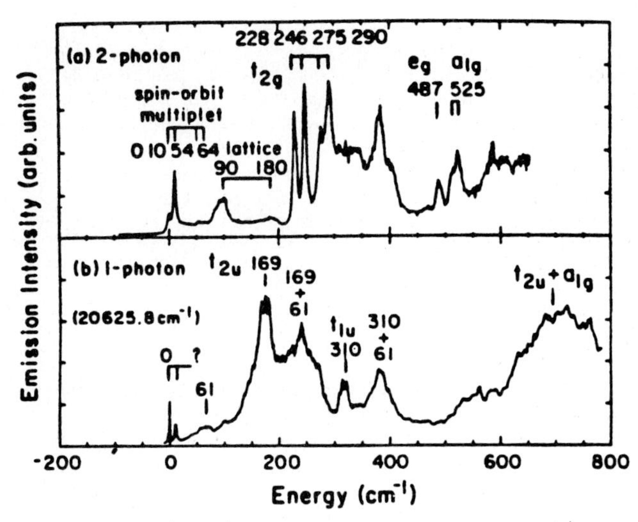

Fig. 2. The $^4A_{2g} \rightarrow {}^4T_{2g}$ transition in $Cs_2GeF_6 Mn^{4+}$. a. Two-photon spectrum at 10K, b. One-photon spectrum at 23K. Origin at 20625.8

i.e., the usual Herzberg-Teller coupling. In a crystal, the lattice modes can induce transitions by this mechanism and this can result in spectral intensity where one would rather not have it. Figure 2 illustrates this point and several others. It shows the first 800 cm^{-1} of the $^4A_{2g} \rightarrow {}^4T_{2g}$ transition. The one-photon spectrum is full of lattice mode structure along with t_{1u} and t_{2u} octahedron modes, and these cover up the interesting structure which is revealed in the two-photon spectrum shown above it. Here one sees all four members of the origin quartet, compressed by a

Jahn-Teller effect (Ham quenching factor); there is a t_{2g} bending mode showing splitting into four components due to coupling into the degenerate electronic state, a progression in a 90 cm^{-1} a_g Cs^+ vibrational mode and several other interesting and informative features.

The other excited states of this system also have informative two-photon spectra.

Two-photon spectra of systems of low symmetry can look just like one-photon spectra. In the case of ruby, for example, we found practically no discernible difference; but two-photon cross-sections could nevertheless be measured and these may be useful quantities.

What are the prospects for wider use of two-photon spectroscopy in this area? We have done preliminary work on Cr^{+3} in symmetrical environments such as elpasolites. It would be very worthwhile to analyze the two-photon spectra of the emitting quartet in weak-field chromium systems. This would give us knowledge of the displacements in the even modes which cannot be observed in one-photon spectra and could lead to a better understanding of the radiationless processes which are important for these materials. The origins of the $^4A_{2g} \rightarrow {}^4T_{2g}$ transition in chloride elpasolites are at about 13,000 cm^{-1}; thus to observe two-photon spectra one will need photons between 6500 and 13,000 cm^{-1}. This range can be achieved with from one to three Stokes shifts after R6G dye.

Another area of interest is that of the divalent transition metal ions, for example, $MgO:Ni^{2+}$. The emission begins at 8000 cm^{-1} in this case, and photons between 4000 and 8000 cm^{-1} can be used to scan the $^3A_{2g}$-$^3T_{2g}$ band and most of the $^3A_{2g} \rightarrow {}^3T_{1g}$ band. One-photon processes would begin above 8000 cm^{-1} and prevent further two-photon observations. This example illustrates some important limitations of the method. First, the fluorescence occurs at energies below the photomultiplier range and less sensitive detectors must be used, for example, Ge-based cells. Second, the laser power will be down since three or perhaps four Stokes shifts are needed. Third, the range over which spectroscopy can be done is limited; in this case only a 8000 cm^{-1} scan can be done.

These limitations may someday be removed with the development of true two-photon absorption spectroscopy.

The writer acknowledges that all the real work on this project was done by his collaborators, Ring-Ling Chien, John Berg, Paul Rabinowitz, Bruce Perry and Nick Levinos. Support by the Department of Energy, Office of Basic Energy Sciences is appreciated.

References

1. S. A. Payne, A. B. Goldberg and D. S. McClure: In J. Chem. Phys., 78, (1983) p. 3688
2. S. A. Payne, A. B. Goldberg and D. S. McClure: In J. Chem. Phys., 81, (1984) p. 1529
3. M. C. Downer, C. D. Cordero-Montalvo and H. Crosswhit: Phys. Rev., B28, (1983) p. 4931
4. P. Rabinowitz, B. Perry and N. Levinos: J. Quantum Electronics, to be published
5. R.-L. Chien, J. Berg, D. S. McClure, P. Rabinowitz and B. Perry: Two-Photon Electronic Spectroscopy of $Cs_2GeF_6:Mn^{4+}$, J. Chem. Phys., 84, (1986) p. 4168

Laser Spectroscopy Measurements on Tunable Solid State Laser Materials

R.C. Powell

Physics Department, Oklahoma State University, Stillwater, OK 74078, USA

1. Introduction

Optical spectroscopy plays a key role in determining the potential usefulness of materials for tunable solid state lasers. Along with obtaining information on the standard absorption and emission properties of the material, it is important to characterize the processes controlling the pumping dynamics of the material such as multiphoton absorption, radiationless relaxation, and energy transfer. We are currently involved in several different projects focused on the evaluation of new tunable solid state laser materials and on characterizing the fundamental properties of established laser materials. We describe here the results of some recent measurements of the optical spectroscopic properties of three different crystals doped with transition metal ions: $RbCaF_3:Rh^{2+}$, $Li_4Ge_5O_{12}:Ti^{4+}$, and $BeAl_2O_4:Cr^{3+}$. The most important results include: (1) the observation of tunable optical gain in a material based on a 4d transition metal ion; (2) the observation of tunable optical gain in a charge transfer band in the blue-green spectral region; (3) the measurement of pump band to metastable state radiationless relaxation rate in Cr^{3+}-doped materials; and (4) the characterization of long-range energy transfer among Cr^{3+} ions in alexandrite crystals at low temperatures.

2. Spectroscopy of $RbCaF_3:Rh^{2+}$ Crystals

We have previously described the optical spectroscopic properties of several 4d and 5d transition metal ions in crystals [1] and have suggested the possibility of using the strong, broad band emission for tunable laser operation.[2,3] Recently we performed single-pass gain measurements on $RbCaF_3$ crystals doped with 800 ppm Rh^{2+} ions, grown in the Oklahoma State University Crystal Growth Laboratory.[4] The pump source was the frequency-tripled output of a mode-locked $Y_3Al_5O_{12}:Nd^{3+}$ laser which provided a pulse of about 30 ps in duration and between 0.1 and 1.6 mJ of energy.

The excitation at 354.7 nm produced a broad fluorescence emission between about 695 nm and 730 nm, peaking at approximately 710 nm as shown in Fig. 1. The fluorescence lifetime is measured to be about 9.0 μs. As the pump power is increased, the lifetime suddenly decreases to about 0.13 μs and the emission band narrows to less than half its original width, as shown in Figs. 1 and 2. The pump energy density threshold

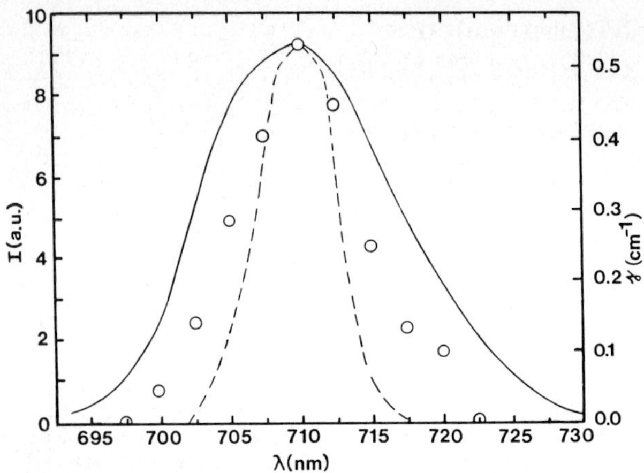

Fig. 1 Fluorescence spectra in arbitrary units and gain of $RbCaF_3:Rh^{2+}$. Solid line is fluorescence below threshold; broken line is fluorescence above threshold; circles are gain.

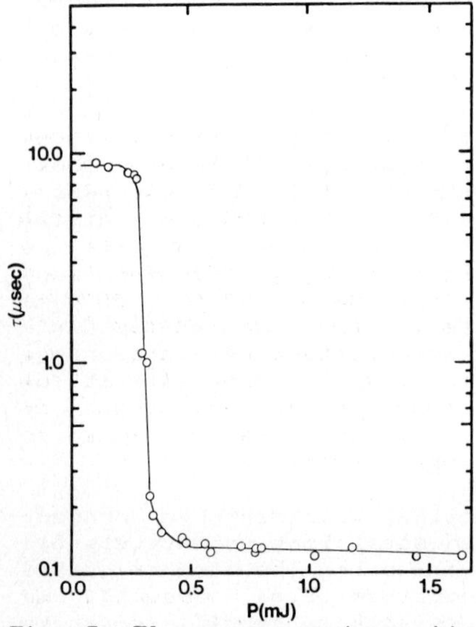

Fig. 2 Fluorescence decay time of $RbCaF_3:Rh^{2+}$ as a function of peak pumping-pulse energy

for these changes occurs at 0.063 Jcm^{-3}. Above this threshold, optical gain was measured by sending a probe beam through the sample and measuring the change in its transmission with the pump beam on and off. The peak gain coefficient was found to be 0.54 cm^{-1} and gain was observed as the probe beam was tuned between 700 and 720 nm as shown in Fig. 1.

These results demonstrate the possibility of using 4d transition metal ions in crystals as tunable laser materials. Rh^{2+}

ions have a $4d^7$ electronic configuration. The excitation band is attributed to a charge-transfer transition and emission can occur from either the Stokes-shifted charge-transfer transition or from d-d transitions after intersystem crossing.[1] The relative importance of these two types of emission depends on the specific environment of the Rh^{2+} ion and this can be altered by thermal or radiation treatments.[1] The band in which gain was observed is attributed to the spin-allowed d-d transition between the spin-orbit split levels of the $^2T_1$ and ^2E states of the $4f^7$ electron configuration. The ideal situation appears to be to use the strong charge transfer band for pumping and maximize intersystem crossing so the longer lived d-d transition can be used for stimulated emission.

The spectral properties of many other 4d and 5d transition metal ions are similar to those of Rh^{2+}.[1] This suggests that this class of ions is potentially useful in solid state laser applications. The major limitation of materials doped with 4d and 5d transition metal ions is the availability of large size, high optical quality crystals. Finding the optimum host crystal for these ions and learning how to control their valance state and intersystem crossing rate are crucial problems in the development of these materials

3. Spectroscopy of $Li_4Ge_5O_{12}$:Ti^{4+} Crystals

The spectroscopic properties of Ti^{4+} ions have been recognized as potentially useful in phosphor applications because of their strong, broad absorption and emission bands.[5-8] The optical properties of this closed shell ion are attributed to charge transfer transitions. These produce strong absorption bands in the ultraviolet spectral region and emission bands appearing in both the blue and red spectral regions, depending on temperature, excitation wavelength, and the host composition. Recently, Loiacono and co-workers at Philips Laboratories have grown several crystals of different compositions of lithium germanium oxide containing various doping concentrations of Ti^{4+}. One of these, $Li_4Ge_5O_{12}$:Ti^{4+} (Ti-LGO), has been shown to exhibit optical properties which are favorable for tunable solid state laser applications.[9]

Using the experimental setup described in the previous section with the Nd-YAG pump laser quadrupled to 266 nm, the fluorescence spectroscopy and optical gain properties of Ti-LGO were measured. The absorption and excitation spectra are shown in Fig. 3. The excitation band for the Ti^{4+} luminescence is located just under the band edge of the host crystal, and thus host absorption complicates the pumping of the active ions. The fluorescence emission is a broad band between about 350 and 600 nm, peaking near 450 nm, as shown in Fig. 4. The intrinsic lifetime is slightly different from sample to sample, but ranges between 7 and 9 μs. As the pump energy is increased above a threshold level of about 0.375 mJ, the fluorescence band width narrows to about one third of its initial value and the fluorescence lifetime decreases to about 32 ns as shown in Figs. 4 and 5. At the same time, a probe beam is observed to have optical gain during pumping. The peak gain coefficient is measured to be 0.42 cm^{-1} with tunable gain measured between 388

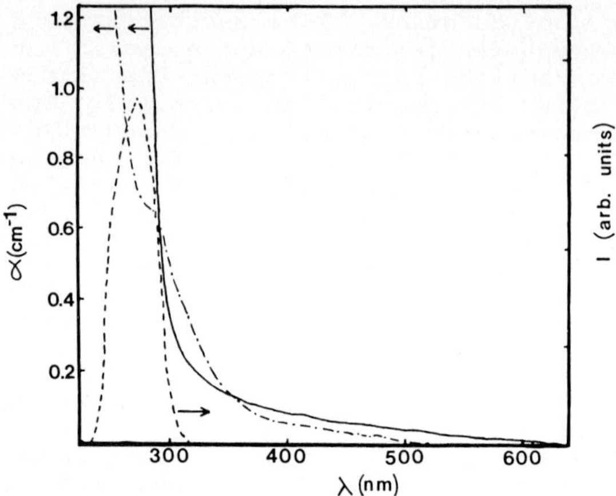

Fig. 3 Absorption and excitation spectra of $Li_4Ge_5O_{12}:Ti^{4+}$. Dashed line is the excitation spectrum of Ti^{4+}; solid line is the absorption spectrum of the doped crystal; dashed and dotted line is the absorption spectrum of the undoped crystal

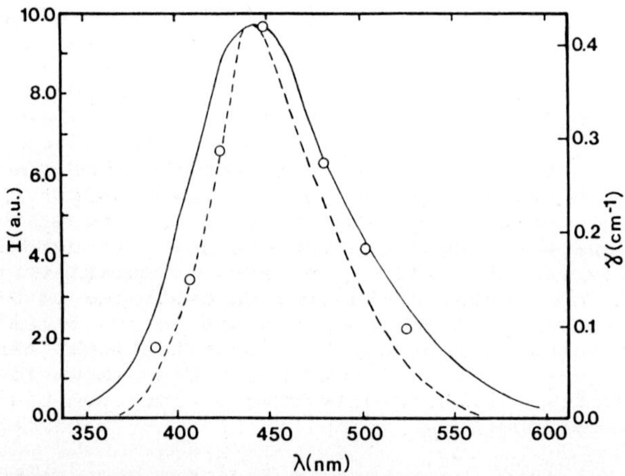

Fig. 4 Fluorescence spectra in arbitrary units and gain of $Li_4Ge_5O_{12}:Ti^{4+}$. Solid line is fluorescence below threshold; broken line is fluorescence above threshold; circles are gain

and 524 nm. (Note that the power scale in Fig. 5 has been corrected for the spectral sensitivity of the power meter, which was not done in ref. [9].) Similar lifetime - shortening results were obtained on samples with two different concentrations of Ti^{4+}. When the sample was aligned with its flat surfaces exactly perpendicular to the probe beam, the gain coefficient increased by over an order of magnitude, which is indicative of feedback due to reflections from the polished faces of the sample.

There are two important aspects to these results on the Philips Ti-LGO crystals. The first is the demonstration of

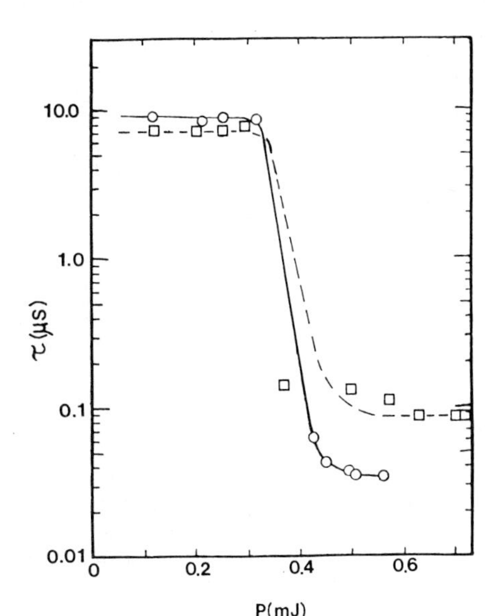

Fig. 5 Fluorescence decay time of $Li_4Ge_5O_{12}:Ti^{4+}$ as a function of peak pumping-pulse energy. Circles are for a Ti concentration of 0.06 % at. and squares are for a Ti concentration of 0.01 % at.

optical gain in a charge transfer transition, and the second is the demonstration of tunable gain in the blue-green spectral region. This indicates that the whole class of phosphor materials based on charge transfer transitions may be useful for solid state lasers. The major problems in using this material in tunable laser applications are the development of reasonable u.v. pumping sources and the development of procedures for repeatedly growing high-quality crystals with the desired optical properties. So far it has not been possible to achieve optical gain except by pumping with a picosecond pulse laser. In addition, the optical properties of the material appear to degrade with age. Current efforts are underway to fully understand the optical properties of Ti^{4+} and how they are altered by changes in the host crystal properties.

4. Laser Spectroscopy Measurements of $BeAl_2O_4:Cr^{3+}$ Crystals

Alexandrite ($BeAl_2O_4:Cr^{3+}$) is already well established as an important tunable solid state laser material.[10] We have been working with researchers at Allied Corporation in characterizing the optical spectroscopic properties of alexandrite crystals relevant to laser operation.[11-13] Results of two-photon absorption and four-wave mixing (FWM) measurements are reported here. They provide additional information on the dynamics of optical pumping of this material.

Figure 6 shows the absorption spectrum of alexandrite obtained at low temperatures. Along with the well-known broad absorption bands the sharp lines appearing in the visible

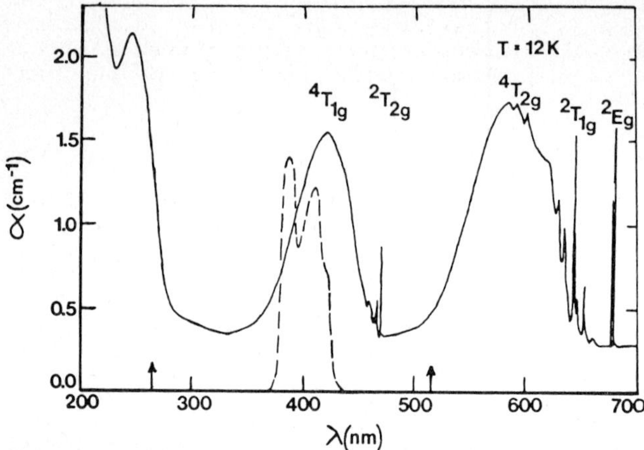

Fig. 6 Solid line represents the absorption spectrum of alexandrite at 12 K. Broken line is the blue fluorescence emission after two-photon excitation at 532 nm as designated by the arrows

spectral region, there is a band on near 250 nm on the absorption edge of the material. This may be due either to a charge transfer transition or to a transition to a high-energy level of the $3d^3$ electron configuration of the Cr^{3+} ion. We used the 30 ps pulse output of our Nd-YAG laser doubled to 532 nm to excite this absorption transition at 266 nm through two-photon absorption. The fluorescence observed after this type of excitation at room temperature consisted of the normal red emission band and a new emission band centered near 400 nm as shown in Fig. 6. The fluorescence lifetime of the 400 nm band is 8 μs at room temperature, while the red emission under these pumping conditions has a rise time of about 20 μs followed by the normal 290 μs decay. Analyzing these decay dynamics indicates that some of the 532 nm excitation is absorbed directly, and the energy decays radiationlessly to the ^2E_g level while some of the rest of the excitation energy is absorbed by two-photon transitions to a level which can decay both radiatively and through radiationless processes to the ^2E_g level.[11] These results provide important information concerning loss mechanisms associated with u.v. pumping and two-photon transitions in alexandrite.

Four-wave mixing measurements were performed using crossed beams from the output of either an argon laser or an argon-pumped dye laser to establish excited state population gratings of Cr^{3+} ions, and using a probe beam from a He-Ne laser to scatter from these gratings.[12] The FWM signal efficiency and decay rate were measured as functions of write beam crossing angle and temperature.[13]

By plotting the FWM signal efficiency versus write beam crossing angle and fitting the experimental data with the predictions of a two-level system model, it is possible to determine the contribution to the laser-induced grating due to the change in the absorption coefficient $\Delta\alpha$ and the contribution due to the change in dispersion Δn. From these results, the dephasing time of the atomic system can be determined through the equation[14]

$$T_2=(2\omega/c)(\Delta n/\Delta\alpha)[\omega-\omega_{12}]^{-1}, \quad (1)$$

where ω and ω_{12} are the laser frequency and the resonant frequency of the atomic transition, respectively. The results of this type of analysis on mirror site and inversion site Cr^{3+} ions in alexandrite and on ruby give values of T_2 ranging from 2.2 to 80 ps. These scale exponentially with the energy difference between the pump band and metastable state for these three types of crystal field environments for Cr^{3+} ions, as shown in Fig. 7. This suggests that the mechanism for dephasing is the radiationless relaxation transition between these levels,[13] and the value of 4.5±3 ps obtained for ruby is consistent with the pump band to metastable state relaxation rate determined by pulse-probe measurements.[15]

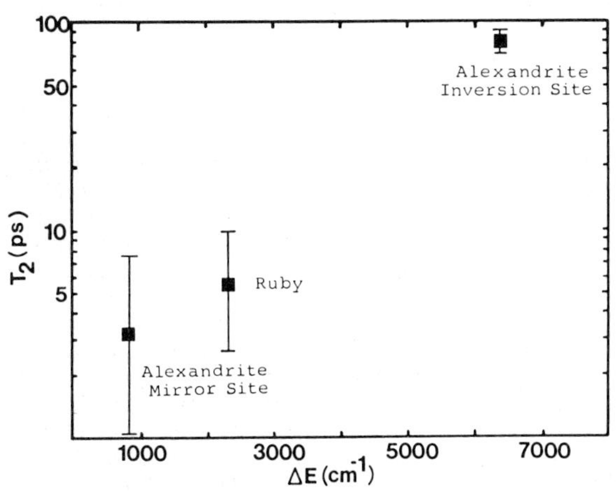

Fig. 7 Variation of dephasing time obtained from analysis of FWM measurements with the energy level splitting of the $^4T_{2g}$ and ^2E_g levels for Cr^{3+} ions in alexandrite and ruby

The decay rate of the FWM signal can be described by[12]

$$k=(2/\tau)+32\pi^2\lambda^{-2}D\sin^2(\theta/2), \quad (2)$$

where the first term represents the fluorescence decay of the Cr^{3+} ions and the second term represents the effects of long range energy transfer between the peak and valley regions of the grating. D is the diffusion coefficient describing the energy migration and the grating spacing depends on λ and θ, the laser wavelength and write beam crossing angle in the crystal, respectively. For Cr^{3+} ions in the mirror sites in alexandrite, the FWM signal decay rate at room temperature is dominated by the fluorescence decay term.[12] However, as

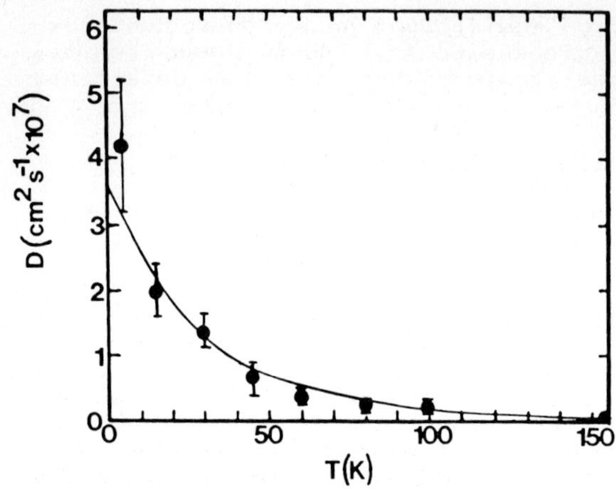

Fig. 8 Temperature dependence of the exciton diffusion coefficient for Cr^{3+} ions in the mirror sites in alexandrite

temperature is lowered below 150 K, a significant angular dependence of the FWM decay rate is observed. This implies the importance of long range energy transfer at these temperatures, and from fitting the equation given above to the experimental data, it is possible to determine the diffusion coefficient for this processes. By 6 K the value of D has increased to about 4.2×10^{-7} cm^2 sec^{-1}, as shown in Fig. 8.

This type of long-range energy transfer is not expected for the low Cr^{3+} concentrations present in this sample. It may be due to the presence of nonuniform distributions of Cr^{3+} ions, since effects such as chromium banding are known to be present in alexandrite. An increase of about a factor of 2.35 in the local density of Cr^{3+} ions over the average concentration will account for the observed magnitude of the energy transfer. The diffusion coefficient for energy migration in the c-direction was found to be an order of magnitude smaller than for migration in the b-direction. These results are consistent with energy transfer through exchange interaction in which the overlap of the electron orbitals decreases exponentially with temperature.[13]

Acknowledgments

This work was supported by the U.S. Army Research Office. The work on $Li_4Ge_5O_{12}:Ti^{4+}$ was performed jointly with Philips Laboratory and the work on $BeAl_2O_4:Cr^{3+}$ was performed jointly with Allied Corp.

References

1. R.C. Powell, R.H. Schweitzer, J.J. Martin, G.E. Venikouas, and C.A. Hunt: J. Chem. Phys. 81, 1178 (1984).
2. R.C. Powell in Proceedings of the Tunable Laser Matrials Workshop, ed. by A. Pinto and J. Paul, June 1983, p. 203.

3. R.C. Powell, in Tunable Solid State Lasers, ed. by P. Hammerling, A.B. Budgor, and A. Pinto (Springer-Verlag, New York, 1985), p. 60.
4. R.C. Powell, G.J. Quarles, J.J. Martin. C.A. Hunt. and W.A. Sibley: Opt. Lett. 10, 212 (1985).
5. F.A. Kroger: Some Aspects of the Luminescence of Solids (Elsevier, New York, 1948).
6. A.J.H. Macke: Phys. Stat. Solidi A 39, 117 (1977); J. Solid State Chem. 18, 337 (1976).
7. G. Blasse and G.J. Dirksen: Chem. Phys. Lett. 62, 19 (1979).
8. G. Blasse: Structure and Bonding (Berlin) 42, 1 (1980).
9. G.M. Loiacono, M.F. Shone, G. Mizell, R.C. Powell, G.J. Quarles, and B. Elouadi: Appl. Lett. 48, 622 (1986).
10. J.C. Walling, O.G. Peterson, H.P. Jenssen, R.C. Morris, and E.W. O'Dell: IEEE J. Quantum Electron. QE-16, 1302 (1980); J.C. Walling, D.F. Heller, H. Samelson, D.J. Harter, J.A. Pete, and R.C. Morris: J. Quantum Electron. QE-21, 1568 (1985).
11. R.C. Powell, L. Xi, X. Gang, G.J. Quarles, and J.C. Walling: Phys. Rev. B 32 2788 (1985).
12. A.M. Ghazzawi, J.K. Tyminski, R.C. Powell, and J.C. Walling: Phys. Rev. B 30, 7182 (1984).
13. A. Suchocki, G.D. Gilliland, and R.C. Powell: to be published.
14. Optical Phase Conjugation, ed. by R.A. Fisher (Academic, New York, 1983).
15. S.K. Gayen, W.B. Wang, V. Petricevic, R. Dorsinville, and R.R. Alfano, Appl. Phys. Lett. 47, 455 (1985).

Spectral Holeburning in Titanium-Doped Sapphire and YAG

R.M. Macfarlane and W. Lenth

IBM Almaden Research Center, 650 Harry Road, San Jose, CA 95120, USA

Abstract: Spectral holeburning in the zero-phonon lines of $Al_2O_3:Ti^{3+}$ (6165Å) and $YAG:Ti^{3+}$ (6502Å) is reported. In $Al_2O_3:Ti^{3+}$, the holes are predominantly transient due to two-level saturation but a shallow persistent hole can be produced which is attributed to photoionization of Ti^{3+}. Substantial hole narrowing was observed in external magnetic fields up to 22 kG. In $YAG:Ti^{3+}$, persistent holes 90 MHz wide were readily observed and the holeburning was shown to be linear in laser power, i.e., a one-photon process. The observation that hole burning is essentially non-permanent in $Al_2O_3:Ti^{3+}$ and permanent in $YAG:Ti^{3+}$ suggests that the photostability of Ti^{3+} ions depends strongly on the host crystal.

1. Introduction

Titanium-doped sapphire ($Al_2O_3:Ti^{3+}$) is a leading contender for a practical, broadly tunable near IR laser [1-4] and other systems such as $Y_3Al_5O_{12}:Ti^{3+}$ [5] and $YAlO_3:Ti^{3+}$ [6] are being actively investigated. These lasers operate on the d – d, $^2E \rightarrow {}^2T_2$ vibronic band of Ti^{3+} around 8000Å. At low temperatures, relatively sharp electronic zero-phonon lines (ZPL) at the origin of the vibronic band are observed in absorption and emission at 6165Å (Al_2O_3) [7] and 6502Å (YAG) [5]. These zero phonon transitions are inhomogeneously broadened, which permits frequency-selective excitation of that small fraction of the Ti^{3+} ions that are resonant with a narrow-bandwidth pump laser. This can result in spectral holeburning, i.e. selective bleaching, in the inhomogeneoulsy broadened absorption line. We have used spectral holeburning at low temperatures to investigate the photostability of the Ti^{3+} ions in Al_2O_3 and YAG. The narrow spectral holes (FWHM <100 MHz) obtained can easily be detected as sharp features on the absorption lines. Thus holeburning permits sensitive detection of the photo-oxidation of Ti^{3+}, for example, and generally enables high-resolution spectroscopic studies to be made of such properties as homogeneous linewidths, saturation parameters, energy transfer and Stark effects.

2. Electronic States

The Ti^{3+} ion has a single 3d electron outside of closed shells and this leads to the simple d-electronic structure of a $^2T_2$ ground state and a ^2E excited state in octahedral coordination. Both sapphire and YAG have trigonal site symmetry which, together with spin-orbit coupling, leads to a splitting of $^2T_2$ into three Kramers' doublets and ^2E into two doublets. In the ground state, a dynamic Jahn-Teller effect leads to a reduction of the $^2T_2$ splittings compared to those expected in the absence of phonon

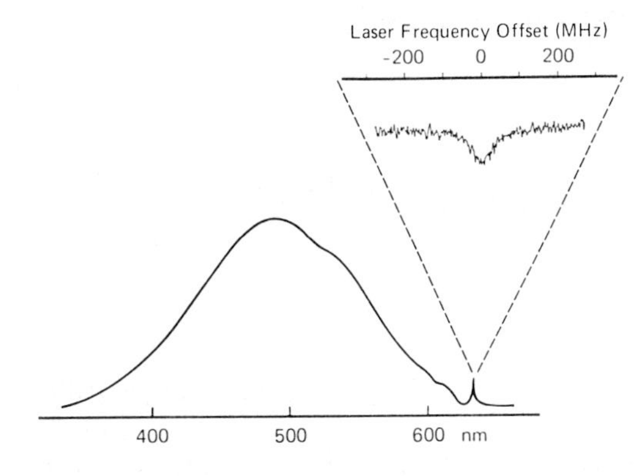

Fig. 1. Schematic illustration of the absorption spectrum of Ti^{3+} showing the broad phonon sideband, the ZPL and a hole burned in the ZPL.

coupling [8]. In the excited state a static Jahn-Teller effect produces a splitting of some 2000 cm^{-1} [9]. The 3d states couple strongly to phonons, which results in absorption and fluorescence spectra that are dominated by broad vibronic bands (Fig.1). The broad fluorescence band enables tunable laser operation of Ti^{3+} materials. At low temperatures sharp zero-phonon transitions between the ^2E and $^2T_2$ levels are observed (Fig.1). Here we will be concerned only with the inhomogeneously broadened ZPL of optical transitions from the lowest component of $^2T_2$ to the lowest component of ^2E.

3. Al_2O_3:Ti^{3+}

In this system, the Ti^{3+} site symmetry is C_{3v} and the $^2T_2$ levels are at 0, 38 and 107 cm^{-1} [7,8]. Absorption to the lowest ^2E level shows a crystal-field induced electric dipole ZPL at 6165Å with a width of 8 cm^{-1}. The fluorescence decay time of the ^2E level at 2 K is 3.9 μsec [4]. Using two single-frequency cw dye lasers; one of fixed frequency to weakly saturate the 6165Å transition and the other of variable frequency to probe the saturation by monitoring the fluorescence excitation spectrum, we observed transient saturation holeburning (Fig. 1). These holes showed essentially complete recovery on removal of the pump beam (but see below). Presumably these saturation holes decay with the excited state lifetime of 3.8 μsec but our time resolution only enabled us to put an upper limit of 100 μsec on the hole lifetime. The cw saturation power density was ~300 W/cm^2 consistent with the oscillator strength of the ZPL of ~ 10^{-8}. The holewidth at 1.6 K and zero magnetic field was 60 MHz, but this narrowed considerably when an external magnetic field (H_0) was applied parallel to the c-axis of the crystal. At H_0 = 20kG, a holewidth of 12 MHz was observed (Fig. 2) at least 3 MHz of which was due to laser frequency jitter of the pump and probe lasers. As in the case of ruby (Al_2O_3:Cr^{3+}) [10, 11], the holewidth is thought to be determined by the fluctuating magnetic field at the Ti^{3+} site due to ^{27}Al nuclear spins. An external field produces an orientation of these spins through the Ti^{3+} moment and slows down the ^{27}Al nuclear spin-flip rate resulting in a narrowing of the saturation holes. After exposure to saturating light for approximately 5 min, a very shallow (~2% of the original saturation hole) hole persisted (Fig. 3). This interesting result is attributed to photoionization of Ti^{3+} which apparently occurs with low efficiency under our excitation conditions. This effect was much more clearly shown by YAG:Ti^{3+}.

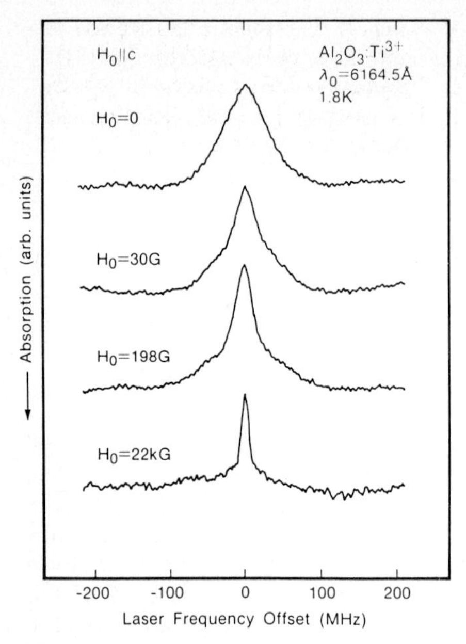

Fig. 2. Transient spectral holes burned in the 6165Å line of $Al_2O_3:Ti^{3+}$ as a function of magnetic field H_0.

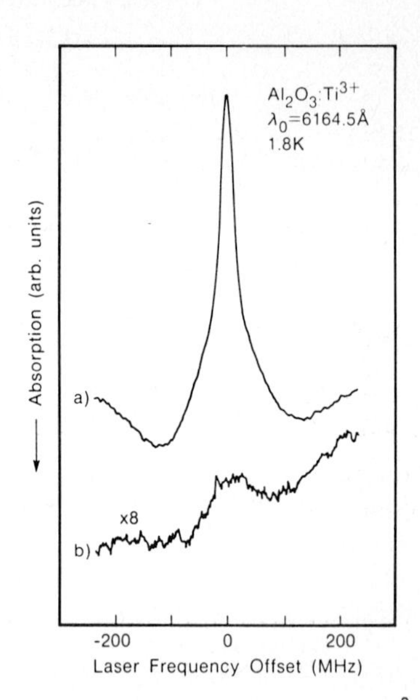

Fig. 3. Transient hole at 6165Å in $Al_2O_3:Ti^{3+}$ (upper) and shallow persistent hole after 5 min irradiation at ~100 W/cm^2 (lower).

4. YAG:Ti^{3+}

In YAG ($Y_3Al_5O_{12}$), Ti^{3+} substitutes for Al^{3+} in sites of C_{3i} symmetry. Thus, the ZPL at 6502Å is magnetic dipole in character and much weaker than the vibronically induced multiphonon sideband [5]. In contrast to sapphire, persistent spectral holes a few percent deep were readily observed at 1.6 K with exposures of ~10s at ~100 W/cm^2. Holes were burned by irradiating in the ZPL at a fixed laser frequency, and subsequently probed by scanning the laser frequency and measuring the fluorescence excitation spectrum. The bleaching was linear in laser power as shown by the observation of constant hole area at constant exposure (power × time) when the power was varied over approximately two orders of magnitude (Fig. 4). Holeburning is thus a single-photon process, rather than a gated two-step process involving excited state absorption such as observed recently in BaClF:Sm^{2+} [12] and $LiGa_5O_8:Co^{2+}$ [13].

Holewidths of 90 MHz were measured in the 500 GHz wide inhomogeneously broadened ZPL. The origin of this width and its behavior with magnetic fields have not yet been determined. Holes could be erased by irradiating at adjacent frequencies in the ZPL or into the broad phonon sideband but irradiation at frequencies below the ZPL led to only a small amount of erasure (Fig. 5). This suggests that Ti^{3+} absorption is bleached by the photoionization process

$$Ti^{3+} \xrightarrow{h\nu} Ti^{4+} + e \tag{1}$$

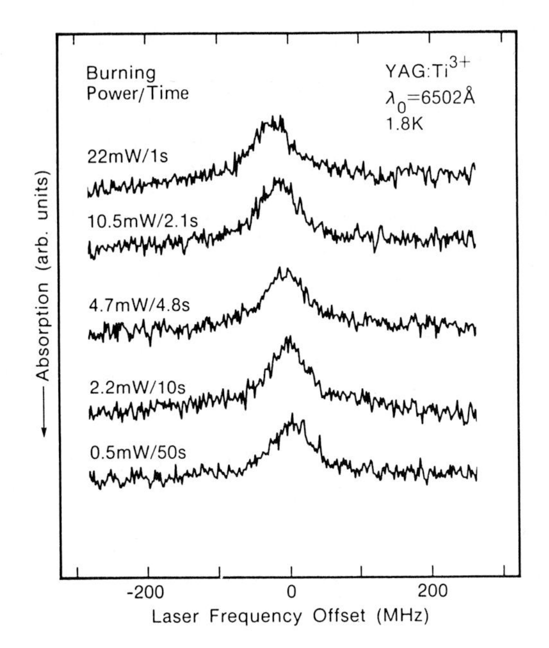

Fig. 4. Persistent holes burned in the ZPL of YAG:Ti^{3+} for laser powers varied by a factor of 40 but at constant fluence showing the linearity of the process.

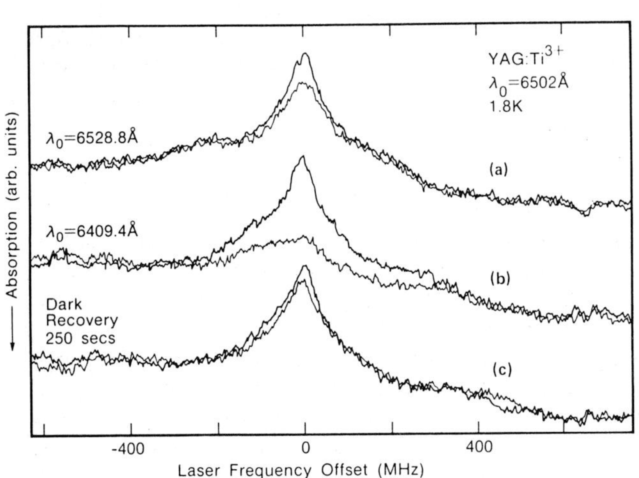

Fig. 5. Erasure of the holes in YAG:Ti^{3+} using different wavelength: (a) below the ZPL;(b) above the ZPL; (c) recovery in the dark. In all cases, the upper trace is the hole before erasure and the lower after erasure.

with the photoreleased electrons being trapped by Ti^{4+} ions which are also present in the crystal. These become reduced to Ti^{3+} and absorb at a different frequency from that of the hole. Certainly other traps could also participate in the photo-induced oxidation of Ti^{3+}. The linear nature of the holeburning (see Fig. 4) suggests that ionization occurs from the ^{2}E state directly, without a second excitation step to the conduction band. For this to occur suitable traps must be accessible from ^{2}E. Of course, the long lifetime of the holes (which were observed not to decay over times ~hr) and the long burning times makes it possible to observe photoionization even when the branching efficiency for ionization from the excited state is very small. For long exposures the hole depth saturates at ~ 10% and some additional hole broadening was observed. This saturation may reflect an equilibrium between burning and erasing in the sense that those Ti^{4+} ions that have been initially produced by photoionization can now serve as electron traps for further photo-induced oxidation of Ti^{3+}. The availability of electron traps (for example Ti^{4+} ions) near a given Ti^{3+} ion will also limit the burning. Generally the occurrence of permanent spectral holeburning by photoionization indicates the presence of such traps.

5. Conclusion

Spectral holeburning, both transient due to two-level saturation and persistent due to photoionization, has been observed in Ti^{3+} laser materials. The different hole burning behavior observed for $Al_2O_3:Ti^{3+}$ and $YAG:Ti^{3+}$ demonstrates that the photostability of Ti^{3+} ions can be very dependent on the host crystal. Holeburning and erasure can be competing processes which reach an equilibrium that depends on the wavelength of the exciting light and the concentration, distribution and nature of the electron traps in the host material. In this context, studies of materials grown under different conditions and with different concentrations of Ti^{4+} or other controlled traps would be worthwhile. In addition to providing a sensitive measure of photoionization, spectral holeburning provides a new and powerful method for high-resolution spectroscopic studies of laser materials.

Acknowledgements: We would like to thank P. Albers and G. Huber, University of Hamburg, and M. R. Kokta, Union Carbide Corporation, for kindly providing samples of $YAG:Ti^{3+}$ and $Al_2O_3:Ti^{3+}$, respectively.

References

1. P. F. Moulton: In Tunable Solid State Lasers, ed. by P. Hammerling, A. Budgor and A. Pinto (Springer-Verlag, Berlin 1985) p.4-10
2. P. Lacovara, L. Esterowitz, R. Allen: Opt. Lett. 10, 273 (1985)
3. G. F. Albrecht, J. M. Eggleston, J. J. Ewing: Opt. Commun. 52, 401 (1985)
4. P. F. Moulton: J. Opt. Soc. Am. B3, 125 (1986)
5. P. A. Albers: Ph.D. Thesis, University of Hamburg (1985)
6. See papers by K. L. Schepler and P. F. Moulton in this volume
7. B. F. Gachter and J. A. Koningstein: J. Chem. Phys. 60, 2003 (1974)
8. R. M. Macfarlane, J. Y. Wong, M. D. Sturge: Phys. Rev. 166, 250 (1968)
9. D. S. McClure: J. Chem. Phys. 36, 2757 (1962)
10. L. Q. Lambert: Phys. Rev. B7, 1834 (1973)
11. A. Szabo: Phys. Rev. B11, 4512 (1975)
12. A. Winnacker, R. M. Shelby, R. M. Macfarlane: Opt. Lett. 10, 350 (1985)
13. R. M. Macfarlane, J. C. Vial: Phys. Rev. B, July (1986)

Part II

Chromium Spectroscopy

Chromium Activated Crystals as Tunable Laser Media – What Makes Them Special?*

J.A. Caird

University of California, Lawrence Livermore National Laboratory,
P.O. Box 5508, Livermore, CA 94550, USA

1. Abstract

The trivalent chromium ion has been successfully used as the optical activator in a large number of tunable crystalline laser materials. Some of the unique properties of its $3d^3$ electron configuration which have led to this widespread success are reviewed. It is found that a combination of positive attributes has led to a higher probability of success with the use of the Cr^{3+} ion than with other transition metal ions. Some negative factors associated with the use of other transition metal ions are identified as areas in need of attention if they are to be used. Optimization of the performance of chromium doped laser materials under flashlamp pumping is also discussed.

2. Introduction

The use of the Cr^{3+} ion as the optical activator in crystalline laser materials will be discussed in terms of a comparison between Cr^{3+} and other transition metal ions of the same row in the periodic table. It is implicit for some of the arguments made here that the potential for efficient flashlamp pumping of the materials is desirable, although to date this has been achieved in only a few cases. In this context, it is appropriate to begin the discussion with a review of some of the major advantages and difficulties associated with the use of solid state transition metal ion lasers in general.

Because of the parity-forbidden nature of transitions within the $3d^n$ configuration of transition metal ions, the excited state radiative lifetimes, τ_r, can be long enough ($\tau_r \gtrsim 100$ μs) to provide for efficient flashlamp pumping with reasonable lamp shot life ($>10^8$ shots). The strong interaction between the electronic energy levels and the crystal vibrations (vibronic interaction) leads to broad absorption and emission bands. This allows for efficient absorption of flashlamp pump light, along with broadly tunable emission. Furthermore, some of the near-infrared emission wavelengths are difficult to generate by any other means. The concomitant low emission cross-sections reduce the potential for parasitic oscillation and shortening of the excited state lifetime by amplified spontaneous emission (ASE). This will allow energy to be stored in the medium at high density, and increase the achievable output energy per laser aperture.

*Work performed under the auspices of the U.S. Department of Energy by Lawrence Livermore National Laboratory under Contract No. W-7405-ENG-48.

In addition to possessing good quantum electronic properties, tunable solid state lasers employing transition metal ions as optical activators may also exhibit good physical properties. These include high stability against photo-degradation and high physical durability of the materials. These are significant advantages over many solid state laser materials activated by color centers.

Of course, the above advantages do not come without a price. The broad emission bandwidths create a difficult tradeoff between emission cross-section and radiative lifetime, which will be discussed in more detail toward the end of this review. If the lifetime is long enough for flash-lamp pumping then the emission cross-section is low, leading to low gain and high saturation fluences. Low gain implies the necessity to develop low loss material to maintain good output efficiency. High saturation fluences increase the potential for optical damage at pulse intensities required for efficient extraction of the stored energy. Other deleterious effects of the strong vibronic interaction include increased potential for broadband excited state absorption and nonradiative relaxation in many ions.

The development of tunable lasers based on the Cr^{3+} ion began with alexandrite[1], $BeAl_2O_4$:Cr and was rapidly extended to numerous other crystalline media.[2] The extraordinary success achieved to date with the Cr^{3+} ion is readily apparent in Table 1, where all of the currently known tunable solid state laser materials are listed. It is seen that the number of materials employing Cr^{3+} exceeds the number of materials employing all other transition metals combined. It is worthwhile noting that with the exception of titanium-doped sapphire, $Al_2O_3:Ti^{3+}$, virtually all of the materials based on other transition metal ions have only been used at cryogenic temperatures. On the other hand, only one of the Cr^{3+} based materials ($ZnWO_4$) requires cryogenic cooling in order to produce cw laser action. This is largely due to the lower nonradiative relaxation rates exhibited by Cr^{3+} at room temperature.

Table 1. A listing of the (reported) tunable solid state laser materials based on activation by transition metal ions as of June 1986

Cr^{3+} Lasers	Others
$BeAl_2O_4:Cr^{3+}$	$MgF_2:Ni^{2+}$
$Be_3Al_2(SiO_3)_6:Cr^{3+}$	$MnF_2:Ni^{2+}$
$Gd_3(Sc,Ga)_5O_{12}:Cr^{3+}$	$MgO:Ni^{2+}$
$Gd_3Sc_2Al_3O_{12}:Cr^{3+}$	$KMgF_3:Ni^{2+}$
$Gd_3Ga_5O_{12}:Cr^{3+}$	$CaY_2Mg_2Ge_3O_{12}:Ni^{2+}$
$La_3Lu_2Ga_3O_{12}:Cr^{3+}$	$MgF_2:Co^{2+}$
$Y_3Ga_5O_{12}:Cr^{3+}$	$ZnF_2:Co^{2+}$
$Y_3Sc_2Ga_3O_{12}:Cr^{3+}$	$KMgF_3:Co^{2+}$
$KZnF_3:Cr^{3+}$	$KZnF_3;Co^{2+}$
$ZnWO_4:Cr^{3+}$	$MgF_2:V^{2+}$
$SrAlF_5:Cr^{3+}$	$CsCaF_3:V^{2+}$
$Na_3Ga_2Li_3F_{12}:Cr^{3+}$	$Al_2O_3:Ti^{3+}$
$ScBO_3:Cr^{3+}$	

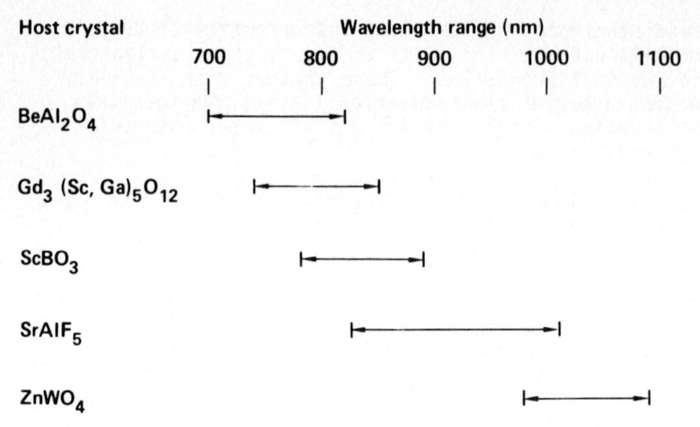

Figure 1. Reported laser tuning ranges for Cr^{3+} ions in representative crystals.

Because the strength of the crystal field has a large effect on the positions of the energy levels, the observed tuning ranges of the various Cr^{3+} doped crystals extends continuously from about 700 nm to nearly 1100 nm. Figure 1 depicts the tuning ranges of representative materials based on Cr^{3+} which have been demonstrated to date.

The broadly successful use of the Cr^{3+} ion is not due to a single factor, but can be attributed to a fortuitous combination of several unique properties of the ion's $3d^3$ electron configuration. These include:

1. The presence of three spin-allowed pump bands which allows efficient flashlamp pumping.
2. Chemical stability--Cr^{3+} is resistant to both oxidation and reduction.
3. The possibility of simultaneous occupation of crystallographic sites with different anion coordination numbers (leading to undesirable spectral characteristics) is minimized by a high octahedral crystal field stabilization energy (CFSE). This is especially important in crystals with both octahedral and tetrahedral cation sites (e.g. garnets).
4. Electron pairing does not produce a low-spin first excited state up to relatively high crystal field strengths, so the dominant emission is usually spin allowed.
5. The relatively large crystal field splittings available stabilize the first excited state against nonradiative relaxation.
6. The positions of the second and third spin-allowed absorption bands result in a gap in the excited state absorption spectrum at the emission wavelengths.

These points will be discussed in more detail in the following section.

3. Discussion of the Special Attributes of Trivalent Chromium

3.1 Pump Bands

If flashlamp pumping is to be employed, then it is important to have broad absorption bands above the metastable state to obtain high lasing efficiency. The strongest absorption bands will be ones for which the spin

Table 2. An enumeration of the spin-allowed absorption (pump) bands for the common transition metal ions.

Ion(s)	Configuration	Spin allowed pump bands	Total
Ti^{3+}, V^{4+}	$3d^1$	$^2E(e)$	1
Ti^{2+}, V^{3+}	$3d^2$	$^3T_2(t_2e), ^3T_1(t_2e), ^3A_2(e^2)$	3
V^{2+}, Cr^{3+}, Mn^{4+}	$3d^3$	$^4T_2(t_2^2e), ^4T_1(t_2^2e), ^4T_1(t_2e^2)$	3
Cr^{2+}, Mn^{3+}	$3d^4$	$^5T_2(t_2^2e^2)$	1
Mn^{2+}, Fe^{3+}	$3d^5$	-	0
Fe^{2+}, Co^{3+}	$3d^6$	$^5E(t_2^3e^3)$	1
Co^{2+}	$3d^7$	$^4T_2(t_2^4e^3), ^4T_1(t_2^4e^3), ^4A_2(t_2^3e^4)$	3
Ni^{2+}	$3d^8$	$^3T_2(t_2^5e^3), ^3T_1(t_2^5e^3), ^3T_1(t_2^4e^4)$	3
Cu^{2+}	$3d^9$	$^2T_2(t_2^5e^4)$	1

quantum number does not change (i.e. spin-allowed transitions). The spin-allowed absorption bands of the various $3d^n$ electron configurations are enumerated in Table 2. It is seen the Cr^{3+} ion with three spin-allowed pump bands is one of a group of ions with the maximum number. In Fig. 2, for example, it is seen that the absorption bands of $SrAlF_5$:Cr span the entire range of visible and near ultraviolet wavelengths.

Comparing Tables 1 and 2 it can be seen that Ti^{3+} is the only ion with fewer spin-allowed pump bands that has been employed as a laser crystal activator. It should be noted however that the absorption band of Al_2O_3:Ti^{3+} (titanium-sapphire) is quite broad due to strong Jahn-Teller splitting of the ^2E level, which improves its absorption efficiency considerably.

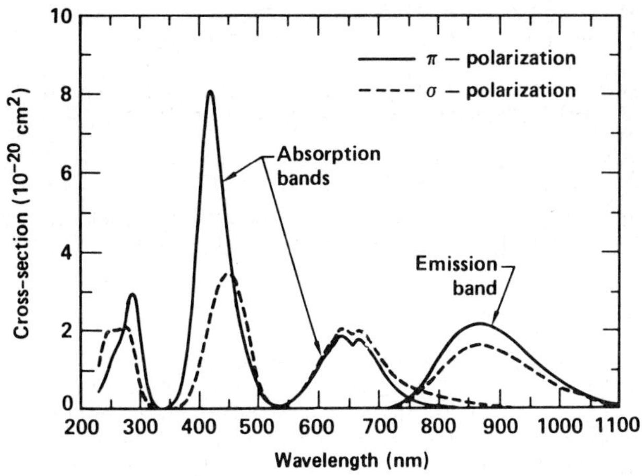

Figure 2. The absorption spectra of $SrAlF_5$:Cr showing three bands spanning the visible and near ultraviolet wavelengths.

3.2 Chemical Stability

In many instances it is possible for transition metals to enter crystal lattices in more than one valence state. The problem can be especially difficult if the desired valence state is different from that of the natural site in the host crystal. The Cr^{3+} ion has been used to produce laser emission in $KZnF_3$ and $ZnWO_4$, where it substitutes for the divalent Zn^{2+} ion. Such behavior has not been demonstrated in any other transition metal ion to date.

In some cases alternate valence states create problems associated with the production of undesirable absorption bands. In other cases it simply reduces the number density of the desired ionic species. Ti^{3+}, for example, is readily oxidized to Ti^{4+}, thus reducing the available concentration of the desired lasing species. Special measures must be taken (e.g. heat treatment in a reducing environment) to stabilize the Ti^{3+} valence state.

The relative stability of the various common transition metal ions against oxidation and reduction is shown in Table 3. Ions not appearing in Table 3 are uncommon because their valence is particularly difficult to achieve. The relative stabilities against oxidation and reduction for the various ions were determined by examining the standard reduction potentials of the electrochemical series.[3] It is seen that Cr^{3+} is one of only four ions that is resistant to both oxidation and reduction.

Table 3. A comparison of the relative stability against chemical oxidation and reduction for common transition metal ions.

Ion	Configuration	Stability against oxidation	Stability against reduction
Ti^{3+}	$3d^1$	Low	High
V^{4+}	$3d^1$	Moderate	Low
Ti^{2+}	$3d^2$	Low	High
V^{3+}	$3d^2$	Moderate	Moderate
V^{2+}	$3d^3$	Low	High
Cr^{3+}	$3d^3$	High	High
Mn^{4+}	$3d^3$	High	Low
Cr^{2+}	$3d^4$	Low	High
Mn^{3+}	$3d^4$	Moderate	Low
Mn^{2+}	$3d^5$	High	High
Fe^{3+}	$3d^5$	High	Low
Fe^{2+}	$3d^6$	High	High
Co^{3+}	$3d^6$	High	Low
Co^{2+}	$3d^7$	High	High
Ni^{2+}	$3d^8$	High	Moderate
Cu^{2+}	$3d^9$	High	Low

3.3 Stability in Octahedral Coordination

In crystals, transition metal cations are most often bonded to four (tetrahedral) or six (octahedral) neighboring anions. Octahedral anion coordination is important for several reasons. First, octahedral coordination produces significantly higher crystal field splittings (i.e. energy gaps between excited states) than tetrahedral coordination. For equal anion charges and metal-ligand distances, for example, the ratio of the tetrahedral to octahedral crystal field strength is given by[4]

$$Dq(tet)/Dq(oct) = 4/9 . \quad (1)$$

Higher crystal field splitting generally results in a greater stability of the metastable excited states against nonradiative relaxation.

Secondly, octahedral coordination generally induces much smaller transition oscillator strengths between states of the $3d^n$ configurations. This is because these ion sites often exhibit inversion symmetry (or near-inversion symmetry). Tetrahedral sites, on the other hand, lack inversion symmetry in the extreme, leading to greater mixing of electronic configurations with opposite parity, and consequently produce much higher transition oscillator strengths. High oscillator strengths lead to short radiative lifetimes for the metastable states and therefore make storage of flashlamp pump energy more difficult. In octahedral coordination radiative lifetimes greater than 100 μs are common.

There are numerous crystals which possess both octahedral and tetrahedral cation sites. Moreover, it is not unusual for the same ion, or for ions with the same valence to be located on such differing sites. The generic formula for an oxide garnet crystal, for example, is $A_3B_2C_3O_{12}$, where A is an 8-fold coordinated (dodecahedral) site, B is a 6-fold coordinated (octahedral) site, and C is a 4-fold coordinated (tetrahedral) site. In gadolinium gallium garnet (GGG), $Gd_3Ga_5O_{12}$, the gallium ions reside on both the octahedral and tetrahedral sites. In gadolinium scandium gallium garnet (GSGG), most of the octahedral sites are occupied by scandium ions which have the same +3 valence as gallium.

As a dopant in garnet crystals Cr^{3+} occupies only the octahedral sites. If there were Cr^{3+} ions on the tetrahedral sites an absorption band would appear at longer wavelengths than the normal Cr^{3+} bands due to the reduced crystal field strength at these sites. Such longer wavelength absorption would most likely overlap the Stokes shifted emission band of the normal octahedrally coordinated Cr^{3+} ions. Furthermore the intensity of this absorption would be high due to the lack of inversion symmetry. Laser action would therefore be inhibited or quite possibly prevented altogether by the increased losses at the normal laser wavelengths. Thus, if Cr^{3+} had even a slight tendency to occupy tetrahedral sites, it is very likely that none of the six oxide garnets in Table 1 would exhibit laser action. (It should be noted that tetrahedrally coordinated Cr^{3+} has been reported in $LiAl_5O_8$.)[5]

The strong preference of Cr^{3+} for octahedral coordination can be understood qualitatively through the model depicted in Fig. 3. In octahedral coordination the d-orbitals which are 5-fold degenerate in the free ion split into low-energy, triply-degenerate t_{2g} orbitals, and high energy, doubly-degenerate e_g orbitals. The magnitude of the splitting is (by definition) 10Dq(oct). Relative to the center of gravity of the configuration, then, the t_{2g} levels lie at -4dq(oct), and the e_g levels lie at +6Dq(oct). The lowest energy state for the $3d^3$ electrons of the Cr^{3+} ion is obtained with one electron in each of the t_{2g} orbitals with parallel spins. The crystal field stabilization energy (CFSE) of this configuration would therefore be -4Dq(oct) x 3 electrons = -12Dq(oct) relative to the center of gravity.

For a tetrahedral site, on the other hand, the e-orbitals lie below the t_2 orbitals. In order to avoid spin pairing (and consequent electrostatic repulsion energy), therefore, one of the $3d^3$ electrons must reside in an upper t_2 orbital. The CFSE of this configuration is

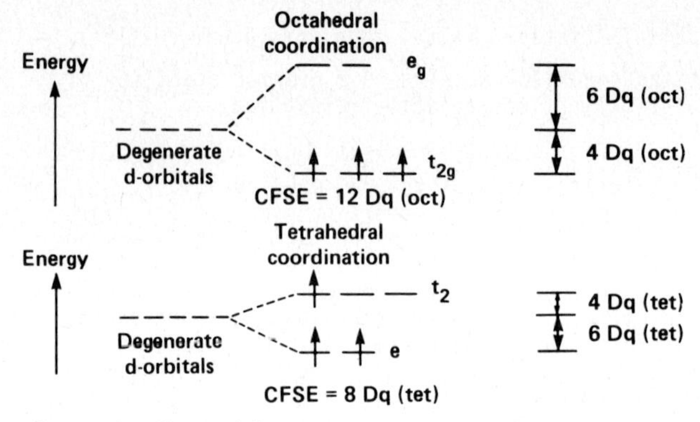

Figure 3. The model used to describe the high octahedral crystal field stabilization energy (CFSE) of the Cr^{3+} ion.

-6Dq(tet) x 2 electrons + 4Dq(tet) x 1 electron = -8Dq(tet). Taking Eq. 1 into account, therefore, the net excess stabilization energy for octahedral coordination over tetrahedral coordination of Cr^{3+} is approximately -8.5Dq(oct). For typical crystals this amounts to an energy of about 1 eV, which is a substantial fraction of the binding energy per cation.

By a similar process it can readily be seen that no other $3d^n$ configuration can have an excess CFSE higher than Cr^{3+}. Data for a number of ions is given in Table 4.[6] It is seen that Cr^{3+} has by far the highest excess CFSE of any of the transition metal ions listed. The high spin state of the $3d^5$ configuration has no net preference at all for octahedral coordination because one electron resides in each of the 5 orbitals in either coordination. It is interesting to note, for example, that magnetite, Fe_3O_4, is an inverse spinel, where the Fe^{3+} ions occupy both octahedral and tetrahedral sites, while the Fe^{2+} ion resides on the other octahedral site.[7] It should be noted that the low CFSE of Ti^{3+} should not be a problem in crystals without tetrahedral sites, such as Al_2O_3.

3.4 Absence of Low Spin First Excited States in Moderate Crystal Fields

There are a number of transition metal ions which can exhibit low spin first excited states in only moderate strength crystal fields. Once the

Table 4. Crystal field stabilization energies (kJ mol^{-1}) estimated for transition metal oxides.[6]

Ion		Octahedral stabilization	Tetrahedral stabilization	Excess octahedral stabilization
Ti^{3+}	d^1	87.6	58.7	28.9
V^{3+}	d^2	160.5	106.8	53.6
Cr^{3+}	d^3	225.0	67.0	158.0
Mn^{3+}	d^4	135.8	40.2	95.5
Fe^{3+}	d^5	0	0	0
Mn^{2+}	d^5	0	0	0
Fe^{2+}	d^6	49.9	33.1	16.8
Co^{2+}	d^7	93.0	62.0	31.0
Ni^{2+}	d^8	122.3	36.0	86.3
Cu^{2+}	d^9	90.5	26.8	63.7

crystal field exceeds a certain critical value, then, the downward fluorescence transition becomes spin forbidden, with a significant reduction in emission oscillator strength. Thus, in order to maintain a reasonably high cross-section for stimulated emission, the crystal field strength must be kept below this critical value. Unfortunately, this leads to considerably longer wavelength emission and, in general, less stability against nonradiative relaxation. A case in point is the Co^{2+} ion which has been shown to exhibit laser emission in a number of low crystal field host materials, but only at cryogenic temperatures where nonradiative relaxation is minimized.

The conditions which produce low spin first excited states in moderate crystal fields can be understood qualitatively by considering the $3d^7$ configuration of the Co^{2+} ion as depicted in Fig. 4. At low crystal fields (i.e. small Dq) the lowest energy levels are high spin states. The lowest energy low spin state is also shown in the figure. For very small crystal fields, the energy of this low spin state is higher than that of

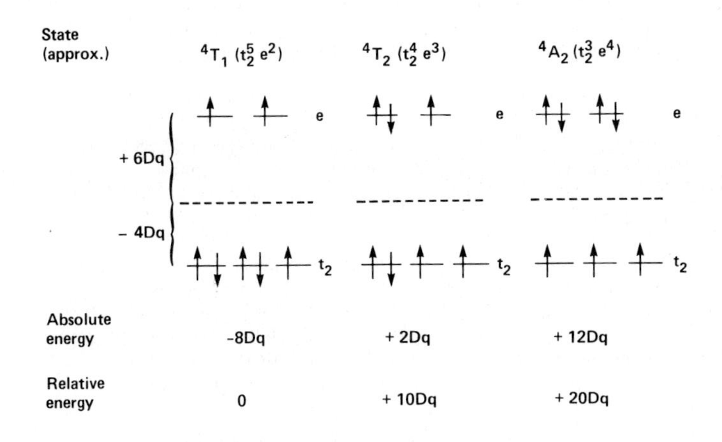

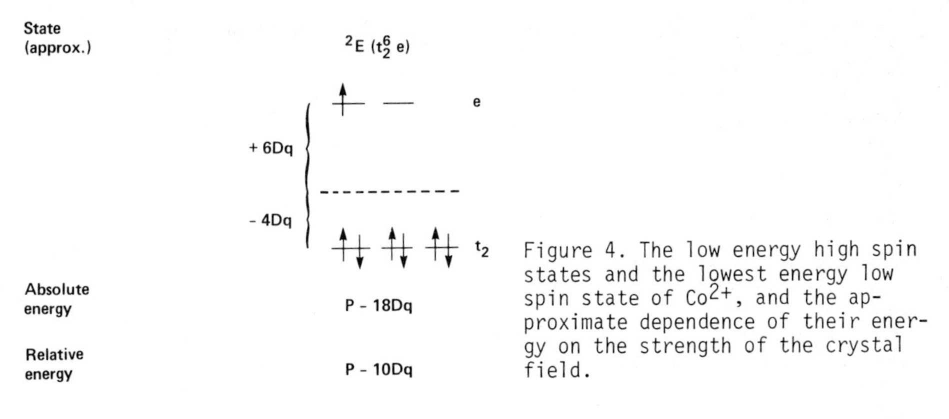

Figure 4. The low energy high spin states and the lowest energy low spin state of Co^{2+}, and the approximate dependence of their energy on the strength of the crystal field.

the high spin states due to the additional electrostatic repulsion energy of the third pair of electrons filling the t_{2g} orbitals. As the crystal field increases, the energy of the low spin state decreases relative to the ground state because of the additional electron in the t_{2g} orbital.

Let P represent the additional "spin pairing" energy of the low spin $^2E(t^6e)$ state. The energy separation between the $^2E(t^6e)$ and $^4T_2(t^4e^3)$ levels is P - 20Dq as shown in Fig. 4. For small crystal fields (10Dq< P/2), the first excited state is $^4T_2(t^4e^3)$. For crystal fields in the intermediate range, P/2 < 10Dq < P, the first excited state becomes the low spin $^2E(t^6e)$ state. For high crystal fields, 10Dq > P/2, the low spin state would actually become the ground state, but the transition between the first excited state and the ground state would still remain spin forbidden. Thus, only for crystal field strengths less than the critical value, 10Dq = P/2, is the transition between the first excited state and the ground state spin allowed, and of reasonably high oscillator strength. The $3d^4$, $3d^5$, $3d^6$, and $3d^7$ configurations all exhibit similar behavior as can be seen in their Tanabe-Sugano diagrams[8] in Fig. 5.

3.5 Availability of Higher Crystal Field Splittings

In contrast to the situation described above, trivalent chromium does not exhibit a low spin first excited state until moderately high crystal fields are reached. Tunable laser emission has been observed on the spin allowed $^4T_2$ to $^4A_2$ transition at much shorter, near infrared wavelengths. The effects of nonradiative relaxation in Cr^{3+} are reduced, and room-temperature operation can usually be attained.

Arguments completely analogous to that given above show that the first excited state of the $3d^3$ configuration of Cr^{3+} does not drop in energy relative to the ground state. Thus, the crystal field strength at which the first excited state becomes a low spin state is approximately doubled (i.e. the critical field strength is now given by 10Dq $\sim$ P), as is clearly seen in Fig. 6. At higher crystal fields Cr^{3+} exhibits the familiar spin-forbidden, R-line emission, which is responsible for 3-level narrowband laser action in ruby. It should be noted that the spin-forbidden transitions in the $3d^4$, $3d^5$, $3d^6$, and $3d^7$ configurations would be broad rather than narrowband, thus leading to much smaller cross-sections for stimulated emission at any given wavelength.

Examination of the Tanabe-Sugano diagrams for the $3d^2$ and $3d^8$ configurations show that they exhibit behavior similar to trivalent chromium in this regard. Furthermore, the $3d^1$ and $3d^9$ configurations are even better because for these simple "one-electron" configurations there is no possibility of spin pairing, and therefore there is only one state of the spin. All of these configurations should exhibit generally higher resistance to nonradiative relaxation in the higher crystal field environments.

3.6 Reduced Excited State Absorption

To date there is very little in the way of direct measurements of the effects of excited state absorption (ESA). However, in most of the $3d^n$ ions ($2 < n < 8$) there are a large number of energy levels which produce such absorption (upward transitions) from the metastable upper laser level. These reduce the available gain and efficiency of any laser based on these ions and, if strong enough can inhibit laser action altogether.

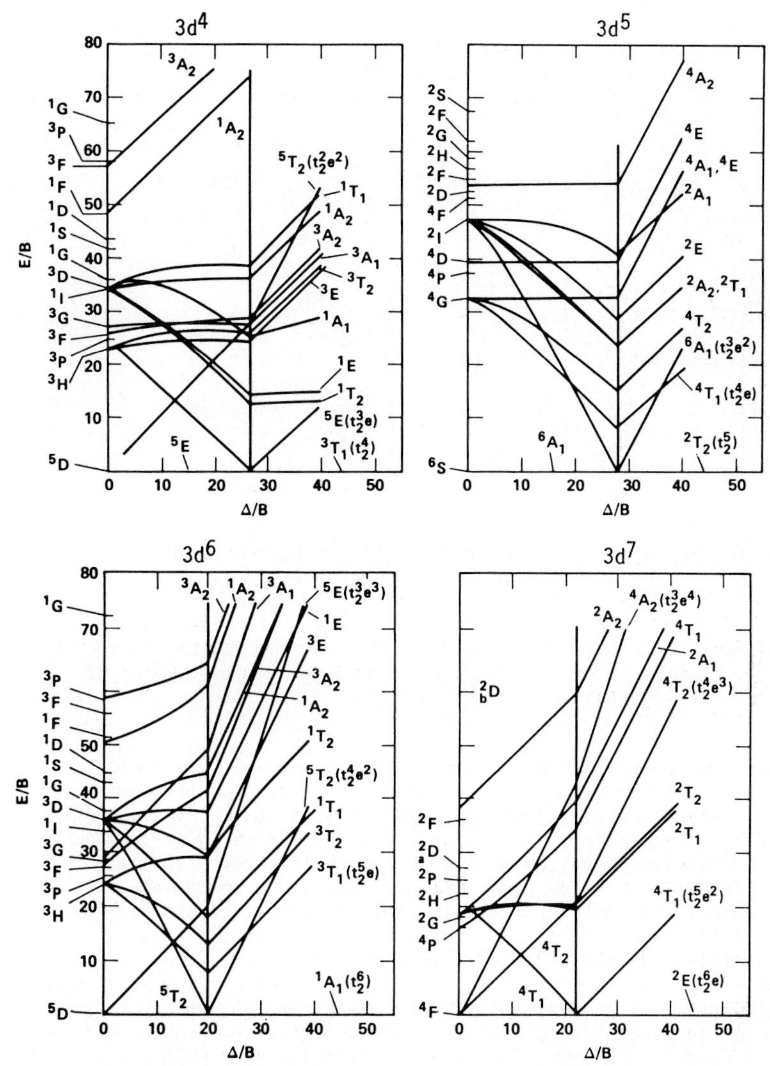

Figure 5. The Tanabe-Sugano diagrams for the $3d^4$, $3d^5$, $3d^6$, and $3d^7$ configurations showing low spin states which drop in energy as the crystal field increases.[8]

Note that for n = 1, and n = 9, this is not a problem since there is only one excited state of the configuration (neglecting minor splittings).

Upward transitions which are spin allowed are generally the strongest, and therefore are of most concern. For the $3d^3$ configuration of Cr^{3+} ESA is reduced because upward transitions at the laser emission wavelength fall in a gap between the two higher-lying states of the same spin, as illustrated in Fig. 6. Furthermore, since the first and second excited states have the same number of e_g electrons, there should be less of a shift in the position of the configuration coordinate potential energy

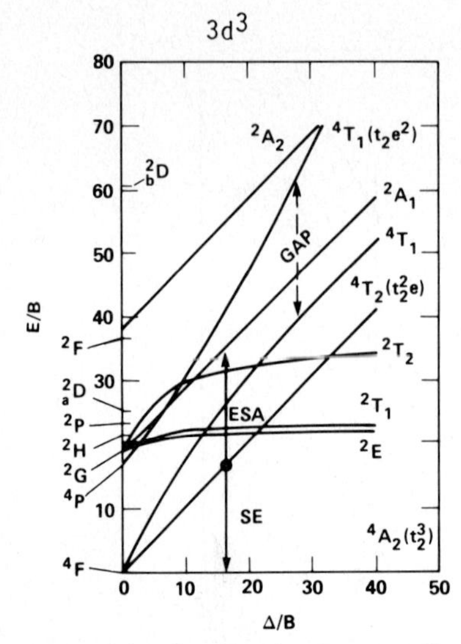

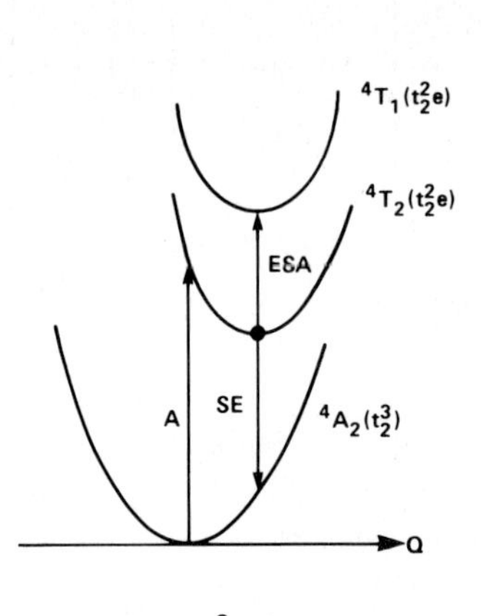

Figure 6. The Tanabe-Sugano diagram for the $3d^3$ configuration shows that the low spin states do not drop in energy as the crystal field increases. Excited state absorption from the $^4T_2$ level falls in the gap between the second and third states of the same spin. The diagram on the right indicates that the ESA to the first $^4T_1$ excited state should be narrow (with a small Stokes shift) because it has the same number of e_g-electrons as the $^4T_2$ state.

minimum. Thus, the lower energy ESA band should be narrow, with less chance of overlapping the emission band. However, mixing of the two $^4T_1$ bands will broaden this absorption, especially at lower crystal field strengths. Measurements of excited state absorption in Cr^{3+} by Andrews, et al.[9] confirm that the emission band falls between the two spin-allowed ESA bands, but the lower ESA band was found to be much broader than expected in GSGG:Cr. Because the excited state absorption is nonvanishing in the region between the two $^4T_1$ ESA bands, it is still of significant concern for Cr^{3+} lasers.

It is also important to point out that excited state absorption bands will not have a significant effect if they are sufficiently weak. This is apparently the case for $MgF_2:Co^{2+}$ ($3d^7$) which exhibits very high lasing efficiency at cryogenic temperatures.[10]

4. Summary Comparison of $3d^n$ Ion Properties

An overall scorecard comparing the properties of the various common 3d transition metal ions in light of the foregoing discussions is displayed in Table 5. The qualitative grades given in the table are generally self-explanatory. Question marks (?) indicate some uncertainty, ambiguity or unavailability of data. It is clear from the overall ratings that the $3d^3$ configuration of the Cr^{3+} ion has the highest probability for successful usage. The ion exhibits a number of positive attributes, so the discovery of a large number of tunable lasers based on it is understandable.

Table 5. An overall comparison (scorecard) of the attributes of the common transition metal ions discussed in Section 3.

Configuration	Ion	No. of pump bands	Chemical stability	Octahedral stabilization	Absence of low spin first excited states	Large crystal field splitting	Reduced ESA
$3d^1$	Ti^{3+}	1	–	–	++	++	++
$3d^1$	V^{4+}	1	–	–	++	++	++
$3d^2$	Ti^{2+}	3	–	+	+	+	–
$3d^2$	V^{3+}	3	?	+	+	+	–
$3d^3$	V^{2+}	3	–	++	+	–	?
$3d^3$	Cr^{3+}	3	+	++	+	+	+
$3d^3$	Mn^{4+}	3	–	++	+	+	+
$3d^4$	Cr^{2+}	1	–	++	–	–	+
$3d^4$	Mn^{3+}	1	–	++	–	–	+
$3d^5$	Mn^{2+}	0	+	–	–	–	–
$3d^5$	Fe^{3+}	0	–	–	–	–	–
$3d^6$	Fe^{2+}	1	+	–	–	–	+
$3d^6$	Co^{3+}	1	–	–	–	–	+
$3d^7$	Co^{2+}	3	+	+	–	–	?
$3d^8$	Ni^{2+}	3	+	+	+	?	?
$3d^9$	Cu^{2+}	1	–	+	++	?	++

However, it could also be said that the success to date with Cr^{3+} is to some extent simply a result of improved statistics. Table 5 is also good for pointing out areas of special concern if other ions are to be used successfully. If an adequate job of valence stablization is done, and if tetrahedral coordination can be avoided (as in Al_2O_3) then the Ti^{3+} ion looks very attractive. Similarly, if chemically stable conditions can be attained, the Mn^{4+} and Cu^{2+} ions also look interesting. On the other hand there is clearly not much you can do with Fe^{3+}.

5. The Tradeoff Between Emission Cross Section and Excited State Lifetime

For flashlamp pumped energy-storage lasers it is desirable to have the excited state lifetime as long as possible consistent with a reasonably high cross-section for stimulated emission. Short lifetimes require short flashlamp pulsewidths to produce efficient operation, resulting in reduced shot life of the lamps. Low emission cross-sections imply high saturation fluences. If the saturation fluence is too high the material can undergo optical damage at fluences required for efficient energy extraction. As a rough guideline lifetimes greater than about 100 μs, and emission cross-sections greater than 2×10^{-20} cm^2 are desirable.

Unfortunately, in an isotropic material the product of the peak emission cross-section, σ_L, and radiative lifetime, τ_r, are constrained by the relation

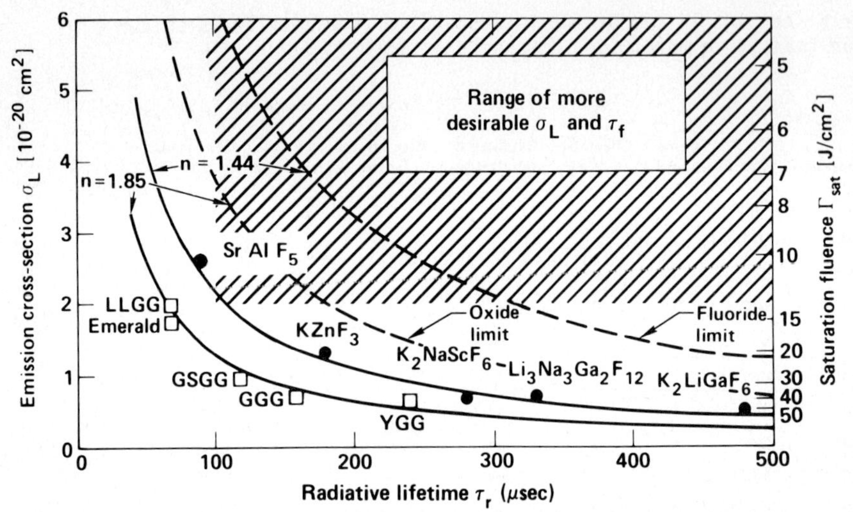

Figure 7. Illustrating the difficult tradeoff between radiative lifetime and emission cross-section. Curves were generated using Eqs. (2) and (6) with λ_L = 750 nm, $\Delta\nu$ = 1700 cm^{-1}, and n = 1.85 (typical of oxides) or n = 1.44 (typical of fluorides). Solid curves are for isotropic media while dashed curves are for idealized anisotropic media with emission concentrated in one polarization.

$$\sigma_L \tau_r = \frac{\lambda_L^2}{8\pi\ n^2 \Delta\nu} \quad , \tag{2}$$

where λ_L is the peak emission wavelength, n is the refractive index of the material, and $\Delta\nu$ is the effective bandwidth of the fluorescence emission. This relation is plotted in Fig. 7 for values of λ_L = 750 nm, and $\Delta\nu$ = 1700 cm^{-1} which are typical of Cr^{3+} emission.[11] The solid curves in the figure are for n = 1.44 (typical of fluoride materials) and n = 1.85 (typical of oxide materials). Also plotted are lifetimes and cross-sections estimated on the basis of spectroscopic measurements of chromium emission in a number of hosts. It can be seen that it is quite difficult to achieve an acceptable tradeoff in an isotropic material.

It has been recognized by a number of researchers, although perhaps not widely appreciated, that the tradeoff can improve considerably in anisotropic media. In a uniaxial crystal, for example, the emission cross-sections, σ_π and σ_σ, for π- and σ- polarizations, respectively, can be treated independently according to

$$\sigma_\pi = \frac{\lambda_L^2}{8\pi\ n^2 \Delta\nu}\ A_\pi \tag{3}$$

and

$$\sigma_\sigma = \frac{\lambda_L^2}{8\pi\ n^2 \Delta\nu}\ A_\sigma \quad , \tag{4}$$

where A_π and A_σ are the spontaneous emission rates for the two polarizations. The total radiative decay rate is given by

$$\frac{1}{\tau_r} = \frac{1}{3} A_\pi + \frac{2}{3} A_\sigma \quad . \tag{5}$$

The optimum situation occurs when the emission rate for π-polarization is substantially higher than that for σ-polarization. If the σ-polarized emission rate can be neglected (3) and (5) can be combined to yield

$$\sigma_\pi \tau_r \approx 3 \cdot \frac{\lambda_L^2}{8\pi \, n^2 \Delta\nu} \quad . \tag{6}$$

Thus, as much as a factor of three increase in the σ-τ product can be had if the emission dipoles are appropriately aligned. The dashed curves in Fig. 7 were obtained by multiplying the solid curve values by 3, thereby indicating the approximate limiting values which can be achieved in highly anisotropic media.

It is well known that the biaxial crystal alexandrite, $BeAl_2O_4$:Cr, exhibits just such a high anisotropy, with the emission rate in one polarization exceeding that of each of the other two by a factor of 10.[1] However, the advantage is substantially reduced in this case by the sharing of population between the ^2E and $^4T_2$ levels which lowers the effective emission cross-section.

6. Conclusion

The Cr^{3+} ion has been used as the optical activator in numerous crystalline laser materials. It has been shown that this success can be attributed to a number of positive attributes of its $3d^3$ electron configuration. While excited state absorption is reduced due to the positions of the spin-allowed ESA bands, its nonvanishing effect on laser emission is still of significant concern, and may limit achievable efficiency in many materials. A number of other transition metal ions show promise if their weak points can be overcome. Finally, improvements in performance of flashlamp pumped materials are possible through the alignment of emission dipoles in anisotropic media.

References

1. J. C. Walling, O. G. Peterson, H. P. Jenssen, R. C. Morris, and E. W. O'Dell, IEEE J. Quantum Electron., QE-16, 1302 (1980).

2. B. Struve, G. Huber, V. V. Laptev, I. A. Shcherbakov, and E. V. Zharikov, Appl. Phys. B, 30, 117 (1983).

3. J. F. Hunsberger, in the CRC Handbook of Chemistry and Physics, 56th Edition, R. C. Weast, Ed., CRC Press, Cleveland, Ohio, D-141 (1975).

4. F. A. Cotton, Chemical Applications of Group Theory, Wiley - Interscience, New York, 278 (1971).

5. D. T. Sviridov, and R.K. Sviridova, J. Appl. Spectros., 34, 431 (1980).

6. J. D. Dunitz and L. E. Orgel, in Advances in Inorganic Chemistry and Radiochemistry, Vol. 2, H. J. Emeleus and A. G. Sharpe, Eds., Academic Press, New York, 30 (1960).

7. A. R. West, Solid State Chemistry and its Applications, John Wiley and Sons, New York, 314 (1984).

8. Y. Tanabe, and S. Sugano, J. Phys.. Soc. Jap., 9, 753 and 766 (1954).

9. L. J. Andrews, S. M. Hitelman, M. Kokta, and D. Gabbe, J. Chem. Phys., 84, 5229 (1986).

10. P. F. Moulton, IEEE J. Quantum Electron., QE-21, 1582 (1985).

11. W. F. Krupke, presented at the NASA Workshop on Tunable Solid State Lasers for Remote Sensing, Menlo Park, CA, October 1-3, 1984.

Radiative and Nonradiative Energy Transfer Between Cr^{3+} and Nd^{3+} in GSGG

G. Armagan and B. Di Bartolo

Department of Physics, Boston College, Chestnut Hill, MA 02167, USA

We have studied the effect of temperature on the radiative and nonradiative energy transfer processes between Cr^{3+} and Nd^{3+} in GSGG. The radiative energy transfer rate was found to decrease slowly with increasing temperature. The nonradiative energy transfer rate was calculated as a function of temperature from the Cr lifetime data and found to increase with increasing temperature. The integrated intensity of the Nd emission in the GSGG: Cr, Nd sample, when the system is pumped in the Nd absorption bands, is constant with temperature (and so is the Nd lifetime); however, when the system is pumped in the Cr absorption bands, it has a different temperature dependence. A model, that takes into account the variations of the energy transfer process with temperature is presented to explain these phenomena and the temperature dependence of the Nd emission in the GSGG: Cr, Nd sample.

1. INTRODUCTION

An area of laser research that is witnessing a very intense activity at the present time is that of solid-state laser-type materials. A new crystal, Gadolinium Scandium Gallium Garnet, called GSGG, is a valuable host for Cr and Nd ions. GSGG: Cr [1,2] and GSGG: Cr, Nd [3] are among the most attractive laser materials of recent years. It has been shown that co-doping of GSGG with Cr and Nd increases the overall efficiency of the Nd emission due to the spin allowed nature of the Cr transition and the good overlap between the Cr emission and the Nd absorption [4,5]; this shows the importance of the role played by the energy transfer process in this material.

Most of the studies on the GSGG: Cr, Nd system have been concerned with its laser performance and no thorough examination has been made of the effects that temperature has on the energy transfer processes between Cr and Nd. We have made such a study and we present here our results.

2. EXPERIMENTAL

The absorption spectra of all the samples were obtained by using a Cary 14 spectrophotometer. The luminescence spectra were obtained by exciting each sample with the light from either a Sylvania DVY 650W tungsten halogen lamp or a He-Ne laser. The luminescence was observed at 90 degrees to the direction of excitation, chopped and focused onto the entrance slit of a Model 2051 McPherson 1-m scanning monochromator. The signal was detected by an RCA 7102 (S-1) photomultiplier tube cooled by crushed dry ice and amplified by a P.A.R. lock-in amplifier.

The pulsed luminescence measurements were made by using either E.G. and G. FX-12 flash tubes or a Molectron Dye Laser Model DL-12 pumped by a Molectron UV-12 N_2 laser.

The excitation measurements were performed by selectively pumping the samples by means of a Model 82-410 Jarrell-Ash monochromator and detecting the luminescence output through an interference filter. A Sylvania DVY 650W tungsten hologen lamp was used as exciting source.

For measurements at room temperature and below, the sample was mounted in a Janis Model 8DT cryostat. The sample temperature was varied from 77 to 300K by using an exchange gas technique. Measurements above room temperatures were made by placing the sample in a quartz furnace and varying the current passing through a Ni-Cr ribbon wound around the furnace. Thermocouples were used to measure the temperature from 77 to 300K (copper-constantan) and above 300K (iron-constantan).

The following samples (dimension ≈ $1cm^3$) were examined:

1) GSGG: Nd (conc.=$1.8x10^{20}cm^{-3}$)
2) GSGG: Cr (conc.=$.8x10^{20}cm^{-3}$)
3) GSGG: Cr, Nd (Cr conc.=$2x10^{20}cm^{-3}$, Nd conc.=$2x10^{20}cm^{-3}$)

3. EXPERIMENTAL RESULTS

3.1 GSGG: Cr

The absorption spectrum of Cr in GSGG: Cr shows the typical broad bands of Cr, $^4A_2 \rightarrow {}^4T_1$ and $^4A_2 \rightarrow {}^4T_2$ centered at 460nm and 645nm (and 73nm and 90nm wide) at 300K, respectively.

The luminescence spectrum of GSGG: Cr ($^4T_2 \rightarrow {}^4A_2$ transition) shows a band centered at 760nm at 300 and 740nm at 77K. The width of this band is 122nm at 300 and 96nm at 77K. The integrated intensity (in number of photons and arbitrary units) of the Cr emission in the temperature range 77-580K is shown in Fig. 1: it increases slightly from 77 to 450K and then decreases considerably with increasing temperature.

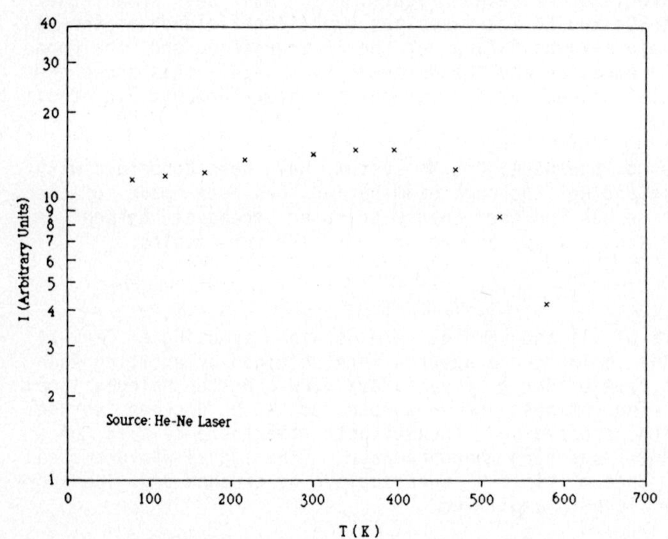

Fig. 1. Temperature dependence of the integrated intensity of the Cr emission in GSGG: Cr

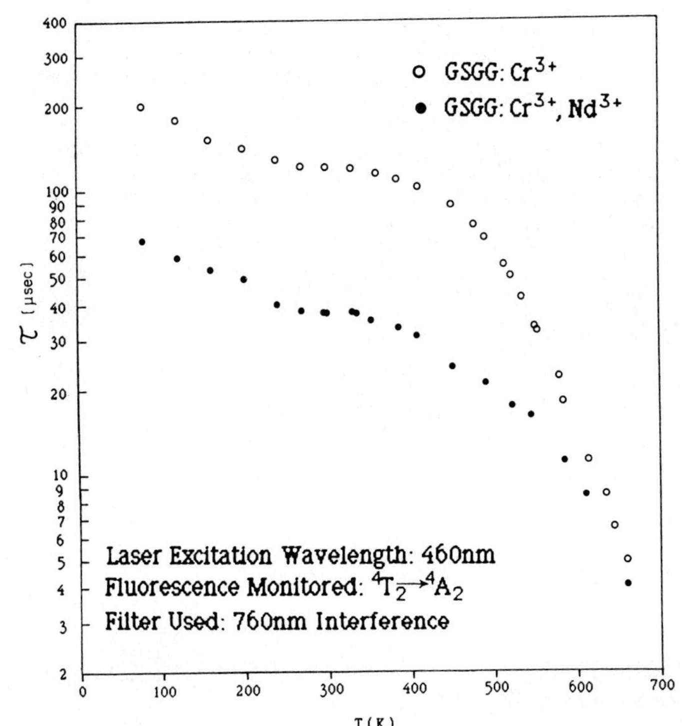

Fig. 2. Temperature dependence of the Cr lifetimes in GSGG: Cr and GSGG: Cr, Nd

The excitation spectrum of the Cr 760nm luminescence at 300K shows the broad $^4T_2$ and $^4T_1$ absorption bands of Cr and correlates well with the absorption spectrum of Cr.

The decay pattern of the Cr 760nm luminescence was found to be exponential in the temperature range 77-660K. The corresponding lifetime is reported in Fig. 2; it decreases with increasing temperature (it is 197μsec at 77K, 120μsec at 300K and 4.9μsec at 660K).

3.2 GSGG: Nd

The absorption spectrum of Nd in the spectral region where Cr emits consists of three structured bands centered at 745, 805 and 880nm, and corresponding to transitions from the ground state to various excited states (see Fig. 3).

The luminescence of Nd in the 1.06μm region corresponds to a transition from the $^4F_{3/2}$ state to the $^4I_{11/2}$ state. The integrated intensity (in number of photons) of this emission in the temperature range 77-650K is practically constant.

The excitation spectrum of the 1.06μm Nd luminescence was obtained at 300K and correlates well with the absorption spectrum of Nd.

The decay pattern of the 1.06μm Nd luminescence was found to be purely exponential in the temperature range 77-700K. The corresponding lifetime

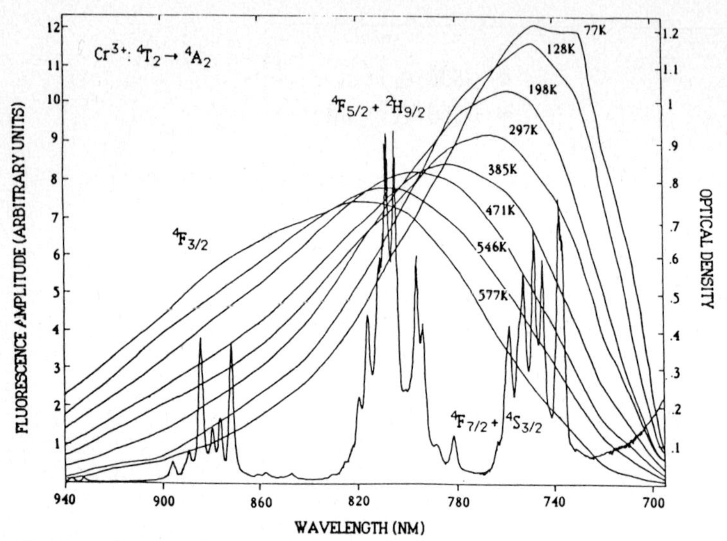

Fig. 3. The overlap between the normalized Cr emission at different temperatures and the Nd absorption at 300K

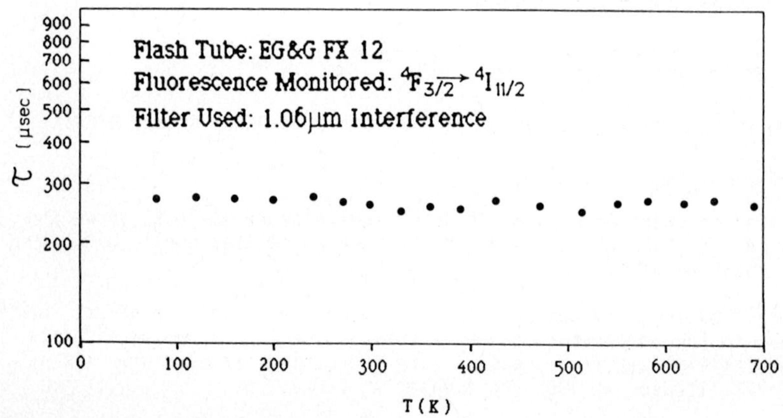

Fig.4. Temperature dependence of the Nd lifetime in GSGG: Nd

is reported in Fig. 4; it is essentially temperature independent, and equal to 260μsec.

3.3 GSGG: Cr, Nd

The absorption spectrum of GSGG: Cr, Nd obtained at 300K is simply the superposition of the absorption spectra of GSGG: Cr and GSGG: Nd.

The luminescence spectrum of Cr in this sample furnishes direct evidence of radiative energy transfer from Cr to Nd: the Cr emission is partly missing in correspondence to the absorption bands of Nd.

The decay pattern of the Cr 760nm luminescence was found to be essentially exponential in the temperature range 77-660K. The corresponding lifetime is shorter than that of Cr in GSGG: Cr on account of the nonradiative energy transfer process (it is 67μsec at 77K, 37.5μsec at 300K and 4μsec at 660K; see Fig. 2).

The temperature dependence of the integrated intensity of the Nd emission in the 1.06μm region, when exciting the system in the Cr absorption bands, is shown in Fig. 5; it is practically constant from 77 to 450K and then decreases considerably with increasing temperature.

The excitation spectrum of the Nd 1.06μm emission in GSGG: Cr, Nd at 300K is shown in Fig. 6; it contains the absorption bands of Cr together with the absorption bands of Nd.

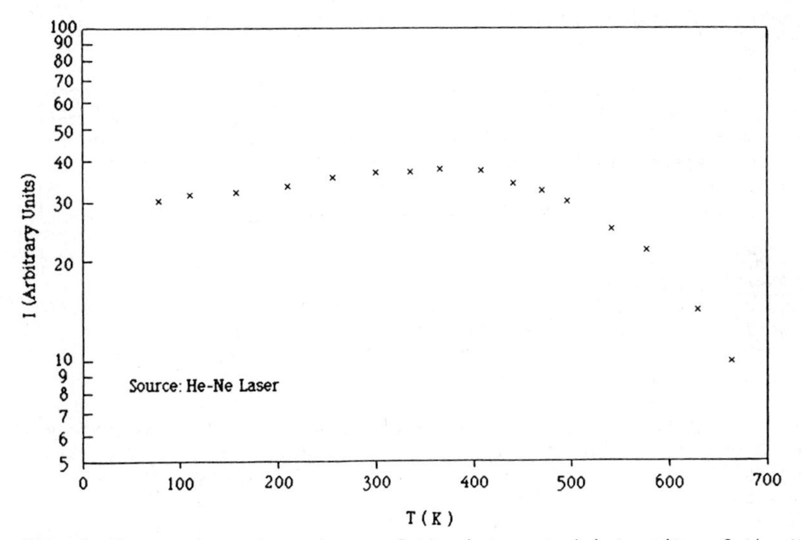

Fig. 5. Temperature dependence of the integrated intensity of the Nd emission in GSGG: Cr, Nd

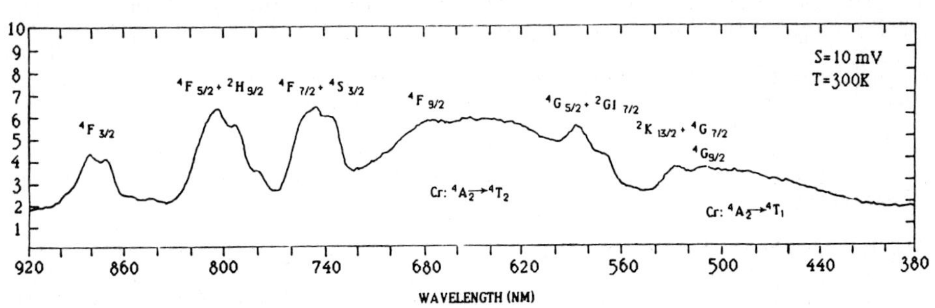

Fig. 6. Excitation spectrum of Nd in GSGG: Cr, Nd (1.06μm fluorescence monitored)

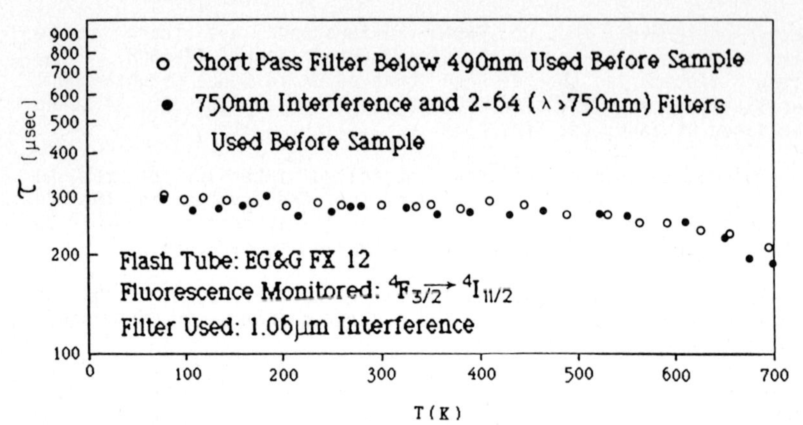

Fig. 7. Temperature dependence of the Nd lifetime in GSGG: Cr, Nd

The decay pattern of this emission was measured under two different excitation conditions (see Fig. 7). The experimental points (o) were obtained by exciting the Cr ions as well as the Nd ions, the experimental points (•) were obtained by exciting the Nd ions only. In the former case, the decay pattern presents a rise at the end of the exciting pulse followed by a decay; in the latter case no rise was observed.

4. DISCUSSION OF RESULTS

The lifetime and the integrated intensity of the Cr emission in the GSGG: Cr sample were found to have different temperature dependences (see Figs. 1 and 2). This is most probably due to the fact that the radiative transition probability of Cr in this system is temperature dependent.

The overlap between the normalized Cr emission at different temperatures and the Nd absorption at 300K is shown in Fig. 3. The Cr luminescence band shifts towards longer wavelengths with increasing temperature. The data in Fig. 3 allow the evaluation of the overlap integral between the Cr emission and the Nd absorption, apart from a constant value. The overlap integral may be taken as a measure of the radiative energy transfer rate; on this basis, the radiative energy transfer rate was found to decrease slowly with increasing temperature (see Fig. 8).

The nonradiative energy transfer rate was calculated from the Cr lifetime data in GSGG: Cr and GSGG: Cr, Nd using the following equation:

$$p(\text{transfer rate})=\tau^{-1}\ (\text{lifetime of Cr in GSGG: Cr, Nd}) - \tau^{-1}\ (\text{lifetime of Cr in GSGG: Cr}).$$

The temperature dependence of this rate is shown in Fig. 9. The nonradiative energy transfer rate was found to increase with increasing temperature.

Let us consider now the simplified model of energy transfer between two different systems of ions N and n (ions N and n represent Cr and Nd, respectively). The assumption is made that the radiationless decays from the absorption bands 3 and 3' to the metastable levels 2 and 2', respectively, are extremely fast; this condition is satisfied in the Cr-Nd

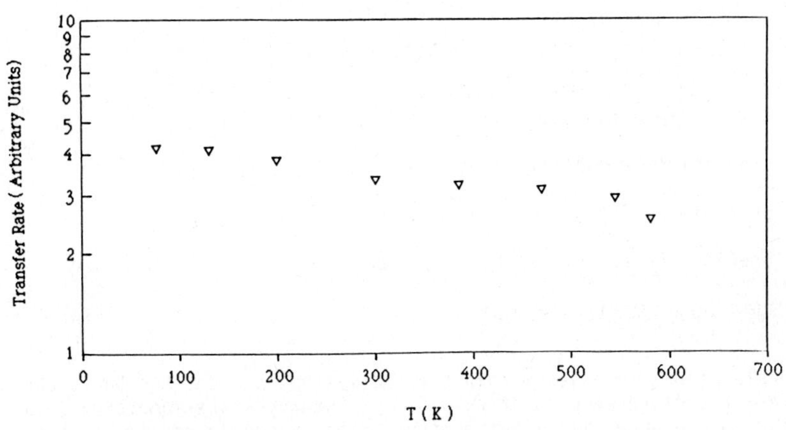

Fig. 8. Temperature dependence of the radiative energy transfer rate from Cr to Nd in GSGG: Cr, Nd

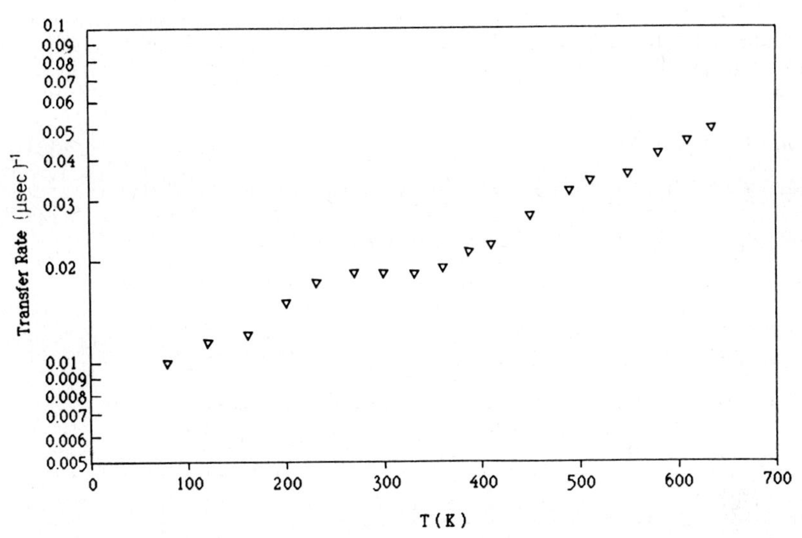

Fig. 9. Temperature dependence of the nonradiative energy transfer rate from Cr to Nd in GSGG: Cr, Nd

case. The rate equations for the populations of the various levels can be written as follows:

$$
\begin{aligned}
\dot{N}_2 &= N_1 w - N_2 P_{21} + n_2 N_1 P_{2'2}/N - N_2 n_1 P_{22'}/n, \\
N_1 + N_2 &= N, \\
\dot{n}_2 &= n_1 w' - n_2 P'_{21} - n_2 N_1 P_{2'2}/N + N_2 n_1 P_{22'}/n, \\
n_1 + n_2 &= n,
\end{aligned}
\qquad (1)
$$

where

N_i = population of state i,

n_i = population of state i',

P_{ij} = transition probability $i \to j$,

P'_{ij} = transition probability $i' \to j'$,

$P_{22'}$ = probability for $N \to n$ energy transfer,

$P_{2'2}$ = probability for $n \to N$ energy transfer, and

w, w' = pumping probabilities.

The fast relaxation processes within the Nd system prevent the Nd to Cr backtransfer; therefore we can set $p_{2'2}=0$. In steady state conditions, as given by the sample under illumination from a steady source of light, eqs. (1) give, with the assumption that $N_1 \approx N$ and $n_1 \approx n$:

$$N_1 w - N_2 P_{21} - N_2 P_{22'} = 0$$
$$n_1 w' - n_2 P'_{21} + N_2 P_{22'} = 0, \qquad (2)$$

or

$$N_2 = Nw/(w + P_2) \approx Nw\tau, \qquad (3)$$

$$n_2 = N_2 P_{22'}/(w' + P'_{21}) + nw'/(w' + P'_{21}) \approx N_2(P_{22'}/P'_{21}) + n(w'/P'_{21})$$
$$= Nw\tau(P_{22'}/P'_{21}) + nw'\tau' = Nw\tau\tau' P_{22'} + nw'\tau', \qquad (4)$$

where

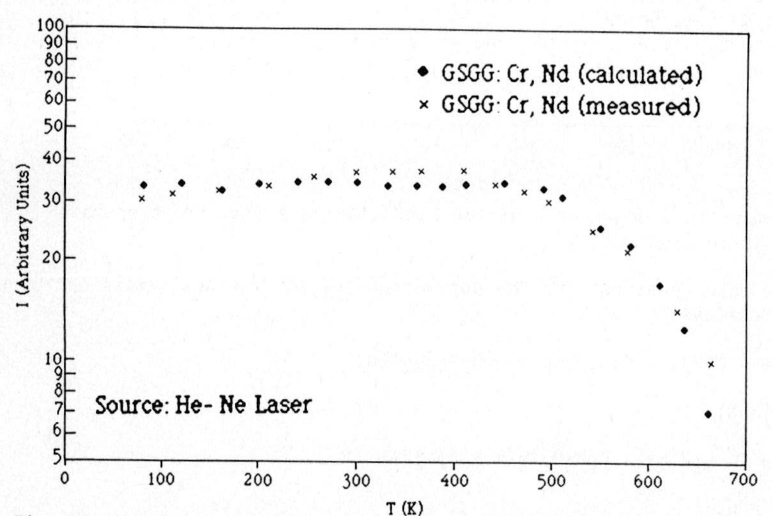

Fig. 10. Match between the calculated and measured integrated emission intensities of Nd in GSGG: Cr, Nd

$P_2 = P_{21} + P_{22'}$,

τ = lifetime of state 2 = $P_2^{-1} = (P_{21} + P_{22'})^{-1}$,

τ' = lifetime of state 2' = $(P'_{21})^{-1}$.

The integrated intensity of the Nd emission (in number of photons) in the GSGG: Nd sample, when pumping the Cr ions only, is given by (4) with w'=0. A match between the calculated and measured intensities of this emission is shown in Fig. 10. Without distorting this match, the maximum allowed radiative energy transfer rate in the sample used was found to be 10% at 77K, 5% at 300K and 2% at 580K.

ACKNOWLEDGEMENTS

The authors wish to thank T. Allik, A. Pinto, M. Shinn, and M. Kokta for useful discussions and the courtesy of lending us the samples used in these experiments.

REFERENCES

1. B. Struve, G. Huber, V.V. Laptev, I.A. Shcherbakov and E.V. Zharikov, Appl. Phys. B30, 117 (1983).
2. E.V. Zharikov, N.N. Il'ichev, S.P. Kalitin, V.V. Laptev, A.A. Malyutin, V.V. Osiko, V.G. Ostroumov, P.P. Pashinin, A.M. Prokhorov, V.A. Smirnov, A.F. Umyskov and I.A. Shcherbakov, Sov. J. Quantum Electron. 13, 9 (1983).
3. D. Pruss, G. Huber, A. Beimowski, V.V. Laptev, I.A. Shcherbakov and Y.V. Zharikov, Appl. Phys. B28, 355 (1982).
4. E.V. Zharikov, N.N. Il'ichev, V.V. Laptev, A.A. Malyutin, V.G. Ostroumov, P.P. Pashinin, A.S. Pimenov, V.A. Smirnov and I.A. Shcherbakov, Sov. J. Quantum Electron. 13(1), 82 (1983).
5. E.V. Zharikov, V.A. Zhitnyuk, G.M. Zverev, S.P. Kalitin, I.I. Kuratev, V.V. Laptev, A.M. Onishchenko, V.V. Osiko, V.A. Pashkov, A.S. Pimenov, A.M. Prokhorov, V.A. Smirnov, M.F. Stel'makh, A.V. Shestakov and I.A. Shcherbakov, Sov. J. Quantum Electron. 12(12), 1652 (1982).

Excited-State Absorption and Photoconductivity of GSGG:Cr and GSAG:Cr

L.J. Andrews

GTE Laboratories Incorporated, 40 Sylvan Road, Waltham, MA 02254, USA

1. Introduction

The garnets $Gd_3Sc_2(GaO_4)_3$ (GSGG) and $Gd_3Sc_2(AlO_4)_3$ (GSAG) have been shown to be promising host materials for tunable, near-infrared lasers when doped with Cr^{3+} [1], and to be the most efficient, flashlamp-pumped Nd^{3+} laser active media when codoped with Cr^{3+} and Nd^{3+} [2]. Both of these notable successes have as their origin vibronically broadened electronic transitions between ligand field states of the Cr^{3+} ion which are responsible for its excellent sensitizer and tunable laser properties. In addition to these transitions associated with the ground state, there is a set of excited state absorption (ESA) transitions associated with the Cr^{3+} fluorescent level which must be understood in order to fully model the performance of both types of lasers. Recently, the ESA spectra of GSGG:Cr and GSAG:Cr were measured [3] and shown to consist principally of two forbidden ligand field transitions at 9,000 and 19,400 cm^{-1} and an electric dipole-allowed transition ($\sigma \sim 10^{-17}$ cm^2) at 32,000 cm^{-1}.

Fortuitously, the uv and visible ESA bands for the most part do not coincide with ground state pump bands and therefore do not excessively compete for pump photons, but they are a source of thermal loading for the crystal. The infrared ESA transition was found to be extraordinarily broad, to the point of truncating the long wavelength tuning range of the Cr^{3+} laser. Its peak cross-section is a factor of 30 smaller than the peak Nd^{3+} fluorescence cross-section at 1.06 µm in GSGG [4], so the infrared band has negligible influence on codoped laser operation.

In this paper, two topics related to the ESA of the garnets will be addressed. First, it will be shown that a single configuration coordinate model can adequately replicate the ground state absorption and fluorescence between ligand field levels, but that the same model fails by a wide margin to explain the ESA bandwidths. It is argued that a degenerate Jahn-Teller active vibration in the upper quartet states must be taken into account in order to understand the ESA bandwidths. Second, the existence of an intense uv ESA band of charge-transfer (or bound to conduction band) origin, could lead to sequential two-photon ionization of Cr^{3+}. Accordingly, photoconductivity measurements were carried out on GSGG:Cr and GSAG:Cr, but instead of the expected quadratic power dependence, these measurements revealed a very large linear photoconductivity in GSGG:Cr which will be described.

2. Configurational Coordinate Model

2.1 Single a_{1g} Coordinate

It is possible to construct a set of single coordinate potential curves which replicate accurately the positions and approximately the bandwidths of electronic transitions between the Cr^{3+} orbitally nondegenerate ground state $^4A_{2g}$ and the orbitally triply degenerate upper states $^4T_{2g}$, $^4T_{1g}a$, and $^4T_{1g}b$ [5]. In the simple model used here to construct such potential curves, the total state energy is taken to be the sum of the ligand field electronic energy and a harmonic potential energy as shown in Fig. 1:

$$E_{total} = E_{el}(\Delta r) + 3\ M\omega^2\ \Delta r^2. \tag{1}$$

The form of the harmonic potential is appropriate for an a_{1g} vibration of the CrO_6 cluster where M is the mass of one oxygen, $\hbar\omega$ is the vibrational energy and Δr is the Cr-O internuclear distance. The variation in state electronic energy is calculated from the Tanabe-Sugano matrices (1x1 for $^4T_{2g}$ and 2x2 for $^4T_1a,b$) under the approximation that Dq varies as Δr^{-5} and B is independent of Dq. This functional dependence is correct only for the totally symmetric a_{1g} mode. The vibrational energy $\hbar\omega$ was varied to produce the best fit to spectra, which occurs for $\hbar\omega$ = 365 cm^{-1}, and therefore bears no direct relationship to the spectroscopic value of the a_{1g} energy which has yet to be determined for GSGG:Cr. Figure 1 shows the resulting potential curves for the Cr^{3+} quartet states, and the vertical arrows indicate the three ground state ab-

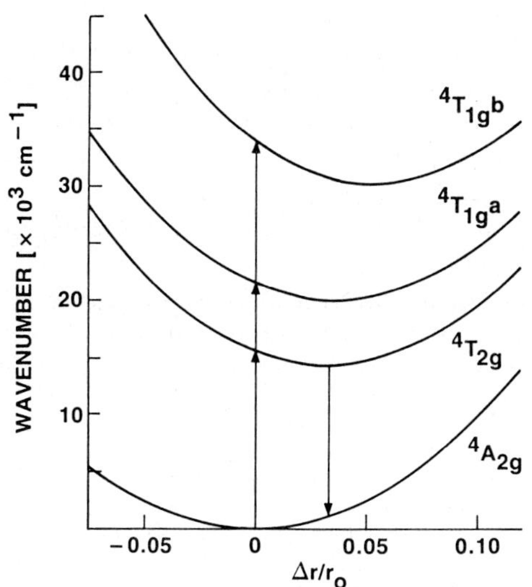

Figure 1. Single configuration coordinate potential energy curves for the four quartet states of a d^3 ion constructed using ligand field parameters appropriate for GSGG:Cr, Dq=1562 cm^{-1}, B=594.9 cm^{-1} and hω=365 cm^{-1}

sorptions and the fluorescence. In the semiclassical approximation, the lengths of these arrows determine the first moments m_1 of the bandshapes as shown in (2) and the second moments can be calculated from (3):

$$m_1 = \Omega_o + S\hbar\omega , \tag{2}$$

$$m_2 = S(\hbar\omega)^2 \coth (\hbar\omega/2kT) . \tag{3}$$

These moments in turn determine a Gaussian representation of the bandshape as shown in (4):

$$G(\bar{\nu}) = (\sqrt{2\pi}\, m_2)^{-1} \exp\left[-(\bar{\nu}-m_1)^2/2m_2^2\right] . \tag{4}$$

Figure 2 shows the absorption of GSGG:Cr and the fit generated from the potential curves in Fig. 1. It is clear that the Stokes shift and the $^4T_{2g}$ – $^4A_{2g}$ transition bandwidths are reasonably well reproduced by this model, but that the $^4T_{1g}a$ bandwidth is somewhat broader than predicted and the $^4T_{1g}b$ band is shown to lie above the band edge of the GSGG host crystal. The moments of these transitions are collected in Table 1.

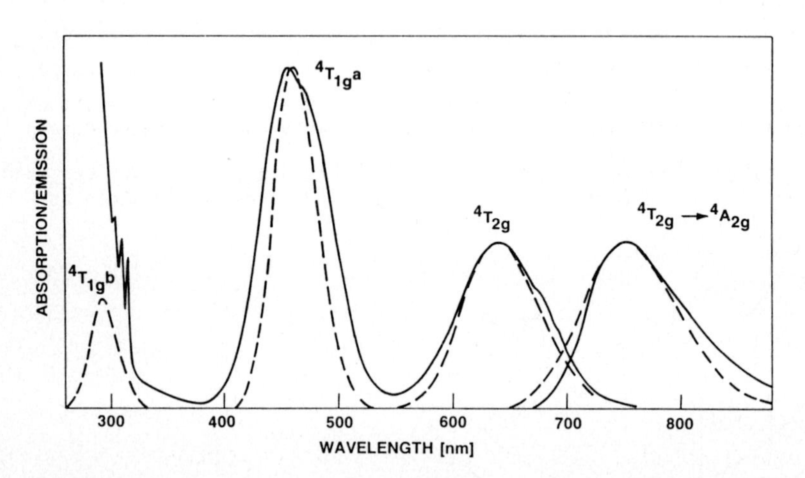

Figure 2. GSGG:Cr ground state absorption and fluorescence spectra (solid lines) and spectra calculated from Fig. 1 potential curves (dashed lines)

The predicted ESA bands are shown in Fig. 3 where they are compared with Gaussian fits to the observed bands as reported in [3]. Figure 3 clearly illustrates the deficiency of a single coordinate model in predicting ESA bandwidths and, to a lesser extent, ESA band positions. The calculated values for $2m_2^{1/2}$ are only 6 and 43% of the measured values for the infrared and visible ESA bands, respectively, and the calculated values for m_1 are 64 and 85% of the peak positions. In the next section, it is argued that coupling to a degenerate vibration through the Jahn-Teller effect is the likely origin of this poor performance.

Table 1. Bandwidths and calculated second moments of GSGG:Cr fluorescence, ground state absorption, and excited state absorption (peak positions in parenthesis); the observed bandwidths are the fullwidths at exp(-0.5) peak intensity

	Bandwidth	$2m_2^{1/2}$
$^4T_{2g} \rightarrow {}^4A_{2g}$	1474 cm^{-1} (13,280 cm^{-1})	1476 cm^{-1} (13,280)
$^4T_{2g} \leftarrow {}^4A_{2g}$	1940 (15,625)	1640
$^4T_{1g}a \leftarrow {}^4A_{2g}$	2760 (21,650)	1800
$^4T_{1g}b \leftarrow {}^4A_{2g}$		2870 (34,150)
$^4T_{1g}a \leftarrow {}^4T_{2g}$	3,200 (9,000)	190 (5,800)
$^4T_{1g}b \leftarrow {}^4T_{2g}$	2,600 (19,400)	1110 (16,400)

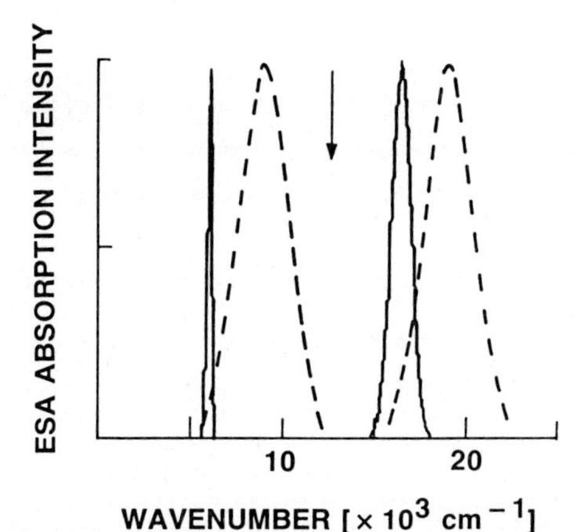

Figure 3. GSGG:Cr calculated ESA spectrum (solid lines) and Gaussian fit to observed spectrum (dashed lines)

2.2 The a_{1g} and e_g Model

Several years ago, high-resolution, low-temperature spectra of the Cr^{3+} ion in O_h and pseudo-O_h symmetries appeared from several groups [6-8], and these studies all demonstrated that the $^4T_{2g}$ state undergoes linear coupling to both a_{1g} and e_g vibrations. This phenomenon is a manifestation of the Jahn-Teller effect and causes a distortion of the $^4T_{2g}$ potential surface along three coordinates, a_{1g} and the Θ and ε components of e_g. If for the moment we neglect the a_{1g} coupling, then e_g alone causes the $^4T_{2g}$ to split into three orbital

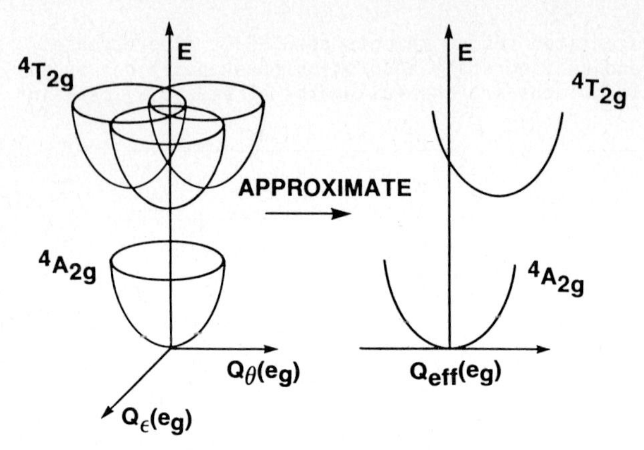

Figure 4. Schematic representation of the $T_{2g} \times e_g$ Jahn-Teller distortion and its approximate mapping onto an e_{eff} coordinate

components, symmetrically displaced in e_Θ and e_ε space as shown in Fig. 4. If we note that the observed progressional intensities of the e_g mode usually conform to a Perkarian distribution characteristic of a nondegenerate mode, we can approximately map the e_Θ and e_ε components onto an effective mode labeled e_{eff} also shown in Fig. 4. This will allow us to make an argument for the origins of the ESA bandwidths while remaining in a three-dimensional space, i.e., a_{1g}, e_{eff} and energy.

Consider the circumstance that the $^4T_{2g}$ state couples only to e_{eff} whereas the ^4T_1a state couples only to a_{1g}. In this situation, a set of potential curves such as shown in Fig. 5 can be drawn. The essential feature of Fig. 5 is that while the upper states are equally displaced from the ground state, their displacement is along different coordinates and so they are also strongly displaced with respect to each other. This leads to a broad ESA transition between them. Of course, a single coordinate model cannot accommodate this complexity and must predict a narrow ESA transition. The success of the simpler model in dealing with transitions involving the ground state derives from the circumstance that no account is taken of the fact that the identity of the coupled coordinate is state dependent.

While it is not realistic to assume that a state will couple with only one vibrational coordinate, it is quite likely that coupling strengths to various coordinates will be state dependent, leading to differing distortions along the available modes. By an extension of the diagram of Fig. 5 it is clear that this situation will also cause broadening beyond that predicted from a single coordinate model.

This qualitative sketch of the proposed broadening mechanism of ESA transitions in Cr^{3+} can be made quantitative through high-resolution spectroscopic studies of the various quartet states which seeks to identify both the symmetry and coupling strengths of the active modes. While traditional one-photon spectroscopy may be adequate for $^4T_{2g}$, the recently developed high resolution

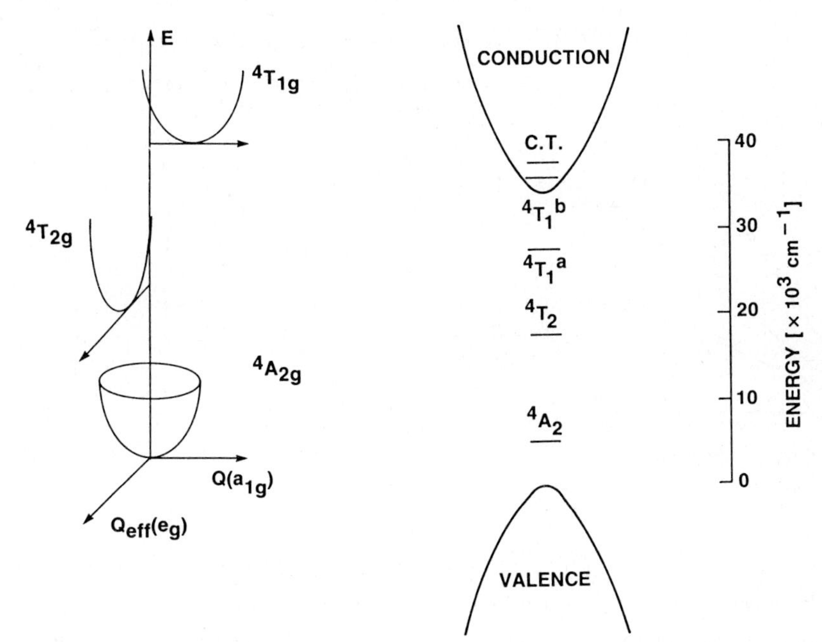

Figure 5. Illustrative double configuration coordinate diagram showing $^4T_{2g}$ coupling to e_{eff} and $^4T_{1g}a$ coupling to a_{1g}

Figure 6. Speculative position of the Cr^{3+} state energies within the GSGG bandgap showing the Cr^{3+} uv ESA band within the host conduction band

two-photon spectroscopy [9] seems particularly well suited for the higher quartet states, ^4T_1a, b, which rarely show resolved vibronic progressions in one-photon spectroscopy.

3. Photoconductivity

The bandgaps for both GSGG and GSAG lie at approximately 32,000 cm^{-1} and the Cr^{3+} ground state is expected to lie somewhere within the gap. If we arbitrarily locate it low in the gap, as shown in Fig. 6, then it is still possible that the energy of the broad uv transition uncovered in the ESA spectrum will lie near or in the conduction band as illustrated in Fig. 6. Should this situation prevail, then charge created by this uv transition could have appreciable mobility and cause detectable photoconductivity. Thus subbandgap radiation could populate the conduction band by a sequential two-photon process involving the metastable $^4T_{2g}$ state,and with this model in mind, photoconductivity measurements were made on GSGG:Cr and GSAG:Cr.

For the photoconductivity measurements, gold electrodes were deposited directly onto polished 0.2 or 1.0 mm thick samples, and either an Ar^+ laser or a 150 watt Xe arc and a f/4 double monochromator was used for excitation. The results for excitation with Ar^+ ion laser lines are shown in Fig. 7 and three features of these data are to be noted. First, the photoconductivity has a

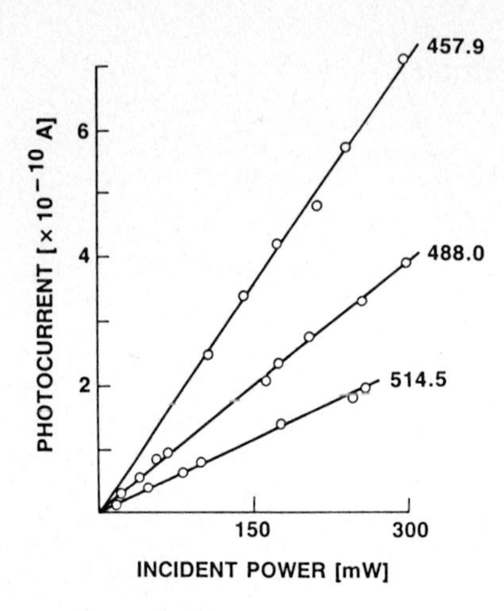

Figure 7. Photocurrent vs incident power for GSGG:Cr and Ar^+ laser excitation

linear power dependence for all the lines over a variation in power of a factor of ten. More extensive measurements not included in Fig. 7 show that this linearity persists as the power is lowered another factor of ten. Second, the wavelength dependence of the photoconductivity is a poor match to ground state absorption cross-section of Cr^{3+} in GSGG, so the effect apparently does not arise from Cr^{3+} absorption. Third, the linear photoconductivity is very large, up to 1 namp can be measured with no apparent crystal damage or onset of irreversibility. Typical currents for photoconducting insulators are usually far less [10].

The magnitude of the photoconductivity made it possible to easily dispense with laser excitation and use a lamp/monochromator to measure the photoconductivity excitation spectrum, and this is shown in Fig. 8. The onset of photoconductivity occurs just below 600 nm and places the ground state of the photoionizing center about in the middle of the gap. The photoconductivity rises monotonically toward shorter wavelengths except for an inflection near 450 nm

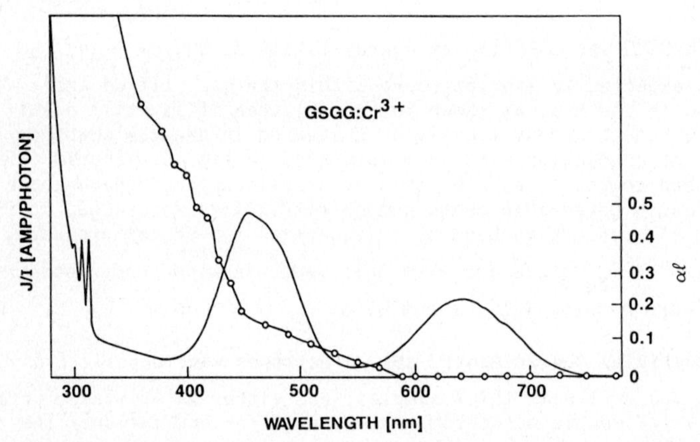

Figure 8. Photocurrent excitation spectrum (amps/incident photon) and ground state absorption spectrum for GSGG:Cr

which appears to be associated with Cr^{3+} inner filtering and reaches a peak near 36,000 cm^{-1} (not shown in Fig. 8). This spectrum bears no resemblance to the Cr^{3+} absorption, confirming the Ar^+ laser data, and therefore must be caused by another center not obvious in absorption.

Similar measurements were performed with GSAG:Cr, but with quite different results. The photoconductivity in GSAG:Cr was 10^3 less than in GSGG:Cr under Ar^+ laser excitation. With our current photoconductivity cell, it was just possible to measure the photoconductivity and demonstrate that there is a wavelength dependence, the current increases by roughly a factor of four over the range 514.5 to 457.9 nm.

We speculate that the large difference in photoconductivity between GSAG:Cr and GSGG:Cr may be related to the photoionization of color centers associated with oxide vacancies in the latter material. At the GSGG melting point, Ga_2O_3 (or some other gallium oxide species) has nonnegligible vapor pressure and can be lost from the melt, thereby causing the formation of nonstoichiometric, oxide-deficient crystals. In the case of GSAG, this process is mitigated due to the lower volatility of Al_2O_3.

Finally, we mention that because of the large linear photoconductivity present in GSGG:Cr, it was not possible to detect two-photon ionization. More sensitive measurements on GSAG:Cr, however, may have a chance to uncover this effect.

Acknowledgements

It is a pleasure to thank Prof. R. Bartram (U. Connecticut) for helpful discussion of the Jahn-Teller effect and in particular for suggesting the transformation shown in Fig. 4.

1. B. Struve, G. Huber, V.V. Laptev, I.A. Shcherbakov, and E.V. Zharikov: Appl. Phys. B 30, 117 (1985); B. Struve and G. Huber: J. Appl. Phys. 57, 45 (1985).

2. D. Pruss, G. Huber, A. Beimowski, V.V. Laptev, I.A. Shcherbakov, and Y.V. Zharikov: Appl. Phys. B 28, 355 (1982); E. Reed: IEEE J. Quant. Elec. QE-21, 1625 (1985).

3. L.J. Andrews, S.M. Hitelman, M. Kokta, and D. Gabbe: J. Chem. Phys. 84, 5229 (1986).

4. J.A. Caird and M.D. Shinn: 1985 Ann. Rpt. Lawrence Livermore Laboratory, p. 6-77.

5. W.H. Fonger and C.W. Struck: Phys. Rev. B 11, 3251 (1975).

6. R.W. Schwartz: Inorg. Chem. 15, 2817 (1976).

7. H.U. Gudel and T.R. Snellgrove: Inorg. Chem. 17, 1617 (1978).

8. R.B. Wilson and E.I. Solomon: Inorg. Chem. 17, 1729 (1978).

9. D.S. McClure: Tech. Digest Am. Opt. Soc. Topical Meeting Tunable Solid State Lasers, Zigzag, OR, June, 1986, WA1-1.

10. C. Pedrini, F. Rogemond and D.S. McClure: J. Appl. Phys. 59, 1196 (1986).

The Fluorescence Behaviour of Chromium-Doped Oxide Glasses

D.B. Hollis[1], *S. Parke*[1], *and M.J. Payne*[2]

[1]Department of Ceramics, Glasses and Polymers, University of Sheffield, Northumberland Road, Sheffield, S1O 2TZ, UK
[2]Royal Signals and Radar Establishment, St. Andrew's Road, Great Malvern, Worcs, WR14 3PS, UK

1. Introduction

Crystals have been used to make vibronic broad-band lasers since the late 1970s. They include chromium-beryl (Alexandrite) and chromium-G.S.G.G. (Gadolinium Scandium Gallium Garnet).

However, glasses have several potential advantages over crystals. They are cheaper to make, can be produced in larger pieces, and can be cast into unusual shapes such as slab, disc, or rod. They are also non-stoichiometric. Hence a large range of compositions and properties is possible. Also, their disordered structure gives lower ligand fields than in crystals. This allows different electric field parameters at the dopant site, and hence different fluorescence behaviour.

Glasses have been studied previously as possible laser hosts [1,2]. However, only a restricted range of samples has been tested. Therefore a more detailed study of how such parameters as decay lifetime and spectral profile of the fluorescence vary with the glass composition, preparation procedure, and content of impurities was undertaken. Results for silicate glasses are reported here. A few phosphate, borosilicate and fluoride types are included for comparison.

2. The structure of glass

In a crystal, a regular lattice of ions provides sites well defined in their symmetry, and position within the unit cell for the dopant ions. However, glasses are essentially random structures in which small regions (up to 500 Å) may have ordered structure. A network former such as $[SiO_4]^{=}$ or $[PO_4]^{\equiv}$ allows existence of rings, chains, or three-dimensional frameworks. Within these, and to some extent disrupting them, lie the "modifier" cations such as alkali, alkaline earth. Network stabilisers such as aluminium or boron serve to randomize the system further, and tighten up the network to prevent onset of crystallization after melting. Refiners such as Ce_2O_3 or As_2O_3 release oxygen during melting, and hence sweep out gas bubbles, but absorb the remaining oxygen again in the cooling period just prior to pouring and casting. This clarifies the glass. Finally, one finds impurities such as iron and hydroxyl in most silicate glasses. The furnace atmosphere can be oxidizing (and hence donate oxygen to the glass) or reducing (and hence rob oxygen from the glass). The valence states of dopants, or coordinations of modifiers may be affected as a result.

3. Preparation of samples

Four main types of glass were made. The"crown glasses" were analogues of the soda-lime-silica glass. Their formulae in molar percentage were

either 70 SiO_2 20 R″O 10 R'_2O
or 70 SiO_2 10 R″O 20 R'_2O
where R was Li, Na, K
and R″ was Mg Ca, Sr Ba, Pb, Zn, Cd.
A few glasses with other than 70% SiO_2 were prepared.

The batches were made from quartz sand plus carbonates of the alkali and alkaline earth metals. The stabilizer was an extra 2% of Al_2O_3, and the refiner was 0.5% of As_2O_3. The dopant was 0.04% of Cr_2O_3. All batches were pre-mixed until patchiness of colour, and lumps within the mix, were eliminated. They were then fired in 300 g weights in mullite crucibles in gas-fired furnaces at 1450 ±10°C for 7 hours. They were then poured into steel moulds, and annealed at 560°C for 1 hour, then slow cooled for about 18 hours at 30°C per hour.

The "lead crystal" glasses were made according to the formula 12.5% K_2O, 12.5% PbO, 75% SiO_2. The stabilizer was 2% B_2O_3. The refiner was As_2O_3. Additives were NaF, NaCl, and the batch materials were sometimes in the form of hydroxides, sometimes in the form of acid powders (such as H_3BO_3). They were mixed and melted as for the crown glasses.

The phosphate glasses were prepared from $CaH_4(PO_4)_2.H_2O$ to which P_2O_5 or B_2O_3 powders were added. A pre-sinter at 800°C for 5 hours followed by a melt at 1100°C for 1 hour was carried out in alumina crucibles in an electric furnace under air.

The fluoride glass was made in the form of fluoride powders ground under dry air to less than 150 μ sieve size. The items were mixed in a plastic tumbling vessel, then placed in a platinum crucible. This was placed in a silica liner, and heated under dry nitrogen. The schedule used was: demoisturize at 200°C for 1 hour; pre-sinter at 300°C for 1 hour then 600°C for 1 hour; sinter at 800°C for 1 hour; melt at 900°C for 1 hour; cool slowly to 670°C (just above crystallization temperature); then rapidly cool in air to 300°C. Finally anneal at 270°C for ½ hour, then slow cool.

The borosilicate glasses had formulae which were similar to the soda-lime-silica series, but the alkaline earth was replaced by B_2O_3, and the stabilizer was Al_2O_3. No refiners were used. The soda and boron were introduced into the batch either as Na_2CO_3 + H_3BO_3 or as $Na_2B_4O_7$. The alumina was in most cases introduced as $Al(OH)_3$, but Al_2O_3 was used directly for some melts. The reasons for these choices are explained during discussion of the results. These glasses were all melted in a gas fired furnace at 1520 ±10°C and annealed at 600°C. Most of them were melted in neutral atmosphere; a few of them were melted in a reducing atmosphere ($PO_2 \leqslant 10^{-4}$ atm).

The structures of some of these glasses have been well determined in previous studies. Fluoride glasses are chain types, in which the modifier cations lie in high coordinated sites, and the dopant ions and former ions both form XF_6 type octahedra [10,11].

For the metaphosphate glasses, the modifier cations, including in this case the chromium ions, lie in essentially octahedral interstices between chains of connected PO_4 tetrahedra [9, page 462ff]. The aluminophosphates and borophosphates and the silicate types have an altogether more complex far less well understood structure, in which the dopant sites are less clearly defined.

4. Experimental systems

The ultraviolet-visible spectra were recorded on a Perkin Elmer 330 Spectrometer, and the densities of the glasses were found by the Archimedes method. From these data, and a knowledge of the glass compositions, the absorption coefficients at 450 nm (the $^4T_1$ (F) band) and at 650 nm (the $^4T_2$ (F) band) were calculated. Estimates of the percentage of hexavalent chromium present were made from these spectra.

The fluorescence spectral profile was obtained using 200 mW C.W. excitation at 488 nm from a Spectra-Physics 2000 laser. An E.G. & G. Reticon type RL1024S self-scanned silicon diode array, which cut off on the long wave side at about 1.1 μ, detected the fluorescence spectrum. The output of the array was displayed, uncorrected for diode response, on an oscilloscope. Filtering of the excitation out of the fluorescent radiation was achieved with an OY4 filter.

The decay times were measured for long excitation pulses. However, detection was by a single element BPX65 PIN silicon diode with a low-noise preamplifier. Although decay traces could be photographed from the oscilloscope, a Brookdeal boxcar detector system triggered by the pulsed laser system was used to output a much less noisy trace to a pen chart recorder. Each boxcar sample was of width 2 μs.

To obtain long laser pulses, a Liconix 4050 He-Cd laser operating at 3 mW C.W. at 442 nm was used. The beam was electro-optic chopped to give "square wave" modulation of the radiation at 1 KHz. The rise time of the electro-optic chopper was less than 1 μs.

5. Results

Table 1 shows the effect of different modifiers. The modifiers open out the structure of the glass such that large modifiers K, Ba fill large vacancies in the structure, and accept high coordinations. As a result, for such glasses, a considerable proportion of the dopant ends up in small vacancies (usually tetrahedral) and in a high valence state (Cr^{6+}) that has a strong ultraviolet absorption. If small modifiers are used, such as Na, Ca, then the reverse tends to happen. The modifiers fit into smaller vacancies at lower coordinations, and a higher proportion of the dopant goes into the glass as Cr^{3+} in octahedral sites. For the borosilicate glasses, the boron introduces large network irregularities which make all the sites for the dopant ions considerably distorted from octahedral. The effect of boron on the initial fluorescence intensity per unit dopant concentration (I_0/conc) is clear in Table 1. The effect of making a glass in a different phase field of the ternary than the other SLS types is manifest by KBaSi/5*, whose initial intensity is weak. (Essentially, as the melt cools, "domains" form, around which and from which the rest of the glass grows rapidly. Whether the glass develops a quartz, tridymite, cristobalite, or other structure is dependant not only on the percentage of former, but also on the type of modifier.) An "all-former" glass SD54/4 has the largest initial intensity. (It is believed that the glass consists entirely of the compound $Na_2SI_2O_5$ [3]).A case intermediate between the all former glasses and the large modifier glasses is given by Pb3a/6.

The effect of refining, and reducing all the chromium to Cr^{3+} by addition of As_2O_3 is seen in SLS2/4 and SLS2/8 whose yields are higher than those of corresponding unrefined glasses. SLS2/4 and 2/8 were made

Table 1 - Cr^{3+} in Silicate Glasses: Effect of Glass Composition

Glass type (Mol%)			No.	Co-dopant (ions/cc x 10^{19})	Fluorescence $\tau'(\mu s^{\frac{1}{2}})$	I_0/conc
SiO_2 70	CaO 10	Na_2O 20	SLS1a/6	None	3.7	0.22
SiO_2 70	CaO 10	Na_2O 20	SLS2/4	5.0 As_2O_3	3.1	0.34
SiO_2 70	CaO 10	K_2O 20	SLS2/8	5.0 As_2O_3	3.1	0.25
SiO_2 70	B_2O_3 10	Na_2O 20	BS5/4	None	2.9	0.08
SiO_2 70	BaO 10	K_2O 20	KBaSi/5*	None	–	0.05
SiO_2 75	PbO 12.5	K_2O 12.5	Pb3a/6	None	2.2	0.28
SiO_2 67		Na_2O 33	SDS4/4	None	1.7	0.45

I_0 = Initial intensity; Conc. = Dopant ion concentration [1-2 x 10^{19} ions/cc]

For these glasses $I_t = I_o \exp(-k'\sqrt{t})$. In Table 1, $\tau' = 1/k'$.

under oxidizing and reducing atmospheres respectively. It is believed that 2/4 gives higher yield because the iron impurity has been oxidized to Fe^{3+} by the furnace, thus removing an unwanted near infra-red absorption band.

The approximately similar decay times for nearly all the silicate glasses indicate that they have similar structures. The ultraviolet-visible d-d absorption spectra vary little from glass to glass except in the ratio of the Cr^{6+} absorption to those of the Cr^{3+} ions. This confirms the above conclusions, and also shows that there is little variation in ligand field at the chromium site amongst these silicate glasses.

The KBaSi/5*, Pb3a/6 and the SDS4/4 have structures very different from the rest. Consequently the decay times and fluorescence initial intensities are also completely different.

Table 2 shows the effect of co-doping and additives on the fluorescence lifetimes. The changes with additive are small. However, the second group of glasses (SLS1b/5 Series) has decay times considerably shorter than those of the first group. The SLS1b/5 glasses were prepared in an atmosphere much more reducing than those of the first group. The effect of a reducing atmosphere is to rob the melt of oxygen. This would increase the number of oxygen vacancies throughout the structure. These vacancies act as energy traps which absorb the energy from the chromium ions, and later decay non-radiatively. It would also reduce any iron impurity to Fe^{2+} which, in octahedral coordination has a strong absorption in the same near infra-red region as the fluorescence of the chromium.

As previously seen from Table 1, Table 2 shows that addition of arsenic trioxide to the melts improves the initial fluorescence intensity of the glass, by ensuring that all the chromium dopant takes the trivalent form.

Table 3 shows the effect of using hydroxides or hydrated compounds in the batch materials for a series of borosilicate glasses. The fluorescent

Table 2 - Cr^{3+} in silicate glasses: Effect of Co-doping and Additives [Cr^{3+} doped to 2 x 10^{19} ions/cc]

Glass Type (Mol%)	No.	Co-dopant	Additive	Fluorescence $\tau'(\mu s^{\frac{1}{2}})$	I_0 conc.
$70SiO_2$	SLS11a/6	Fe_2O_3	None	3.5	0.18
10CaO	SLS11b/5	MnO_2	None	3.7	0.15
$20Na_2O$	SLS2/4	None	As_2O_3 2.0%	3.2	0.35
	SLS1a/6	None	None	3.7	0.23
$70SiO_2$	SLS1b/5*1	None	None	2.0	0.23
10CaO	SLS1b/5*2	None	NaF 0.2%	1.8	0.25
$20Na_2O$	SLS1b/5*3	None	As_2O_3 0.05%	2.0	0.25
	SLS1b/5*4	None	As_2O_3 0.67%	2.3	0.22

Table 3 - Cr^{3+} in $70SiO_2.10B_2O_3.20Na_2O$ glasses: Effect of hydroxyl, and iron.
Cr^{3+} doped to approximately 2 x 10^{19} ions/cc. Dehybor = $Na_2B_4O_7$.
All except BS8/4 had neutral furnace atmosphere.
Glass BS8/4 had reducing furnace atmosphere.

No.	Co-dopant x 10^{19}	Additive (mol.%)	B_2O_3 compound and Na compound	Fluorescence $\tau'(\mu s^{\frac{1}{2}})$	Intensity I_0/Dopant Conc.
BS5/3	None	1.0 $Al(OH)_3$	Dehybor	2.5	0.026
BS5/4	None	1.0 $Al(OH)_3$	H_3BO_3 + Na_2CO_3	2.9	0.024
BS5/9	None	1.0 $Al(OH)_3$	H_3BO_3 + Na_2CO_3	2.9	0.030
BS5/7	2.0 Fe_2O_3	1.0 $Al(OH)_3$	Dehybor	2.8	0.015
BS5/8	2.0 Fe_2O_3	1.0 $Al(OH)_3$	H_3BO_3 + Na_2CO_3	2.8	0.006
BS4/7	None	0.3 $Al(OH)_3$	Dehybor	2.5	0.001
BS8/4	None	3.0 Al_2O_3	Dehybor	3.2	0.030
BS6/10	None	0.6 NaCl	Dehybor	3.0	0.136

decay times are all similar, the variations being attributable to experimental error bars of about 0.5 $\mu s^{\frac{1}{2}}$ for the long pulse excitation results. However, the fluorescent intensities vary considerably with additive or co-dopant.

Alumina has to be added in some form to prevent phase separation in these glasses. The effect of too little alumina is clear from BS4/7. Addition of even 1.0% $Al(OH)_3$ - which forms 0.5% Al_2O_3 in the melt - as in the BS5/-series improves the yield about 20-30 times. The determination of which phase accepts the chromium in the phase-separated glasses has not been carried out, but its dramatic effect on yield would be worthy of study. Addition of more than 1.0% Al_2O_3 - as in BS8/4 - has little further effect.

Hydroxyl is often cited as the cause of poor fluorescence yield, yet these samples do not show a dependence of yield on use of hydroxides instead of oxides. Dehybor ($Na_2B_4O_7$) serves just as well as H_3BO_3 + Na_2CO_3, and $Al(OH)_3$ serves just as well as Al_2O_3. For that reason, one has to consider other factors, not only in the borosilicate glasses, but in all other types of glass as possible causes of weak fluorescence and non-exponential fluorescence decay.

Table 3 shows one further entry - that for BS6/10 - which has NaCl instead of alumina compounds, and a rather higher percentage of dehybor than the others. Consequently the phase-separation region of the phase diagram is avoided, without paying the price of distorting and tightening the network structure by adding alumina. BS6/10 has by far the highest yield of any of the borosilicate glasses, and is comparable with glasses of the soda-lime-silica type.

It should be said that no refiners or reductants were needed for the borosilicate glasses. They were generally free of "seed" (fine gas bubbles). Their ultraviolet cutoffs were the same for doped or undoped specimens, so showing that very little Cr^{6+} was present.

Addition of iron has a disastrous effect on these glasses. The existence of an Fe^{2+}/Fe^{3+} mixture imparts to even the undoped glasses a faint green colour, and the Fe^{2+} absorption in the near infra-red kills the fluorescence.

Table 4 shows a comparison of glasses with different formers. The fluoride glass with a total of $[ZrF_4] + [LaF_3] + [AlF_3] = 60\%$ former gives the weakest fluorescence. This glass has chains of ZrF_6 octahedra in which

Table 4 - Cr^{3+} in glasses with different formers: Effect of changing type and molar proportion of former.
Cr^{3+} concentration is 0.5 to 2 x 10^{19} ions/cc. Items starred (*) give almost true exponential decay.

No. and Network Type	Formula (mol.%)	Fluorescence Parameters $\tau(\mu s)$	$\tau'(\mu s^{1/2})$	Intensity Per Unit Dopant Concentration
Chain glasses				
F003/7	$50ZrF_4.4.5LaF_3.5.5AlF_3$ $20BaF_2.20NaF$	- (believed about 1 μs)	-	Hardly detectable
CaP/2	$50P_2O_5.50CaO$	11.5*	(3.6)	0.59
CaBP/2	$45P_2O_5.20BaO.35CaO$	16.0*	(5.0)	1.20
Network glasses				
KBaSi/5*	$70SiO_2.10BaO.20K_2O$	-	-	0.05
SLS2/8	$70SiO_2.10CaO.20K_2O$	-	3.1	0.25
SLS1a/6	$70SiO_2.10CaO.20Na_2O$	-	3.7	0.22
BS6/10	$70SiO_2.10BO.20Na_2O$	-	3.0	0.14
Pb3a/6	$70SiO_2.12.5PbO.12.5K_2O$	-	2.2	0.28
"Former only" glasses - see text				
CaP/2		11.5*	(3.6)	0.59
SDS4/4	$100Na_2Si_2O_5$	22-24*	(2.5)	0.45
Crystals				
GSGG5	$Gd_3Sc_2Ga_3O_{12}$ (garnet)	70*	-	1.46

the fluorine anions are hexagonal close packed, and the cations go into octahedral or higher coordinated sites. A near relative (in structure), the phosphate type, is represented by CaP/2 and CaBP/2. These glasses are again chain structured, but epr spectrum indicates that the chromium in the fluoride glass is in a strongly axially distorted site similar to that seen in silicate glasses [4], whereas the chromium in the phosphate glasses is in cubic to slightly rhombically distorted sites [5,6,9 page 584]. The signal at g = 2 [9, page 584] in both the fluoride and phosphate glasses is believed to be due to Cr-Cr exchange coupling. The differing types of distortion of the octahedron affect not only the emission cross-section but also the fluorescence lifetime of the chromium. Both glasses show all Cr^{3+}, no Cr^{6+}, so the lower total time-integrated yield of the fluoride glass has to be attributed to a stronger competition from non-radiative decay paths than is present in the phosphate glass. One possible reason is that the very low ligand field of the fluoride glass results in larger than usual Jahn-Teller distortions of the octahedra by the Cr^{3+} when in its $^4T_2$ (F) first excited electronic state. Possible flattening of the configuration coordinate-energy parabolae may also occur in the fluoride glass. The result would be easier thermal assisted intersystem crossing from $^4T_2$ (F) to $^4A_2$ (F) without fluorescent emission. This surmise is partly supported by fluorescence spectroscopy at different temperatures of the fluoride glass. The coefficient of variation of fluorescence yield with temperature was 2 or 3 times stronger than for a typical soda-lime-silica type over the range 80 K-300 K.

Considering now the silicates, the strong distortion due to the boron in the borosilicate glass clearly reduces the fluorescence yield. The reason why the silicate glass SLS2/8 should give more fluorescent yield than KBaSi/5* is apparent from the phase diagrams for these systems [7]. The K-Ba-Si oxide system is in a different part of the phase diagram than for all the other "SLS" type glasses, including Pb3a/6. Hence for KBaSi/5*, the distortions of the octahedra coordinating the Cr^{3+} could be very different from those to be expected for the other silicate glasses.

The longest time decays and strongest initial fluorescence intensity were obtained for the "formers only" type glasses CaP/2, CaBP/2 and SDS4/4. These are of compositions known from the relevant phase diagrams [7] to form compounds, not eutectics. Further work on "formers only" glasses would be interesting.

The initial fluorescence strength in all the glasses seems to be related to the field strength of the formers. The fluoride glass has the lowest ligand field, and the lowest output. The silicates are intermediate, both in field and in fluorescence intensity. The sodium disilicate and phosphate types are highest in both parameters. The fluorescence decay time generally increases with increasing fluorescence intensity, which suggests that the time-integrated fluorescence quantum efficiency varies greatly over the range of glasses considered, and is highest in the "formers only" types.

The increase in both initial intensity and decay time could be explained by either a reduction in the range of site types in the formers only glasses, and a large spread of site types in such as the fluoride or KBaSi type, or by reduction of the number of oxygen vacancies or unsatisfied oxygen bonds in the structure by removal of alkali and alkaline earth ions, or by prevention of formation of chromium-chromium aggregates in the less viscous, alkali-rich parts of the melt.

Comparison with the garnet crystal shows that the non-radiative loss, not the absorption and emission cross-sections, is the important limitation to fluorescence efficiency in glasses.

6. The origin of the non-exponential fluorescence decay in glasses. An expression $I_t = I_0 \exp(-K'\sqrt{t})$ almost fits the fluorescence decay typical of most of these glasses. It conforms to a dipole-dipole coupled process between the chromium ions and dipoles randomly distributed throughout the glass, or else to a random distribution of site distortions [8]. The form of the decay equations for long pulse excitation is $I_t = I_0(1 + K'\sqrt{t}) \exp(-K'\sqrt{t})$.

Our glasses show a decay which is not purely exponential, nor does it conform exactly to the above expressions. This would suggest that either the range of distortions of the octahedra is limited, or that the dipoles involved in any non-radiative process are distributed in a non-random manner about the chromium sites.

The "formers only" glasses are interesting in that their decays are much closer to exponential than the mixed modifier-former types. This could be due to a greater degree of regularity in these glasses - i.e. an approach towards a disordered mixed crystal system rather than a true random glass.

7. Conclusions

Our results indicate that the glasses have finite initial fluorescence intensities which compare favourably with those at present obtainable from garnets. However, the garnets have much longer fluorescence decay times of about 70 to 100 μs, whereas the glasses have decay times of 1 to 10 μs.

The short fluorescence times of the glasses, and their strongly non-exponential behaviour, is due to a range of distortions of the CrO_6 octahedra and/or dipole-dipole coupling of energy from the chromium ions to such non-radiative centres as oxygen vacancies.

The modifiers and formers of the glass structure have a strong effect on the fluorescence parameters. So also do impurities which absorb in the spectral region 650-1150 nm in which the chromium fluoresces. However, the effect of hydroxy, and of minor dopants and stabilizers/refiners is small except in cases where the Cr^{6+}/Cr^{3+} ratio is affected.

Future study of "formers only" glasses should give further improvement in the fluorescence parameters, since these appear to have the highest initial fluorescence intensities and the longest decay times after pulsed excitation.

Acknowledgements

This work has been funded by the Ministry of Defence under the auspices of the RSRE, Malvern, Worcestershire, for the past three years and is expected to continue into the future. It is the first attempt in the UK to develop a glass host tunable laser.

We would like to thank Professor M. Cable for many useful suggestions, and Mr J. Smedley for help in preparation of glasses. (Both are from Sheffield University.) Many thanks also to Mr H. Evans and Mr N. Lowde of

RSRE, Malvern, for preparing experimental equipment and assisting with testing of glass samples.

Finally, D.B. Hollis gratefully acknowledges the effort made at short notice by the D.C.V.D., and travel and finance departments of the Ministry of Defence and Sheffield University to make funding available for the conference.

References

1. L.J. Andrews, A. Lempicki, B. McCollum: J. Chem. Phys. 74(10) 5526-5538 (1981).

2. E.Snitzer: Bull Amer.Ceram. Soc. 52 516-525 (1973).

3. C.H.L. Goodman: Phys. Chem. Glasses 26(1) 1-10 (Feb. 1985).

4. (a) F. Durville, B. Champagnon, G. Boulon: J. Phys. Chem. Sol. 46(6) 701-707 (1985).
   (b) F. Durville, et al.: Phys. Chem. Glasses 25(5) 126-133 (Oct. 1984).
   (c) Gan Fuxi, Deng He, Liu Huiming: J. Non. cryst. Sol. 52, 135-141 (1982).

5. D.L. Griscom: J. Non. cryst. Sol. 40 211-272 (1980).

6. N.R. Yafaev: Proc. 4th All Union Conf. USSR, "The Glassy State", Leningrad, USSR, 1965. Vol.7 "The Structure of Glass" ed. E.A. Porai-Koshits. Translated Consultants Bureau of New York, 1966. pp. 65-67. Application of the Electron Paramagnetic Resonance of Transition-Metal Ions to the Study of the Structure of Glass.

7. Phase Diagrams for Ceramists. American Ceramic Society Volumes I-V inclusive (1964, 1967, 1975, 1981, 1983). [Relevant phase diagrams are scattered throughout these volumes.]

8. D.L. Huber: In Coherence and Energy Transfer in Glasses. eds. P. Fleury, B. Golding. (Plenum Press, 1984). pp. 125-141: Theoretical Studies of Optical Energy Transfer in Glasses and Other Disorderd Systems.

9. J. Wong, C.A. Angell: Glass Structure by Spectroscopy. (Marcel Dekker Inc., New York 1976).

10. (a) H. Inoue, H. Hasegawa, I. Yasui: Phys. Chem. Glasses 26(3) 74-81 (June 1985).
    (b) Y. Kawamoto: ibid 25(3) 88-91 (June 1984).
    (c) R.M. Almeida, J.D. Mackenzie: ibid 26(5) 189-191 (October 1985).

11. P.W. France, S.F. Carter, J.M. Parker: Phys. Chem. Glasses 27(1) 32-41 (Feb. 1986).

Part III

Crystal Growth

Chromium-Doped Garnet Hosts: Crystal Chemistry Development and Properties

E.V. Zharikov

General Physics Institute of USSR Academy of Sciences,
38 Vavilov Street, SU-117942 Moscow, USSR

1. INTRODUCTION

The single crystals of chromium-doped rare-earth garnets are used now in tunable solid state lasers, as well as in the lasers operating on a number of fixed wavelengths [1], and they compete successfully with other known active media of solid state lasers. Already initial results on tunable laser action of Cr^{3+} ions in gadolinium-scandium-gallium garnet (GSGG) and other rare-earth gallium garnets (REGG) obtained by both selective excitation [2], and lamp pumping [3] indicates the great promise of these crystals. Lasers based on single crystals of chromium-doped REGG with neodymium demonstrate higher efficiencies than singly doped neodymium lasers [4-6]. The slope efficiency of Nd, Cr-GSGG-laser achieves 6-7% for both long pulse and free-running mode operation [7,8], and 3% for Q-switch operation [7,9]. The calculated maximum efficiency of lamp pumped Q-switched Nd, Cr-GSGG-laser is 11% [10].

The present paper considers some crystal chemistry aspects of chromium-doped REGG, and thermomechanical properties of these crystals. This class of crystals belongs to the garnet structure [11], which due to its three different sublattices (Fig. 1), enables the introduction of more than 50% of all elements from the Periodic Table. On this account one can change the garnet composition within extensive limits in order to purposefully change their physical properties with exceeding care. Different aspects of development of garnet crystals with given properties are discussed in [1,13-17].

The Cr^{3+} ions, which have wide and intense absorption bands in the visible, have attracted researcher attention for a long time. The first solid state laser, the ruby laser, developed more than 25 years ago [18], used Cr^{3+} as active ions in sapphire. New interest in chromium ions was aroused by successful operation of the alexandrite tunable laser [19]. During the investigation of a class of REGG single crystals [20] the connection between intense broad-band luminescence of Cr^{3+} ions and efficient Cr^{3+}-sensitization of Nd^{3+} ions was demonstrated. On this account one can develop both the tunable Cr^{3+}-laser and laser with Cr^{3+}-sensitization based on the same garnet host [1].

2. CRYSTAL CHEMISTRY DEVELOPMENT

Chromium ions were repeatedly tried to be used as the sensitizer in various laser hosts including yttrium aluminum garnet crystals doped with neodymium (YAG-Nd) [21]. The chromium dopant provides a much more efficient absorption within the emission spectral range of lamp pump sources. But the inversion population increase of upper laser level of activator was quite small in all cases. The cause of this is the low

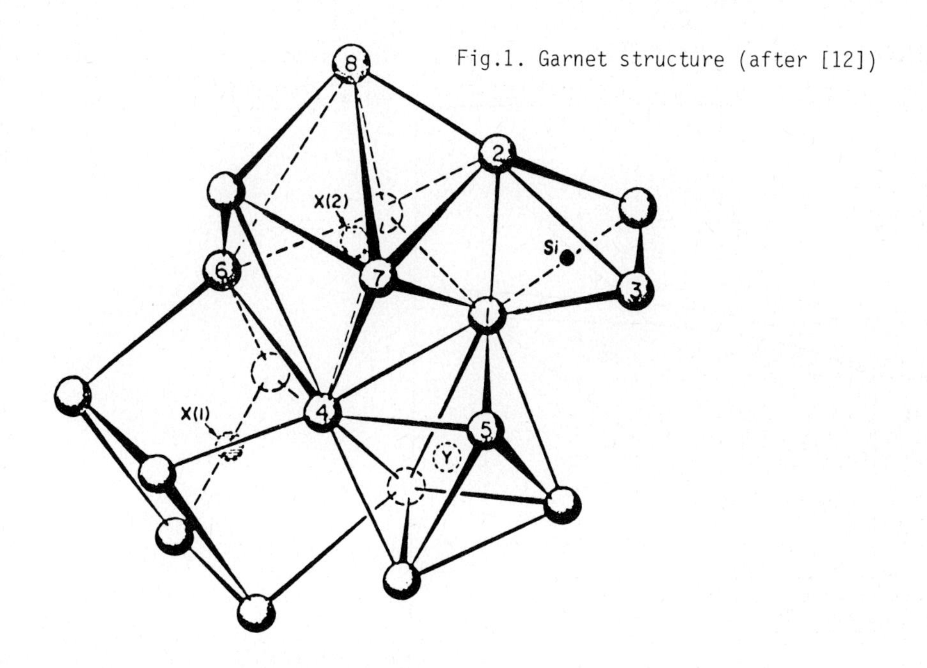

Fig.1. Garnet structure (after [12])

efficiency of energy transfer from Cr^{3+} ions to Nd^{3+}.

In [22] it was concluded that high rates of energy transfer of $Cr^{3+} \rightarrow Nd^{3+}$ and large sensitizing efficiencies are reached when such transfer initiates from the $^4T_2$ state of Cr^{3+}, and, not from ^2E state, as in YAG. The transfer efficiency was connected with the value of the energy gap ΔE between ^2E and $^4T_2$ chromium levels. When ΔE is small, the $^4T_2$ state is thermalized with the ^2E state, which results in the intense broad-band luminescence from $^4T_2$ state [21]. This allows realization of efficient energy transfer to Nd^{3+} ions, as well as to other rare-earth ions. Such ions are Pr^{3+}, Sm^{3+}, Dy^{3+}, Ho^{3+}, Er^{3+}, Tm^{3+}, and Yb^{3+} [23].

The practical opportunity of realizing the needed ΔE value is reached by the fact that the position of $^4T_2$ chromium ion level sharply changes versus crystal field strength (Dq) while the position of ^2E level remains practically constant. Fig. 2 schematically shows the positions of the Cr^{3+} ion levels versus crystal field strength [24]. From this

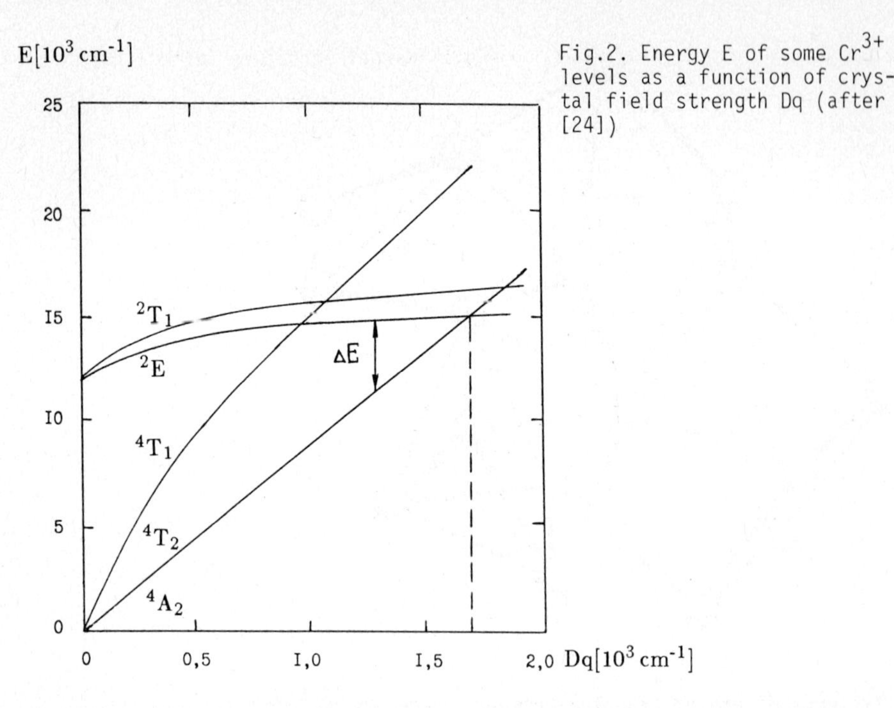

Fig.2. Energy E of some Cr^{3+} levels as a function of crystal field strength Dq (after [24])

figure one can see that the crystal field strength must be about 1500 cm^{-1} in order to reach the small values of ΔE.

The energy gap ΔE linearly depends on crystal field strength for a family of rare-earth garnets created by trivalent cations (Fig. 3).

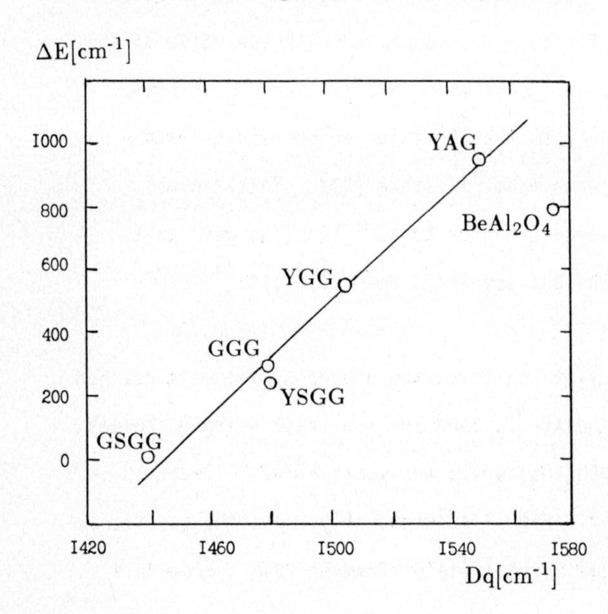

Fig.3. Energy gap ΔE as a function of crystal field strength Dq

Realization of conditions of relatively weak crystal field strength for Cr^{3+} ions and small values of ΔE in garnet crystals can be provided by variation of cation-oxygen distance in the octahedral sites of the structure (d_6).

After establishment of relations between Dq, ΔE, d_6, and a [1] an important dependence $\Delta E = f(a)$ was obtained. This dependence allows estimation of the fitness of garnet hosts for providing efficient Cr^{3+} sensitization when only the lattice parameter is known. From Fig. 4 one can see that REGG have relatively large lattice parameters and are suitable hosts, providing a small value ΔE and allowing wide variations.

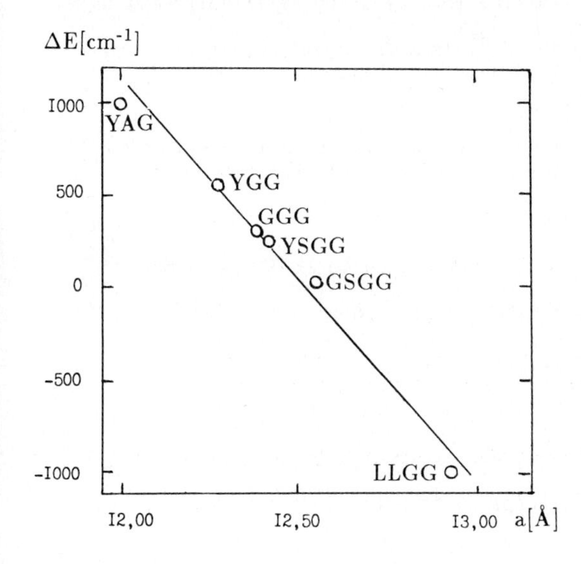

Fig.4. Energy gap ΔE as a function of lattice parameter a of REGG

The growth problems contribute the main difficulties to the development of crystals with dopants, including the garnet crystals. Because of growth difficulties, obtaining the necessary crystals for practical use of sufficient size and required optical quality is not always successful. One of the principal crystal growth requirements is to obtain a value of dopant distribution coefficients close to unity. The distribution coefficient is defined as the ratio of the concentration in the solid to the concentration in the melt ($K_i = C_i^s / C_i^l$).

Let us consider the problem of two practical important dopants--Nd^{3+} and Cr^{3+} ions. It is known [25] that the distribution coefficient of Nd^{3+} in YAG is only 0.18. This is a disadvantage of these crystals because of the change of melt composition during the growth process. In order to obtain crystals with minimal compositional variation, the growth of YAG-Nd is usually carried out with convex interface. Typical pulling rates for these crystals are 0.75 - 1 mm/h [26]. Any increase of dopant distribution coefficient (with its approach to unity) allows, in principle, proportional increase of the pulling rate of the crystal [27].

The rare-earth ions with screened f-shell are under a relatively small influence of the host crystal field, and therefore their distribution coefficients are determined mainly by geometrical factors. Brandle and Barns [28] gave an empirical relation of the dependence of neodymium distribution coefficient in garnets upon lattice parameter:

$$K_{Nd} = 1.029\ a - 12.164 .$$

From this relation it follows that garnets with lattice parameters greater than 12.40 Å will function as a host with high Nd^{3+} distribution coefficient ($K_{Nd} > 0.6$). Such distribution coefficients can be provided in REGG, and will provide an opportunity to obtain large crystals with homogeneous dopant distribution from the melt by using relatively high pulling rates [29]. By choosing the Gd^{3+} ions for dodecahedral sites in REGG, it was found that in case of gadolinium-gallium garnet (GGG) K_{Nd} is 0.62 [30], and in case of GSGG K_{Nd} is 0.75 [28, 30].

In the case of Cr^{3+} ions, the electronic configuration affects their incorporation in the crystal and their distribution coefficient during growth process. The Cr^{3+} ions in garnets only occupy the octahedral sites [31]. This is connected with their tendency to sp^3d^2-hybridize (octahedral type of bonding) due to the specific character of Cr^{3+} electronic structure. Therefore the Cr^{3+} ions enter very readily into crystalline structures containing octahedral sites and demonstrate anomalously high effective distribution coefficients (K_{Cr}) in garnets. So, in YAG $K_{Cr} = 2.4$ [26], whereas in GGG it is of a greater value: $K_{Cr} = 3.3$ [29]. That is the sharp discrepancy to octahedral ion sizes correlation:

$$r_6(Cr^{3+}) = 0.615\ \text{Å};\ r_6(Al^{3+}) = 0.53\ \text{Å};\ r_6(G^{3+}) = 0.62\ \text{Å}\ [32].$$

An anomalously high K_{Cr} value results in an essentially inhomogeneous chromium distribution along the crystal and in the appearance of intensive growth striation. In the case of GGG these inhomogeneities are accompanied with specific "fan-beam"-defects [29, 30].

The specific character of Cr^{3+} entry to Ga^{3+} octahedral sites in GGG and its conjunction with its electronic configuration are explained [13] in terms of the relative stability of the octahedral clusters $[CrO_6]^{9-}$ and $[GaO_6]^{9-}$. It is assumed that the cations forming most stable clusters (in other words, the clusters with maximum energy gain origin) enter in growing crystal more preferably. Appearing cation competition reveals distribution coefficients during the growth.

Hence there is an important conclusion. In order to make the Cr^{3+} distribution coefficient close to unity, it is necessary to fill the octahedral sites in a garnet structure with a host trivalent cation where octahedral cluster $[MeO_6]^{9-}$ stability is close to that of $[CrO_6]^{9-}$. Maintaining similar Me^{3+} and Cr^{3+} ionic radii is desirable. An examination from a viewpoint of the outer electron shell structure carried out in terms of molecular orbital theory [13] reveals that Sc^{3+} ions (electron configuration $3d^0$) form clusters of somewhat greater stability than $Cr^{3+}(3d^3)$ ions, while $Ga^{3+}(3d^{10})$ clusters are less stable as compared to $[ScO_6]^{9-}$ and $[CrO_6]^{9-}$ ones. Based on these reasons, Sc^{3+} ions were introduced into octahedral sites of garnet structure. Thus were developed the scandium-containing garnets (for instance GSGG, YSGG), whose Cr^{3+} distribution coefficients are close to unity.

The octahedral sites for these compositions are sufficiently large ($d_6 \approx 2.1$ Å). This allows the conditions of low crystal field for Cr^{3+} ions to be realized,which results in a small energy gap ΔE between $^4T_2$ and ^2E levels of chromium. (So, in case of GSGG, ΔE is close to zero [1, 33].) At the same time, REGG crystals permit sufficiently large lattice parameters and correspondingly high distribution coefficients of Nd^{3+} and other rare-earth ions.

Note, that the aluminum garnet features are unable to provide either of these conditions because the maximum lattice parameter in a line of Al-garnets is only 12.133 Å (for $Gd_3Al_5O_{12}$). On the average, the size of octahedral sites in Al-garnets is $d_6 \approx 1.94$ Å [34].

In the extensive family of germanium garnets, the means of isomorphic substitution of different kinds can realize a lattice parameter up to 13 Å and more [35, 36] and provide the distances d_6 in the structure up to 2.1-2.2 Å [35].

But Ge-garnets have a few essential disadvantages. These are: substantial evaporation of GeO_2 during the crystal growth, requiring growth of a majority of Ge-garnets by flux methods [35, 37]; and inevitable heterovalent isomorphism in the dodecahedral sites of the structure if the crystal is doped by rare-earth ions. The charge compensation occurring in this case may cause the appearance of different point defects and color centers in the crystals [14].

Of certain interest are the rare-earth scandium-aluminum garnets (RESAG), including gadolinium scandium-aluminum garnet (GSAG) [36, 38]. These garnets have relatively low crystal field for Cr^{3+} ions, as well as sufficiently high Nd^{3+} distribution coefficients. Moreover they have higher heat conductivity as compared with REGG [28]. But the problem of perfect crystal production seems not to be solved yet. GSAG crystals have one order of magnitude greater defect density than YAG and GSGG, and correspondingly lower onset of laser damage [36].

The technology of growing REGG crystals with different compositions is highly developed now [30, 39, 40]. In the case of GSGG there is no difficulty in keeping the flat interface during the growth process, in comparison with aluminum and scandium-aluminum garnets [28, 36]. Near-unity distribution coefficients of the dopants provide the opportunity to grow activated crystals with flat interface without the so-called "core" (see, for instance [41, 42]). This is an important advantage of these crystals from the viewpoint of growth, in comparison with other activated hosts including YAG-Nd crystals.

3. THE PROPERTIES OF GARNETS

Some practically important thermal and mechanical properties of the crystals can be calculated, which when interpreted in terms of material requirements, can be satisfied by choosing garnet composition. The necessary calculations were carried out in [15, 16, 43] based on garnet elastic constants from experimental data obtained from Mandelshtam-Brillouin light-scattering spectra. The corresponding relations are given in references cited above. Data on some mechanical and thermophysical properties of a number of garnet crystals are given in the table. The calculated values are in sufficiently good agreement with experimental data, as shown in Refs. [15, 16, 44]. It is possible through use of such calculations to determine the thermo-mechanical properties of crystals with garnet structure. Moreover the obtained data provide an opportunity to see how these properties evolve through the garnet line.

Table 1. Thermo-mechanical properties of garnets

Crystal	a [Å]	M [a.m.u]	B [GPa]	G [GPa]	E [GPa]	σ	θ [K]	κ [$W \cdot m^{-1} \cdot K^{-1}$]	C_p [$J \cdot g^{-1} \cdot K^{-1}$]	α [$cm^2 \cdot s^{-1}$]
YAlG	12.01	594	187	112	280	0.25	741	12.9	0.629	0.0450
CaGaGeG	12.25	669	165	71	186	0.31	565	6.5	0.627	0.0215
YGaG	12.29	807	175	92	235	0.28	584	8.6	0.514	0.0288
GdGaG	12.38	1012	168	88	225	0.28	512	7.4	0.427	0.0244
GdScGaG-ND,Cr C_{Nd} = 2.2at% C_{Cr} = 1.2at%	12.57	970	158	80	205	0.28	503	7.0	0.448	0.0240
LaNdLuGaG C_{Nd} = 10at%	12.93	1195	154	67	176	0.31	423	5.1	0.379	0.0183

a: lattice parameter; M: molecular weight; B: uniform expansion (compression) modulus; G: shear modulus; E: Young modulus; σ: Poisson coefficient; θ: Debye temperature; κ: thermal conductivity; C_p: heat capacity; α: thermal diffusivity

Some useful dependences, given below, simplify determination of such important characteristics, as Debye temperature θ, melting point T_m, thermal conductivity $\varkappa$, specific heat C_p, and thermal diffusivity α.

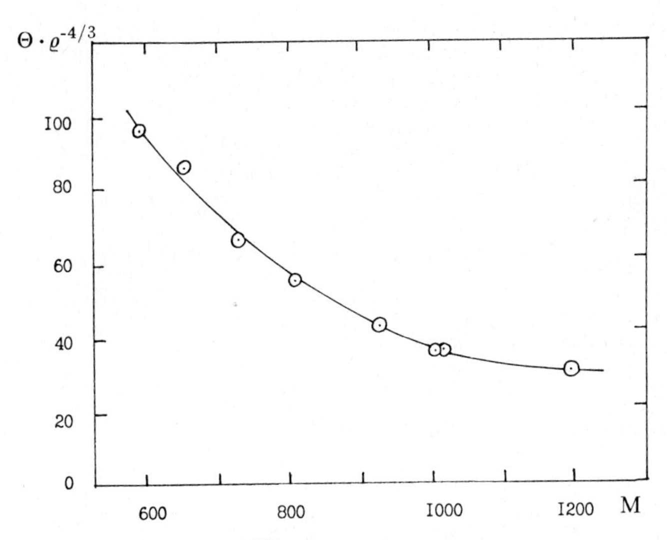

Fig.5. Function $\theta\rho^{-4/3}$ versus molecular weight M of garnets

Note that in the present paper the dependences are only given for nonmagnetic synthetic garnets created by trivalent cations. The analogous dependences for germanium and natural silicate garnets, as well as data for magnetic garnets, are given in [15, 16].

Fig. 5 shows the dependence of θ versus molecular weight and density of garnets. The analytical dependence may be obtained by the formula:

$$\theta = B \cdot a^{-4} \cdot M^{-0.39}$$

where B is the constant, $B = 1.84 \times 10^8$

a is lattice parameter (Å)
M is molecular weight.

The Debye temperature characterizing the crystal phonon spectrum permits prediction of some spectroscopic features of the host, including the opportunity to obtain lasing on rare-earth ion transitions with longer wavelength [15]. Moreover, the knowledge of θ allows one to calculate other important thermo-physical properties of the crystals.

Fig. 6 shows the dependence $T_m = f(\theta \sqrt{M})$. Fig. 7 demonstrates the dependence of thermal conductivity versus θ, and in Fig. 8 the dependence $C_p = f(\theta)$ is given. Fig. 9 depicts the inverse value of thermal diffusivity versus garnet molecular weight. The needed values of corresponding characteristics of crystals with garnet structure may be easily estimated from given drawings. The only data necessary for this purpose are lattice parameter and molecular weight.

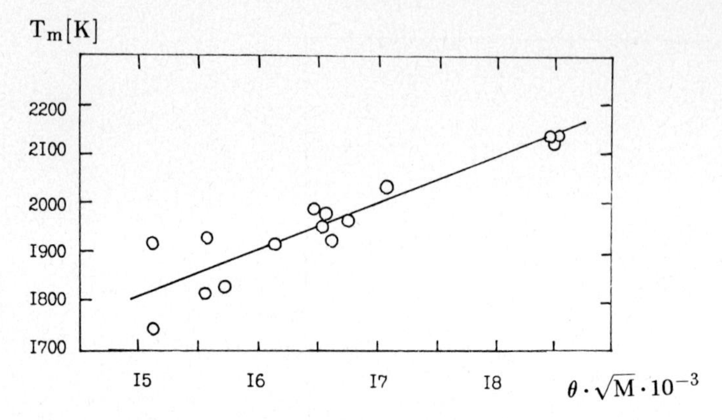

Fig.6. Melting point of garnets as a function of $\theta\sqrt{M}$

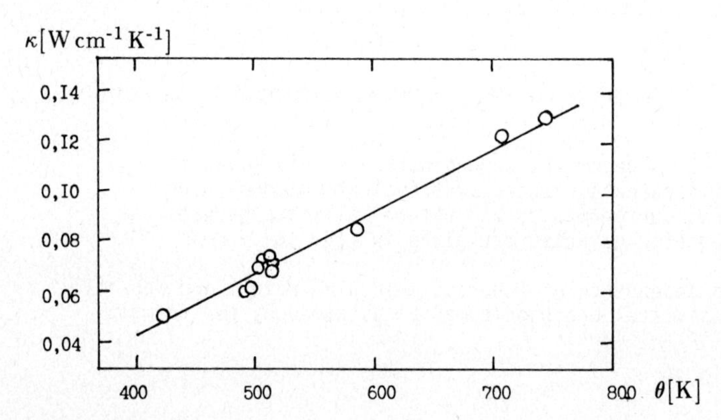

Fig.7. Dependence of thermal conductivity κ on Debye temperature

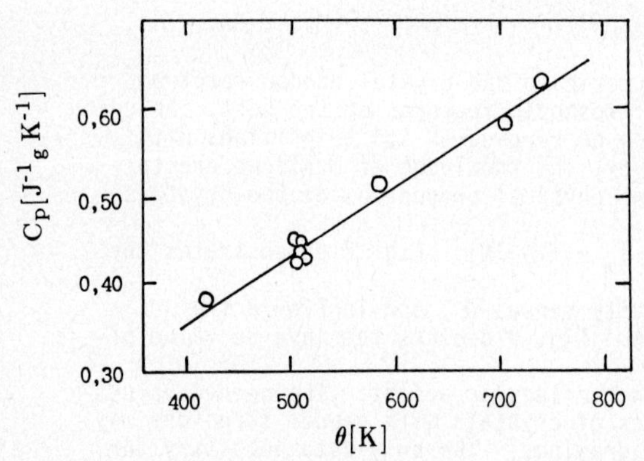

Fig.8. Heat capacity C_p of garnets as a function of θ

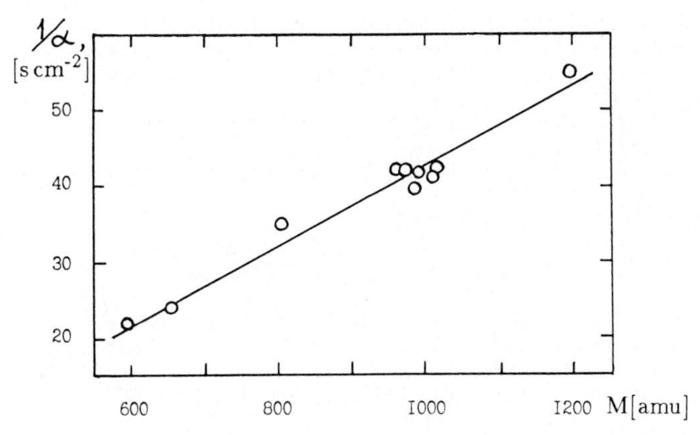

Fig.9. Inverse value of thermal diffusivity $1/\alpha$ versus garnet molecular weight M

REFERENCES

1. E. V. Zharikov, V. V. Osiko, A. M. Prokhorov, and I. A. Scherbakov: Bull. Acad. Sci. USSR, Phys. Ser. 48, 81-94 (1984).
2. B. Struve, G. Huber, V. V. Laptev, I. A. Scherbakov, and E. V. Zharikov: Appl. Phys. B 30, 117-120 (1983).
3. E. V. Zharikov, N. N. Il'ichev, S. P. Kalitin, V. V. Laptev, A. A. Malyutin, V. V. Osiko, V. G. Ostroumov, P. P. Pashinin, A. M. Prokhorov, V. A. Smirnov, A. F. Umyskov, and I. A. Scherbakov: Sov. J. Quantum Electron. 13, No. 9, 1274-1276 (1983).
4. E. V. Zharikov, N. N. Il'ichev, V. V. Laptev, A. A. Malyutin, V. G. Ostroumov, P. P. Pashinin, and I. A. Scherbakov: Sov. J. Quantum Electron. 12, No. 3, 338-341 (1982).
5. D. Pruss, G. Huber, A. Beimowski, V. V. Laptev, I. A. Scherbakov, and E. V. Zharikov: Appl. Phys. B 28, No. 4, 355-358 (1982).
6. E. V. Zharikov, V. A. Zhitnyuk, G. M. Zverev, S. P. Kalitin, I. I. Kuratev, V. V. Laptev, A. M. Onishchenko, V. V. Osiko, V. A. Pashkov, A. S. Pimenov, V. A. Smirnov, M. F. Stel'makh, A. V. Shestakov, and I. A. Scherbakov: Sov. J. Quantum Electron. 12, No. 12, 1652-1653 (1982).
7. A. V. Dobrovol'skii, A. S. Dorkin, M. B. Zhitkova, I. I. Kuratev, V. A. Lipatov, G. I. Kharkhova, M. Yu, Nikitin, V. A. Pashkov, N. S. Ustimenko, E. M. Shvom, and A. V. Shestakov: Bull. Acad. Sci. USSR, Phys. Ser. 48, 101-102 (1984).
8. J. A. Caird, W. F. Krupke, M. D. Shinn, L. K. Smith, and R. E. Wilder: In Conference on Lasers and Electro-Optics (CLEO) Tech. Digest, Paper THR 3, 232 (1985).
9. V. N. Grigir'ev, G. N. Egorov, E. V. Zharikov, V. A. Mikhailov, S. K. Pak, Yu. A. Pinskii, E. I. Shklovskii, and I. A. Scherbakov: General Physics Inst. Report No. 52, Moscow (1986).
10. A. D. Gondra, V. M. Gradov, E. V. Zharikov, Yu. N. Terent'ev, A. A. Scherbakov, and I. A. Scherbakov: General Physics Inst. Report No. 5, Moscow (1986).
11. G. Menzer: Z. Kristallogr, Bd. 69, No. 3, 300-396 (1928).
12. G. A. Novak, G. V. Gibbs: Amer. Mineralog 56, No. 5-6, 791-825 (1971).
13. E. V. Zharikov, V. V. Laptev, A. A. Mayer, and V. V. Osiko: Izv. AN SSSR, Sev. Neorgan. Mater. 20, No. 6, 984-990 (1984).

14. E. V. Zharikov, S. P. Kalitin, V. V. Laptev, A. A. Mayer, and V. V. Osiko: General Physics Inst. Report No. 135, Moscow (1984).
15. V. F. Kitaeva, E. V. Zharikov, and I. L. Chistyi: P. N. Lebedev Inst. Report No. 146, Moscow (1984).
16. V. F. Kitaeva, E. V. Zharikov, and I. L. Chistyi: Phys. Stat. Solidi (a) 92, 475-488 (1985).
17. E. V. Zharikov: Technical Digest, VI All-Union Conf. on Crystal Growth 1, 32 (Yerevan, USSR, 1985).
18. T. H. Maiman: Nature 187, 493-494 (1960).
19. J. C. Walling, O. G. Peterson, H. P. Jenssen, R. C. Morris and E. W. O'Dell: IEEE J. Quantum Electron. QE-16, 1302-1315 (1980).
20. E. V. Zharikov, S. V. Lavrishchev, V. V. Laptev, V. G. Ostroumov, Z. S. Saidov, V. A. Smirnov, and I. A. Scherbakov: Sov. J. Quantum Electron. 14, No. 3, 332-336 (1984).
21. Z. J. Kiss, R. C. Duncan: Appl. Phys. Lett. 5, No. 10, 200-202 (1964).
22. V. V. Osiko, A. M. Prokhorov, and I. A. Scherbakov: Izv. AN SSSR, Serv. Phys. 44, No. 8, 1690-1715 (1980).
23. E. V. Zharikov, S. P. Kalitin, V. V. Laptev, V. V. Osiko, A. M. Prokhorov, V. A. Smirnov, and I. A. Scherbakov: P. N. Lebedev Inst. Report No. 196, Moscow (1983).
24. Y. Tanabe, S. Sugano: J. Phys. Soc. Japan 9, No. 6, 753-779 (1954).
25. R. R. Monchamp: J. Cryst. Growth 11, No. 3, 310-312 (1971).
26. R. R. Monchamp: J. Solid State Chemistry 12, 201-206 (1975).
27. C. D. Brandle, J. C. Vanderleeden: IEEE J. Quantum Electron. QE-10, No. 1, 67-71 (1974).
28. C. D. Brandle, R. L. Barns: J. Cryst. Growth 20, No. 1, 1-5 (1973).
29. E. V. Zharikov, N. N. Il'ichev, V. V. Laptev, A. A. Malyutin, V. G. Ostroumov, P. P. Pashinin, A. S. Pimenov, V. A. Smirnov, and I. A. Scherbakov: Sov. J. Quantum Electron. 13, No. 1, 82-85 (1983).
30. E. V. Zharikov, S. P. Kalitin, V. V. Laptev, A. A. Mayer, V. V. Osiko, V. G. Ostroumov: Technical Digest II All-Union Conference, "The State and Development Prospects of the Single Crystals Production Methods," (Khar'kov, USSR, 1982) p. 57.
31. S. Geller: Z. Kristallogr, Bd. 125, No. 1, 1-47 (1967).
32. Ionic radii are given according to: R. D. Shannon: Acta Crystallogr. 32A, Ser. 1, 751-767 (1976).
33. B. Struve, G. Huber: Appl. Phys., B 36, 195-201 (1985).
34. F. Euler, J. A. Bruce: Acta Cryst. 19, No. 6, 971-978 (1965).
35. B. V. Mill, E. L. Belokoneva, M. A. Simonov, and N. V. Belov: In Problems of Crystalology," 3, (Moscow, 1982) pp. 161-179.
36. M. R. Kokta: Proc. First Intl. Conf. on Tunable Solid State Lasers, La Jolla, CA, June 1984. In Tunable Solid State Lasers, Eds. P. Hammerling, A. B. Budgor, and A. Pinto (Springer-Verlag, Berlin, 1985), pp. 105-114.
37. E. J. Sharp, J. E. Miller, D. J. Horowitz, A. Linz, V. Belruss: J. Appl. Phys. 45, No. 11, 4974-4979 (1974).
38. J. Drube, B. Struve and G. Huber: Opt. Commun. 50, 45-48 (1984).
39. F. J. Bruni: In Crystal Growth, Properties and Applications, C. J. M. Rooijmans, Ed., (Springer-Verlag, Berlin, 1978) 1, pp. 54-70.
40. D. Mateika: In Current Topics in Materials Science 11, ed. by E. Kaldis (Elsevier Science Publishers B.V., Amsterdam-New York-Oxford-Tokyo, 1984) pp. 141-239.

41. A. A. Kaminskii, V. V. Osiko, S. E. Sarkisov, M. I. Timoshechkin, E. V. Zharikov, J. Bohm, P. Reiche and D. Schultze: Phys. Stat. Sol. (a) 49, No. 2, 305-311 (1978).
42. Y. Kuwano: J. Cryst. Growth 47, 353-361 (1982).
43. E. V. Zharikov, A. S. Zolot'ko, V. F. Kitaeva, V. V. Laptev, V. V. Osiko, N. N. Sobolev, and I. A. Sychev: Sov. Phys. Solid State 25, 568-571 (1983).
44. W. F. Krupke, M. D. Shinn, J. E. Marion, J. A. Caird, and S. E. Stokowski: J. Opt. Soc. Am. B3, 102-114 (1986).

Crystal Growth of $ScBO_3:Cr^{3+}$ – A New Near-IR Tunable Laser Crystal

B.H.T. Chai, M. Long, R.C. Morris, and S.T. Lai

Electronic Materials and Devices Laboratory, Allied-Signal Corporation, P.O. Box 1021R, Morristown, NJ 07960, USA

1. INTRODUCTION

Since the discovery of the alexandrite room-temperature tunable laser [1], much effort has been expended in exploring new laser materials and several new room-temperature tunable laser hosts [2] have been discovered. In almost every case Cr^{3+} has been used as the active ion. More recently, there is excitement over the discovery of the $Al_2O_3:Ti^{3+}$ laser [3] because of its broader tuning range, simple energy-level structure and relatively large gain cross-section. But the short lifetime and poor absorption has made Ti^{3+} difficult to pump. Cr^{3+} doped tunable lasers generally have efficient absorption and adequate lifetime but the optical gain is proportionately lower. Fortunately, through understanding of the physics of Cr-doped tunable lasers, it is possible to manipulate the tunability range by varying the crystal field strength of the host crystal. It is also possible to select crystal hosts with low refractive indices as well as proper site distortions, so that the emission cross-section can be increased to a certain extent without significantly degrading the fluorescence lifetime.

We are currently engaged in survey efforts to grow or synthesize and evaluate various Cr-doped crystal hosts. It seems feasible to expect series of Cr-doped crystal hosts which will cover the wavelength range from 700 to 1000 nm with about 50 nm of useful tunability per crystal. The discovery of $ScBO_3:Cr^{3+}$ is one of the results of this effort.

2. ABUNDANCE OF SCANDIUM IN NATURE

One of the interesting consequences of Cr^{3+} tunable laser research has been the discovery of Sc sites in a variety of crystal hosts which are attractive for Cr doping. Since the Sc^{3+} ion in an octahedral site (0.75Å) is bigger than Ga^{3+} (0.62Å) and Al^{3+} (0.53Å), the crystal field for Cr^{3+} (0.62Å) is generally weaker and the fluorescence spectrum is red shifted into the near IR region.

As the interest in Sc bearing compounds increases, the demand for high-purity Sc_2O_3 for material synthesis and crystal growth research is also increasing. Since the current supply of Sc_2O_3 is very limited, the cost has increased dramatically.

In reality, Sc is relatively common in nature. The average earth crust abundance of Sc is 22 ppm [4] as compared to Y (35 ppm), Nd (44 ppm), Ga (17 ppm), In (0.2 ppm), Cu (58 ppm) and Au (0.002 ppm). One problem for Sc mining is the lack of enriched ore, since Sc is normally distributed in various rock-forming minerals, very similar to Ga. Fortunately, some Sc enrichment is found in association with U and Th ores and it can be extracted as by-product. Once the consumption is up and steady, the price of scandium should drop to a level comparable to that of yttrium and rare-earth elements.

3. CRYSTAL CHEMISTRY OF SCANDIUM BORATE

Early work by Levin et. al. [5] showed that $ScBO_3$ has a $CaCO_3$-calite-type structure. It has a triangular rhombohedral symmetry with a space group of $R\bar{3}c$. The structure is analogous to NaCl with Na and Cl ions replaced by Sc and BO_3 ions respectively and the unit cell distorted by compression along the triad axis. The distortion accommodates the planar triangular shape of the BO_3 group. The lattice constants are a = 4.759Å, c = 15.321Å.

Similar to calcite, $ScBO_3$ has a perfect cleavage along the plane $(10\bar{1}1)$. The crystal has a hardness of about 5 and a density of 3.45 gm/cm^3. Optically, $ScBO_3$ is uniaxial negative with ω = 1.872 and ε = 1.780. The birefringence (δ = 0.092) is about half that of calcite but still adequate for polarizer applications.

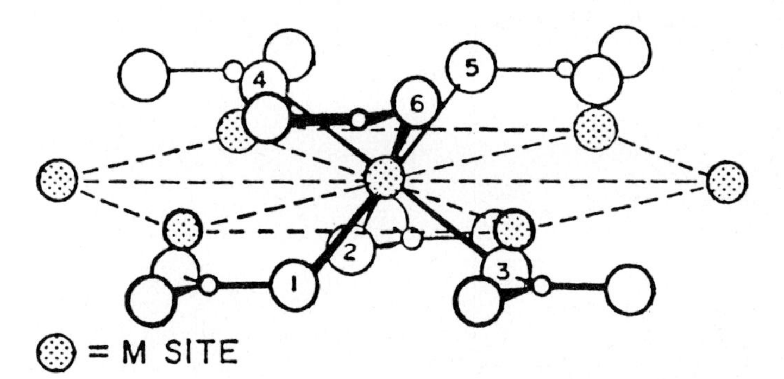

Fig. 1. The structure of $ScBO_3$. The oxygens surrounding the scandium are indicated; they form a near-perfect octahedron.

Structurally, $ScBO_3$ has a single Sc site (Fig. 1) and ScO_6 forms a near-perfect octahedron without distortion. The single site reduces complications for Cr fluorescence and the undistorted site contributes to the relatively long lifetime. The average Sc-O bond length is 2.13Å and the average B-O bond length is 1.35Å. The Sc cation size is 0.75Å as compared to the Cr ion (0.62Å). A large cation site reduces the crystal field strength and thus red-shifts the Cr fluorescence spectrum into the near IR range (peaked at 815 nm).

The calcite structure is just one of the three stable polymorphs of ABO_3 compounds where A = trivalent cation. Apparently, the calcite structure is stable with a cation size from 0.60Å to 0.85Å. Known trivalent cations which will form the calcite structure ABO_3 compounds include Lu, In, Sc, Ti, V, Mn, Cr and Fe. The heavy rare earth elements (from Sm to Lu) and Y with ionic sizes from 0.96Å to 0.85Å form stable borates in the vaterite structure whereas the light rare earth (from La to Nd) borates stabilize in the aragonite structure.

4. SYNTHESIS AND GROWTH OF SCANDIUM BORATE

Scandium borate can readily be synthesized by solid state reaction below 900°C [5]. Single crystals of $ScBO_3$ up to 1 mm in size were first successfully grown from a barium borate flux [6]. The melt was cooled slowly from 1170°C to 740°C at a rate of 20°C per day.

The phase diagram of the binary Sc_2O_3-B_2O_3 system (Fig. 2) was published by Levin [7] in 1967. It is characterized by the existence of a single binary compound, $ScBO_3$, and two immiscible liquids above 1526°C with over 66 mol.% B_2O_3. Scandium borate melts congruently at 1610°C and there are no solid-solid phase transitions. In view of these facts we decided to grow $ScBO_3$ crystal directly from the melt using two approaches.

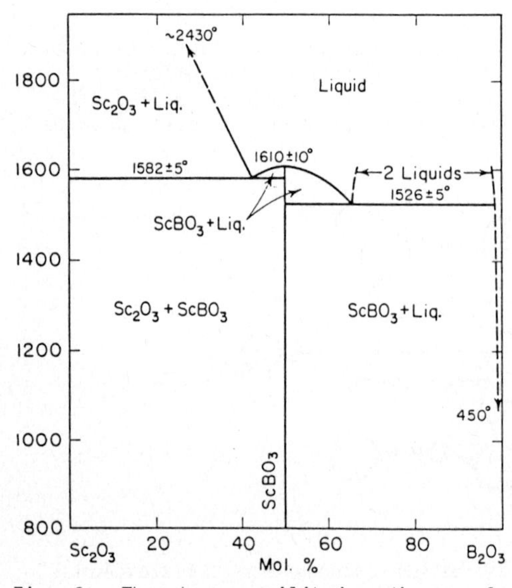

Fig. 2. The phase equilibrium diagram for the system Sc_2O_3 - B_2O_3 [7].

4.1 Kyropoulos Growth of $ScBO_3$

In order to obtain a single crystal seed, we elected to grow the first crystals by slow freezing from the melt. The starting materials included 99.9% Sc_2O_3 (Aesar) and 99.99% Sc_2O_3 (Research Chemicals) supplied by LLNL. High-purity anhydrous B_2O_3 glass was also used. A one inch diameter iridium crucible inside a 3-inch diameter iridium crucible was used to hold the melt. The crucible was heated by RF induction under N_2 + 0.8% O_2 atmosphere.

The melting point measured by a pyrometer was very close to the phase diagram result. We experienced loss of B_2O_3 by evaporation, as evidenced by white smoke and condensation. Therefore excess B_2O_3 (2 ~ 5%) was added to the melt to compensate the loss. By cooling the melt slowly (<10°C per hour), we were able to obtain clear crystals up to 10 mm in dimension which provided seeds for subsequent Kyropoulos and Czochralski growth runs.

4.2 Czochralski Growth of $ScBO_3$

With oriented single crystal seeds in hand, we proceeded to grow by direct pulling from the melt. The starting materials as well as the preparation procedures for the melt were as described in Section 4.1. Here we used a 2-inch Ir crucible inside a 3-inch Ir crucible to hold the melt. The growth apparatus is shown in Fig. 3.

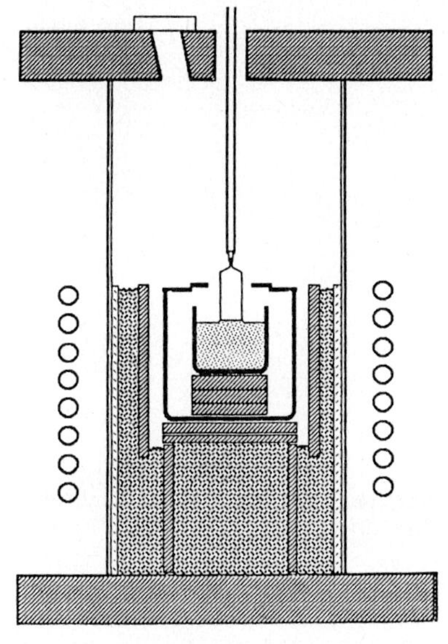

Fig. 3. Illustration of the Czochralski pulling arrangement for $ScBO_3$.

Again the melt was compensated for B_2O_3 loss. The crystal was pulled along the Y-axis so that both the σ and π polarizations were perpendicular to the growth axis. The pulling rate varied from 0.5 to 3 mm per hour, inversely proportional to Cr concentration. The rotation rate ranged from 20 to 40 rpm. Crystals of 15 mm diameter and up to 100 mm long have been successfully pulled.

5. DISCUSSION OF RESULTS

The main difficulty in the growth of $ScBO_3$ is the excessive evaporation and loss of B_2O_3 which changes the melt composition and limits both the total time and the fraction of melt available for the growth. Moreover, we also noticed loss of Sc_2O_3. Since Sc_2O_3 is highly refractory, we speculate that the loss was probably in the form of $ScBO_3$ vapor. An as-grown crystal still in the 2" crucible is shown in Fig. 4. We can see clearly the dendritic material deposited at the lid of the crucible. X-ray powder diffraction shows that it is the cubic form of Sc_2O_3, most likely formed by vapor deposition through the decomposition of $ScBO_3$ vapor.

The most prominent defects in $ScBO_3$ crystals are voids resulting from constitutional supercooling instability, starting at the center of the

Fig. 4. As-grown $ScBO_3$ crystal in a 2" Ir crucible. Notice the deposition of dendritic crystals of Sc_2O_3 at the lid of the crucible.

boule and progressively moving out until the entire boule cross-section is affected. Unstable growth is more serious in highly doped melts. We are able to minimize void formation by the combination of increasing rotation rate and reducing pull rate. We also noticed the correlation of growth stability and the melt composition. Instability becomes more severe toward the late stages of growth where the melt is more B_2O_3 deficient and further away from the congruent composition. We expect reduced vaporization losses to accompany the use of larger melts since the volume-to-surface ratio increases with the melt size.

Despite these difficulties, we have successfully grown high optical quality $ScBO_3$ crystals doped with Cr by both methods. In fact, the first lasing of $ScBO_3$:Cr was achieved using a crystal with 0.4 at% Cr produced by the Kyropoulos method. Using the Czochralski technique, we have successfully grown crystals with 0, 0.3 and 1.0 at% Cr. The measured distribution coefficient of Cr in $ScBO_3$ is 0.86. Lasing was also achieved using Czochralski grown crystals.

6. CONCLUSION

We have demonstrated the growth of $ScBO_3$:Cr from the melt by both the Kyropoulos and Czochralski methods. The congruent nature of the $ScBO_3$ melt and its melting temperature were confirmed. Despite the excessive evaporation of both B_2O_3 and $ScBO_3$, the growth of large high optical quality crystals seems feasible. Lasing was achieved with large near-IR tuning range (778-892 nm), long lifetime (115 μsec) and high slope efficiency (30%) [8]. The combination of these properties makes $ScBO_3$:Cr attractive in many applications.

ACKNOWLEDGEMENT

This work was supported by the LLNL under DOE Contract W-7405-ENG-48.

REFERENCES

1. J.C. Walling, O.G. Peterson, H.P. Jenssen and R.C. Morris, IEEE J. Quantum Electron. vol. QE-16, 1302 (1980).
2. K. Petermann and G. Huber, J. Luminescence, vol. 31 & 32, 71 (1984).
3. P.F. Moulton, J. Opt. Soc. Am. B/vol. 3, 125 (1986).
4. K.K. Turekian, Chemistry of the Earth, Holt, Rinehard and Winston, Inc. 131 pp. (1972).
5. E.M. Levin, R.S. Roth and J.B. Martin, Am. Mineralogist, vol. 46, 1030 (1961)
6. A. Biedl, Am. Mineralogist, vol. 51, 521 (1966).
7. E.M. Levin, J. Am. Ceramic Soc. vol. 50, 53 (1967).
8. S.T. Lai, B.H.T. Chai, M. Long, M.D. Shinn, J.A. Caird, J.E. Marion and T.R. Starver, this issue, (1986).

Growth of Laser-Quality $Ti:Al_2O_3$ Crystals by a Seeded Gradient-Freeze Technique

R.E. Fahey, A.J. Strauss, A. Sanchez, and R.L. Aggarwal

Lincoln Laboratory, Massachusetts Institute of Technology,
244 Wood Street, Lexington, MA 02173, USA

A seeded vertical gradient-freeze technique has been used to grow laser-quality $Ti:Al_2O_3$ single crystals with low residual infrared absorption from charges containing between 0.15 and 1.0 wt.% Ti_2O_3 in tungsten crucibles.

1. Introduction

The demonstration [1] of the wavelength-tunable $Ti:Al_2O_3$ laser has led to renewed interest in the growth of sapphire crystals doped with Ti^{3+} ions, which are responsible for the laser action. Such crystals have previously been grown by the Czochralski and heat-exchanger methods. Laser performance has generally been limited by an infrared absorption band of unknown origin that is present in the laser output wavelength region [2,3].

In this paper we describe the growth of $Ti:Al_2O_3$ crystals with low residual absorption by a technique derived from the gradient-freeze method used previously for the growth of MgF_2 crystals doped with Ni [4]. A vertical temperature gradient is imposed on the growth crucible, so that the molten charge is gradually frozen from the bottom up by reducing the temperature of the furnace. Single crystals of $Ti:Al_2O_3$ have been grown by this technique from charges containing between 0.15 and 1.0 wt.% Ti_2O_3. Values as high as 100 have been obtained for the ratio of the absorption at 490 nm, the peak of the Ti^{3+} laser pump band, to the residual infrared absorption at 900 nm. In laser experiments on samples cut from the gradient-freeze crystals, room-temperature cw operation has been obtained for rods containing between 0.024 and 0.099 wt.% Ti_2O_3 [5].

2. Crystal Growth Procedure

The furnace and tungsten crucible used for crystal growth are shown schematically in Fig. 1. As described previously [4], the furnace is resistance heated by means of a cylindrical tungsten mesh element that is enclosed in a water-cooled, stainless-steel jacket, permitting either vacuum or inert-gas operation. Multilayer molybdenum heat shields are used for thermally insulating the element and the crucible, which is cooled from the bottom by close contact with the water-cooled base of the furnace. To establish the temperature gradient required for crystal growth, nine heat shields are placed across the top of the hot zone, and this zone is insulated along its length by means of a spiral shield which is wound so that the number of molybdenum layers decreases continuously from ten at the top to two at the bottom.

The crucible ends at the bottom in a closed capillary about 5 mm in diameter. The main body of the crucible has an approximate diameter of

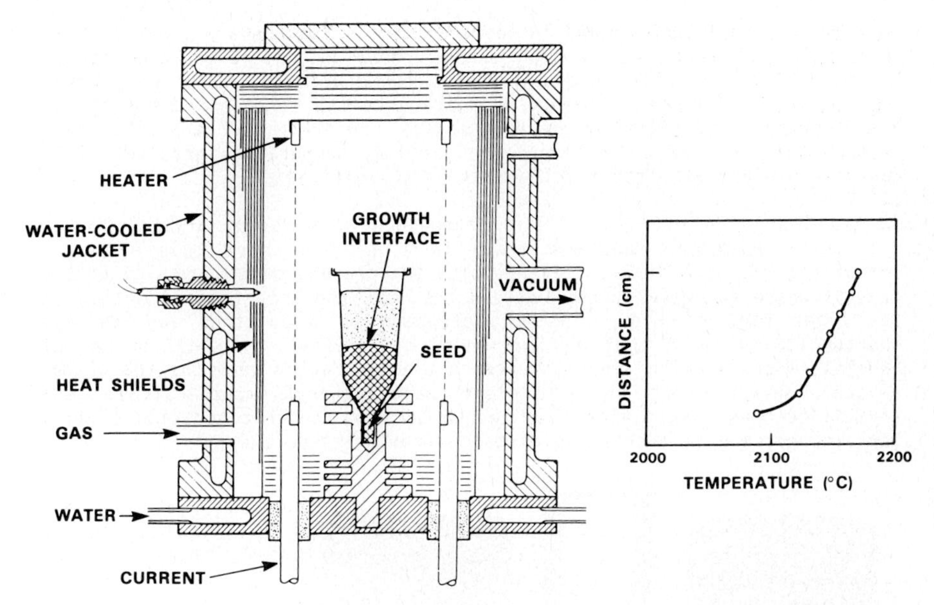

Fig. 1 Schematic diagram of apparatus for gradient-freeze growth of $Ti:Al_2O_3$ crystals

either 2.5 or 5 cm. An oriented sapphire seed is placed in the capillary, and the crucible is loaded with a charge consisting of commercial high-purity sapphire crackle and single-crystal Ti_2O_3. The Ti_2O_3 is prepared in our laboratory by directional solidification of a melt obtained by the fusion of TiO_2 with Ti powder prepared by hydriding Ti crystal bar. A tungsten cap is then spotwelded to the crucible at several points but not sealed.

The crucible is placed in the furnace, which is then evacuated with a mechanical pump to a pressure of about 0.1 Torr, heated sufficiently to melt the charge, backfilled with Ar or He at atmospheric pressure, and gradually cooled to room temperature. The initial cooling rate is ~3 °C/h. After the top of the charge reaches a temperature that is well below the melting point, the rate is increased to ~20 °C/h. According to measurements made by moving a W/W-Re thermocouple along the central axis of an empty 2.5-cm-diameter crucible, during solidification of the molten charge the vertical temperature gradients along the main body of the crucible are ~8 and 15 °C/cm for operation in Ar and He, respectively, corresponding to growth rates of about 4 and 2 mm/h.

3. Experimental results

3.1 Crystal growth

Nine $Ti:Al_2O_3$ single crystals, each weighing about 100 g, have been grown in 2.5-cm-diameter crucibles. In each case the seed was oriented parallel to the c (optic) axis. Five crystals were grown in an Ar atmosphere, the others in He. The as-grown crystals exhibit little strain, so that samples can be cut and polished without cracking. The principal internal defects are small bubbles and inclusions; the inclusions are generally associated with bubbles. The occurrence of these defects is greatly reduced by the use of He, which because of its lower density should be removed more easily from the melt during evacuation of the furnace and during solidification. In addition, He reduces the probability of constitutional supercooling because it increases the vertical temperature gradient along the crucible and also reduces the rate of solidification.

On the left side of Fig. 2 is a photograph of a recent crystal, about 2.5 cm in diameter, that was grown in a He atmosphere from a charge containing 0.5 wt.% Ti_2O_3. Several pits are present on the outside of the crystal where gas was trapped between the melt and the crucible wall. On the right side of Fig. 2 is a photograph of a polished longitudinal section 1.8 cm thick that was cut from this crystal. A small number of bubbles, most of which are clustered in a small region near the top of the crystal, appear to be the only significant internal imperfections. The crystal becomes darker along its length because the concentration of Ti^{3+} ions increases with increasing distance from the seed end.

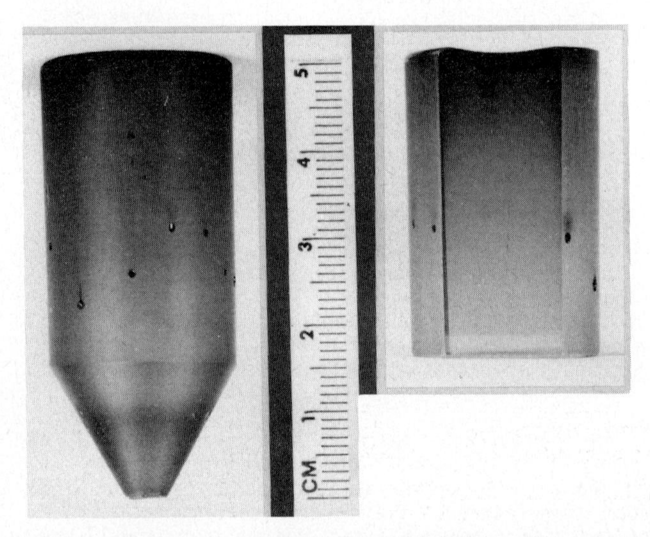

Fig. 2 (Left) Single crystal of $Ti:Al_2O_3$ (Right) Longitudinal section

The growth runs made to date with 5-cm-diameter crucibles have been carried out in an Ar atmosphere. The crystals obtained contain numerous inclusions, making them unsuitable for laser use. We anticipate that laser-quality crystals of the larger diameter can be obtained by growth in a He atmosphere. Such growth is not feasible at present because the increased heat loss resulting from the higher thermal conductivity of He would require operation of the tungsten resistance heater at unacceptably high temperatures. In order to reduce the heat loss to a satisfactory level we are fabricating a new spiral heat shield in which the maximum number of Mo layers will be twenty instead of ten.

3.2 Optical Absorption and Concentration of Ti^{3+} Ions

The Ti^{3+} concentrations in the $Ti:Al_2O_3$ crystals have been monitored by measuring the optical absorption band peaking at 490 nm, which is due to

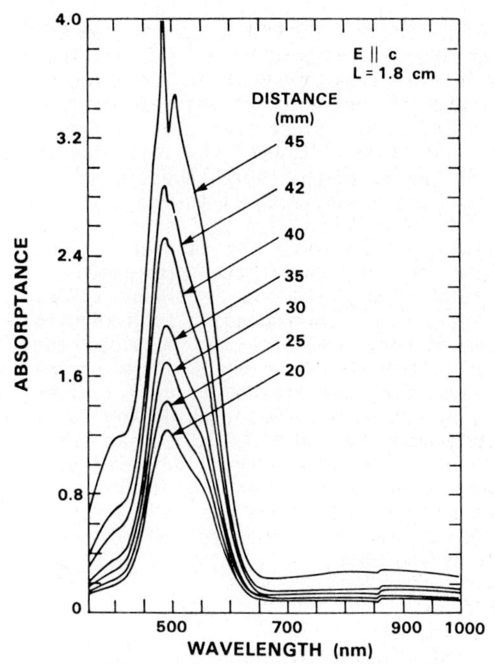

Fig. 3 Absorptance spectra for longitudinal section of Fig. 2, taken at indicated distances from seed end

transitions between the $^2T_2$ ground state and ^2E excited state of the single d electron of the Ti^{3+} ion. In Fig. 3 the absorptance values measured with a spectrophotometer at seven positions along the length of the longitudinal section shown in Fig. 2 are plotted as a function of wavelength from 350 to 1000 nm for light polarized parallel to the c axis. As expected from the change in color observed visually, the absorption band becomes stronger with increasing distance from the seed. The shoulder on the long-wavelength side of the band results from Jahn-Teller splitting of the doubly degenerate ^2E level. The splitting at the peak of the band is an artifact that appears when too little light is transmitted to the spectrophotometer detector.

Data like those of Fig. 3 have been used to determine the optical absorption cross-section for Ti^{3+} ions in Al_2O_3. For samples cut from several of the gradient-freeze crystals, the absorption coefficient (α_{max}) at the peak of the Ti^{3+} band has been plotted against the Ti^{3+} concentration determined from magnetization measurements [6]. The experimental points are fit well by a straight line through the origin whose slope corresponds to a Ti^{3+} cross-section of $9.3 \pm 1.0 \times 10^{-20}$ cm^2. By using this value, which is about 40% higher than the lower limit for the cross-section found by Moulton [1] from x-ray fluorescence data, we obtain the concentration of Ti_2O_3 in weight percent as $(0.032 \pm 0.03)\ \alpha_{max}$.

Having established that α_{max} is proportional to the Ti^{3+} concentration, we can use the data of Fig. 3 to evaluate the distribution coefficient, $k = C_{solid}/C_{liquid}$, for Ti^{3+} in Al_2O_3. According to the theory of normal freezing, the impurity distribution along a crystal grown from the melt is given by the equation $C = kC_0(1-g)^{k-1}$, where C is the concentration in the crystal at the point where a fraction g of the original liquid has frozen and C_0 is the initial concentration in the liquid. Therefore k can be evaluated from a plot of log C, or of the logarithm of a property proportional to C, against log (1-g), since this plot will be a straight line with slope of (k-1). The value of k will also be given by the ratio of C for (1-g) = 1, which is the first-to-freeze composition, to the concentration in the charge, which is presumably equal to C_0.

Figure 4 is a log-log plot of α_{max} vs (1-g) based on the data of Fig. 3. The right-hand ordinate scale gives Ti^{3+} concentrations (expressed in wt.% Ti_2O_3) that correspond to values of α_{max} on the left-hand scale. The highest measured value of α_{max} is 5 cm^{-1}, corresponding to 0.16 wt.% Ti_2O_3, which is believed to be the highest value so far obtained for $Ti:Al_2O_3$. The data are fit well by a straight line whose slope corresponds to k = 0.21. However, extrapolating the line to (1-g) = 1 gives the first-to-freeze concentration as 0.0395 wt.% Ti_2O_3. Taking the ratio of this value to the charge concentration gives k of only 0.079. It seems likely that this value more closely approximates the equilibrium distribution coefficient, while the higher value is an effective distribution coefficient obtained because growth at a finite rate results in the formation at the growth interface of a liquid boundary layer with higher impurity concentration than the bulk of the melt.

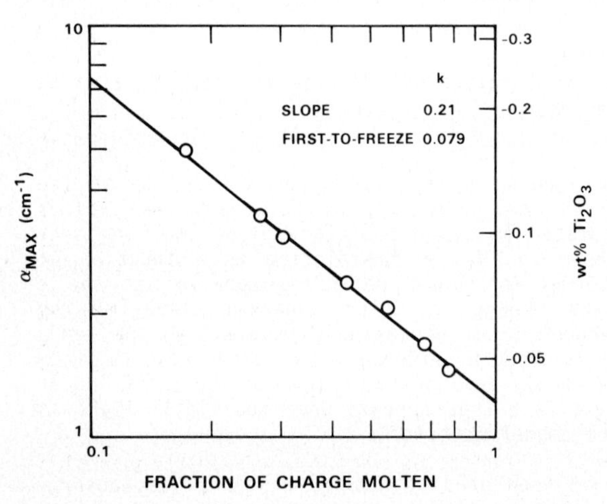

Fig. 4 Peak Ti^{3+} absorption coefficient (α_{max}) and Ti_2O_3 concentration vs fraction of growth charge remaining molten (1-g).

3.3 Residual Infrared Absorption

Routine spectrophotometer measurements, giving data like those plotted in Fig. 3, have shown that for light polarized parallel to the c axis the residual infrared absorption is not more than a few percent of the peak Ti^{3+} absorption for any of the as-grown gradient-freeze crystals. (For light polarized perpendicular to c, the peak absorption is lower and the residual absorption higher, so that light propagation perpendicular to the c axis is used in laser operation.) For two of the crystals, more precise

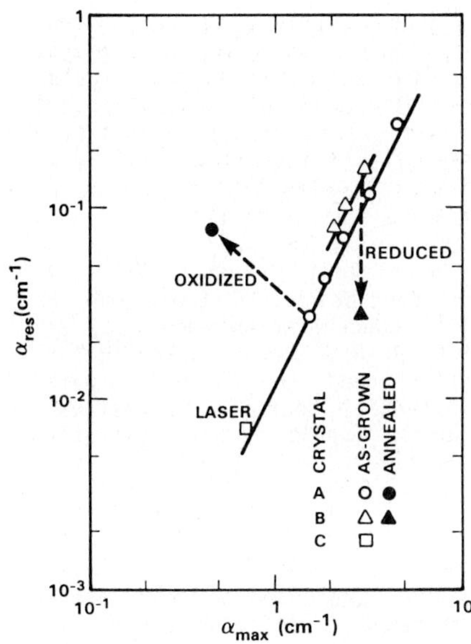

Fig. 5 Residual absorption at 900 nm (α_{res}) vs Ti^{3+} absorption at 490 nm (α_{max}) for $Ti:Al_2O_3$ crystals

measurements have been made to determine the residual absorption at 900 nm, α_{res}, as a function of α_{max}. The results are shown on a log-log plot in Fig. 5. The points represented by open circles were obtained for the crystal of Fig. 3, designated as crystal A, while those represented by open triangles were obtained for a crystal designated as B, which was grown from a charge containing 1 wt.% Ti_2O_3. The data fit two parallel straight lines, one for each crystal, with a slope of 2. For a laser rod cut from a crystal designated as C, α_{max} was found by spectrophotometer measurements, and an upper limit was found for the residual absorption at 770 nm by analyzing the laser threshold and slope efficiency to determine the round-trip cavity loss [5]. The point obtained in this manner falls close to the straight line obtained for crystal A. For this sample, α_{res} is less than one percent of α_{max}.

The linear dependence of α_{res} on the square of α_{max}, that is, on the square of the Ti^{3+} concentration, indicates that the residual absorption is due to an absorbing species composed of two members, one of which might be a Ti^{3+} ion. The possibility that the absorbing species is a Ti^{3+}-Ti^{3+} pair is ruled out because in this case α_{res} would be a single-valued function of α_{max}. This prediction is inconsistent with the difference observed between the sets of data for the as-grown crystals A and B, and even more so with the results shown in Fig. 5 for two annealed samples. One of these, cut from crystal A, was annealed in Ar gas containing 10 ppm O_2. This treatment decreased α_{max} more than a factor of 3 while increasing α_{res} by almost as much. The other sample, from crystal B, was annealed in Ar gas containing about 15 percent H_2, causing α_{res} to decrease by almost a factor of 6 without a significant change in α_{max}.

The data obtained to date suggest that the residual absorption is due to Ti^{3+}-Ti^{4+} pairs, as proposed by JORGENSON [7]. In this case α_{res} will be proportional to $(\alpha_{max})^2$ as long as the Ti^{3+}/Ti^{4+} concentration ratio is constant. According to this model, the difference between the results for as-grown crystals A and B can be attributed to a difference in their Ti^{3+}/Ti^{4+} ratios. For a fixed total concentration of Ti ions, α_{res} will have its maximum value when the Ti^{3+} and Ti^{4+} concentrations are equal, and it will disappear when either Ti^{3+} or Ti^{4+} becomes zero. We believe that for our as-grown crystals the fraction of Ti^{3+} ions is not far from one. In this case, partial oxidation of Ti^{3+} ions to Ti^{4+} ions, while reducing α_{max}, can increase the concentration of Ti^{3+}-Ti^{4+} pairs and therefore increase α_{res}, as observed for the sample that was annealed in Ar-O_2. On the other hand, partial reduction of the Ti^{4+} ions to Ti^{3+} can produce a significant decrease in pair concentration and therefore in α_{res} without a significant increase in Ti^{3+} concentration and a_{max}, as observed for the sample that was annealed in Ar-H_2. We are carrying out additional annealing experiments in order to test the Ti^{3+}-Ti^{4+} pair model. Regardless of mechanism, the substantial decrease in residual absorption achieved by Ar-H_2 annealing should result in significantly improved laser performance.

4. Conclusion

A seeded vertical gradient-freeze technique has been used to grow laser-quality Ti:Al_2O_3 single crystals with low residual infrared absorption from charges containing between 0.15 and 1.0 wt.% Ti_2O_3. By magnetization measurements it has been found that the concentration of Ti_2O_3 in weight percent is given by 0.032 (α_{max}). Evidence has been obtained that the residual absorption is due to Ti^{3+}-Ti^{4+} pairs, but confirmation of this model will require additional data.

Acknowledgements

The authors are grateful to L. J. Belanger, G. D. Silva, M. M. Stuppi, and D. J. Sullivan for technical assistance. This work was sponsored by the Defense Advanced Research Projects Agency.

References

[1] P.F. Moulton: J. Opt. Soc. Am. B 3, 125 (1986).
[2] P. Lacovara, L. Esterowitz, and M. Kokta: IEEE J. Quantum Electron. QE-21, 1614 (1985).
[3] P. Albers, E. Stark, and G. Huber: J. Opt. Soc. Am. B 3, 134 (1986).
[4] T. B. Reed, R. E. Fahey, and P. F. Moulton: J. Crystal Growth 42, 569 (1977).
[5] A. Sanchez, R.E. Fahey, A.J. Strauss, and R.L. Aggarwal: Opt. Lett. 11, 363 (1986); this conference, Paper ThA6.
[6] R. L. Aggarwal, A. Sanchez, R. E. Fahey, and A. J. Strauss: Appl. Phys. Lett. 48, 1345 (1986).
[7] C. K. Jorgenson, quoted in Ref. [1].

Effects of Growth Conditions and Post-Growth Thermal Treatment on the Quality of Titanium-Doped Sapphire

M.R. Kokta

Union Carbide Corporation, 750 S. 32nd Street,
Washougal, WA 98671, USA

Effects of growth conditions, mainly the ambient growth atmosphere, were investigated. It was found that optical absorption in wavelength range 650 to 1000 nm can be correlated with presence of oxygen in the ambient atmospheres, as well as with presence of oxygen dissolved in the liquid phase during the growth. Effects of high temperature thermal treatment on crystals grown were also investigated. Annealing of titanium-doped sapphire crystals in vacuum or in inert atmosphere leads to reducing the Ti^{4+} present in as-grown crystals, and increases apparent absorption of Ti^{3+} at 490 nm, lowers absorption in between 650 to 1000 nm and shifts the UV absorption edge slightly towards shorter wavelength.

1. Introduction

Tunable lasers may find attractive application in remote sensing, medicine and scientific research. Presently, the only available tunable laser in visible range is the dye laser. The solid state lasers would have numerous advantages as compared to dye lasers. Recently a number of solid state materials utilizing transitions from upper level laser levels to a vibrationally excited ground state - vibronic lasers - emerged as potentially useful candidates for tunable lasers. From this class of vibronic lasers, titanium Ti^{3+} ion demonstrated broad range laser operation at room temperature in sapphire (1), (2).

The major problem limiting usefulness of titanium-doped sapphire is an optical loss occurring in the lasing wavelength range of 650 to 1000 nm. This optical loss is caused by broad band absorption extending across the whole lasing wavelength band (3), Fig. 1. While the origin of this absorption cannot be positively assigned, available evidence indicates that it may be related to point defects present in the host crystal - sapphire.

A substantial number of point defects (10^6 to $10^8/cm^3$) exist in an undoped sapphire crystal, and these defects manifest themselves by electrical conductivity and luminescence under short wavelength UV excitation. The presence of point defects in undoped sapphire crystals was extensively studied in the past (4-9). The point defects in sapphire are arising from non-stoichiometry of oxygen and from the presence of impurities in valence states other than 3+. The concentration of defects caused by non-stoichiometry is directly related to crystal growth conditions such as high melt temperature and ambient growth atmosphere. The concentrations of defects caused by impurities is related to quality of raw materials used in crystal growth process.

The Ti^{3+} ions substituted in normal substitutional sites with sixfold symmetry absorb light at 490 nm, Fig. 1. A wavelength corresponding to transition from 2^T ground level to 2^E excited state. The sizes of the voids r_{oct} = 0.60 Å are appreciably smaller than the radius of Ti^{3+} ion $(r_{Ti^{3+}})_6$ =

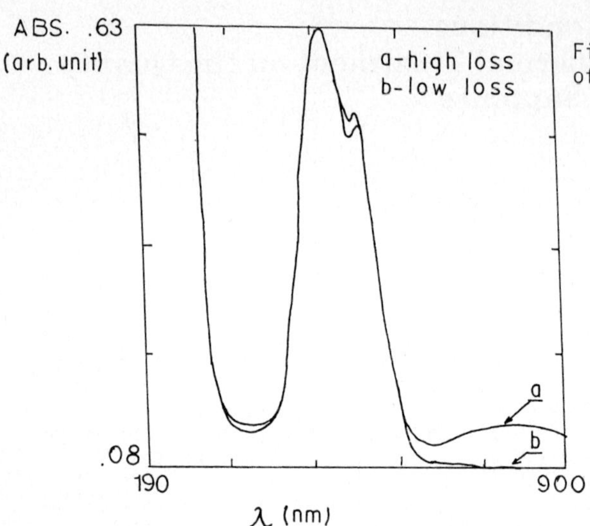

Fig. 1. Optical absorption of titanium-doped sapphire

0.67 Å. With an introduction of the activator (Ti^{3+}), anionic and possibly cationic vacancies are concentrated in its vicinity, which leads to formation of coordination polyhedra with fluctuations in field parameters. These fluctuations affect the crystal field splitting of 3d levels. Changes in separation of pertinent levels result in absorption observed at longer wavelength. Jones et al. (9) describe a reversible oxydo-reduction process involving titanium in sapphire. Such an equilibrium requires a presence of vacancies and may be described by a simple relationship:

$$Al_{(2-4x)}V_{x}^{Al}Ti_{3x}^{4+} \rightleftarrows Al_{)2-4x)}V_{x}^{Al}Ti_{3x}^{3+}O_{(3-3/2x)}V_{3/2x}^{O} + 3/4xO_{2}$$

The incorporation of Ti^{4+} may be expected during the growth process depending on growth conditions such as growth atmosphere, and aluminum oxide dissociation at elevated temperature. The Ti^{4+} ion is a defect-causing impurity and therefore increases in optical loss may be expected with increases of titanium concentration in sapphire. In addition, the large difference between the radii of crystal-forming and activator ions also explains low value of distribution coefficient (0.12), and resulting low solubility of trivalent titanium in sapphire.

2. Experimental

Growth - Growth technique employed was the Czochralski pulling technique. The aluminum oxide containing desired concentration of dopant, Ti_2O_3 was melted in an iridium crucible by means of RF heating under an atmosphere of pure nitrogen or mixture of nitrogen with 1% by volume of hydrogen. Crystals were grown at orientation perpendicular to "C" axis. The pulling rates employed were 0.015"/hour with rotation rate of 5 rpm. The grown crystals were fabricated into samples with dimensions of 7x7x20 mm and 7x7x10 mm in "as grown" states. Some of the boule material was subjected to "post-growth" treatment prior to sample fabrication.

Effects of Growth Atmosphere - In standard growth conditions, the titanium-doped sapphire crystals were grown under a nitrogen atmosphere containing 0.1% by volume of hydrogen. In order to determine whether the adverse long

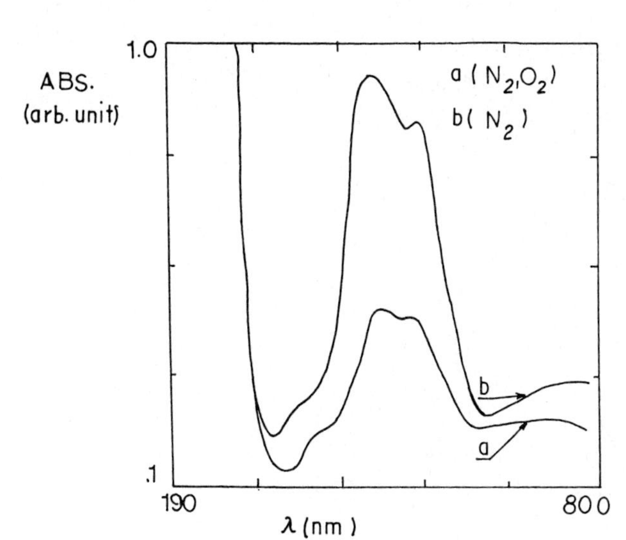

Fig. 2. Absorption in spectrum of titanium-doped sapphire grown under a) nitrogen + 100 ppm (volume) oxygen, b) pure nitrogen

wavelength absorption effects are caused by Ti^{4+} incorporation during the growth, crystal was grown under an atmosphere composed of nitrogen with 100 ppm of oxygen. The optical absorption spectrum of this crystal is shown in Fig. 2 along with spectrum of a crystal grown under the "standard" conditions with identical titanium concentration.

3. Post-Growth Treatment

Improvement in quality of Czochralski-grown $Ti^{3+}:Al_2O_3$ by post-growth treatment was already described previously (3). This treatment involves high-temperature annealing under either vacuum or inert atmosphere. The temperature of the annealing was 1940°C. The soak time at this temperature was a minimum of 40 hours followed by cooling to room temperature at rates from 10°C/hour to 200°C/hour.

In Fig. 3, the effect of post-growth treatment is evident for a crystal grown under "standard" conditions. It is evident that annealing converts Ti^{4+} present in the crystal to Ti^{3+} and thus stronger absorption at 490 nm is seen. This reduction of Ti^{4+} is consistent with improvement of optical quality of the crystal by decreasing point defect concentrations.

Optical Evaluation of Internal Loss - Optical transmission spectrum was taken from each boule grown in the range of 190 to 900 nm. Each boule was evaluated for presence of scattering centers such as bubbles and inclusions. Optical loss measurements were carried out at 830 nm by using a light beam generated by semiconductor laser diode (cw 10 mw output). The output of diode laser was collimated prior to it entering the sample. In order to separate the surface reflection losses from the internal loss, the surface loss of an undoped high purity sapphire was measured at the same wavelength using samples of different length, and extrapolating for the loss at zero length.

Dependence of absorption losses on titanium concentration was tested by growth of a series of crystal with titanium concentration varying from 0.02%

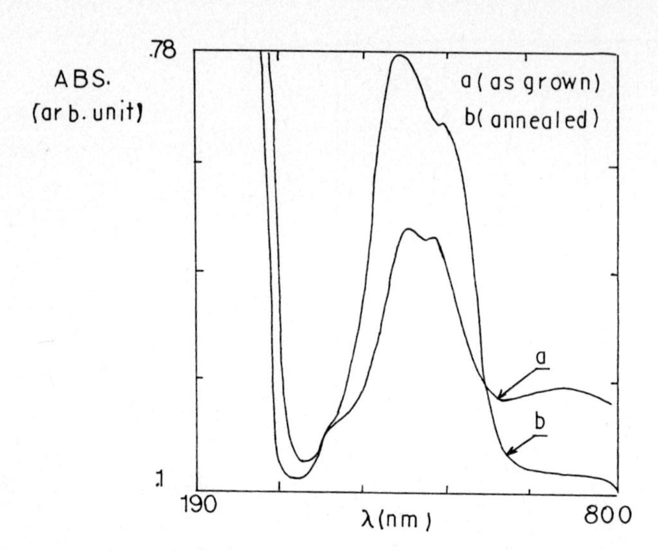

Fig. 3. Optical absorption spectrum of titanium-doped sapphire a) as grown (under N_2 atmosphere), b) annealed (in vacuum)

atomic to 0.13%. The absorption losses were measured in "as grown" as well as in "annealed" samples. Results of these measurements are compiled in Table 1.

Table 1 Absorption losses in % cm^{-1} at 830 nm

Sample	Ti Conc. (%)	As Grown	Annealed
VLD	0.02	0.0	0.0
LD	0.045	2.8	1.3
HD	0.09	12.8	8.3
HD-II	0.105	14.8	8.8
VHD	0.13	24.0	11.8

These results are plotted in Fig. 3. These results clearly show that the optical loss depends on titanium concentration in both "as grown" and "annealed" samples. However, there is significant difference in optical losses for the "annealed" as compared with "as grown" samples, indicating a decrease in point defect concentration.

4. Summary and Discussion

It was found that significant optical losses in wavelength range 650 to 1000 nm exist in "as grown" titanium-doped sapphire crystals, and that these losses are proportional to titanium concentration (Fig. 4) as well as to the presence of oxygen in ambient growth atmosphere. However, it was also found that post-growth annealing of crystals in reducing atmosphere can reduce the optical losses significantly. Various annealing conditions were investigated from which only annealing in vacuum or under reducing conditions resulted in

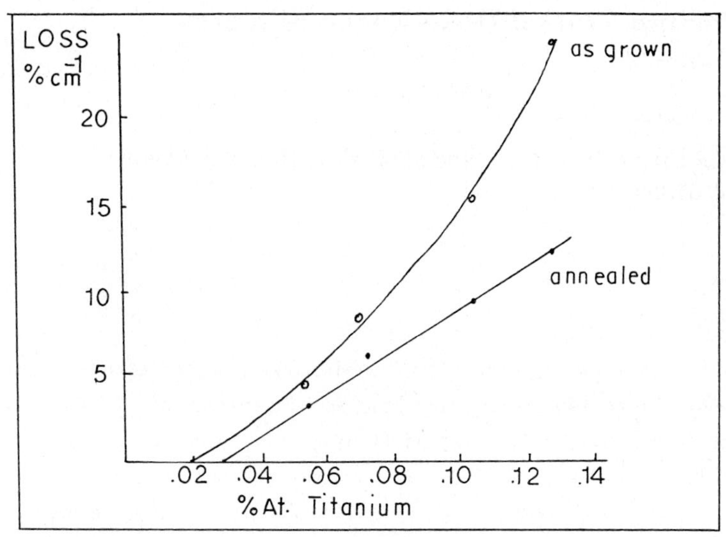

Fig. 4. Absorption loss in titanium-doped sapphire as function of titanium concentration

decreasing optical losses. It was found that final result of annealing crystals grown under atmosphere with varying oxygen concentration yields the same value of optical loss independent of original loss of as grown crystals. The same appears to be true for crystals where dopant (Ti) was introduced in melt in different oxidation states (Ti_2O_3, Ti_3O_5 or TiO_2). Therefore, it is evident that optical quality of titanium-doped crystals depends on their thermal history more than on original growth conditions.

This conclusion is consistent with an assumption that optical loss is generated by structural defects associated with titanium ions, because the equilibrium concentration of such defects is determined by thermal history of the crystal.

References

1. P.F. Moulton, paper WA2 at Conference on Laser and Electro-optics, June 19-22, Anaheim, CA
2. P.F. Moulton, Solid State Res. Rep. Lincoln Lab., Mass. Inst. Tech, pp 15-21 (1982)
3. P. Laçovara, L. Esterowits, and M. Kokta, IEEE J. Quant. Electr., Vol. QE-21, No. 10, 1614 (1985)
4. E.W.J. Mitchell, J.D. Ridgen, P.D. Townsend, Phil. Mag. (8th Series), 5, 10013, (1960)
5. P. Gibbs, I.B. Cutler, J.L. Bates, Bull. Amer. Phys. Soc., 2, 300, (1957)
6. R.R. Dills, Dissertation, Stanford Univ. SU-DMS No. 65-28, 6, (1965)
7. J.J. Mills, J. Phys. Chem. Solids, Vol. 31, 2577-2579, (1970)
8. S.K. Mohapatra, S.K. Tiku, F.A. Kroger, J. Amer. Ceram. Soc., Vol. 62, 50-57, (1979)
9. T.P. Jones, R.L. Coble, G.J. Mogab, J. Amer. Ceram. Soc., Vol. 52, No. 6 331-334 (1969)

Growth of Cerium-Doped Rare Earth Silicates for Tunable Lasers

R.F. Belt and J.A. Catalano

Airtron Division of Litton Industries, Inc., 200 East Hanover Avenue, Morris Plains, NJ 07950, USA

1. Introduction

The rare earth silicates form an interesting isomorphous group of structures for laser hosts. The few preliminary studies of rare earth silicate growth were all performed by the flux method [1-5]. This method utilizes fluxes such as Bi_2O_3, Li_2O-MoO_3, KF, PbO-PbF_2, or combinations of the preceding to grow a particular composition of silicate. The crystals grown from these fluxes were of mm size and useful for X-ray or other structural studies. The earliest growth of Y_2SiO_5 single crystals was done in conjunction with laser research in the Soviet Union. In 1973 it was reported [6-7] that Nd^{3+}:Y_2SiO_5 was lased successfully at 300 K. Rods of mm size were used for these studies.[5] The emission spectrum, threshold, and other laser parameters were highly dependent on the orientation of the crystal axes of the laser rods. The crystal structure was given as orthorhombic and the two site symmetries at the Y^{3+} were slightly different [8]. In 1983, workers in Japan reported [9] the Czochralski growth of Ce^{3+}:Gd_2SiO_5 for a scintillator application. These crystals were 25 mm diameter and 40 mm long. A weak cleavage along (100) was present but the growth direction was not given. The ultraviolet excited luminescence has a broadband emission due to 4f $\rightarrow$ 5d transitions of the Ce^{3+}. The fluorescent lifetime is about 60 ns. The host properties suggest that high-quality single crystals may be useful for tunable laser action. Our efforts report the crystal growth and doping of R_2SiO_5 type compounds.

2. Experimental Procedures

The precursor powders used in these experiments were of 4-9's purity or better. The CeO_2 was purchased from Spex Industries, Inc., the Gd_2O_3 and SiO_2 from the United Mineral Corporation, the Y_2O_3 from Rhone-Poulenc, and the La_2O_3 and Lu_2O_3 from Research Chemicals. These powders were weighed to the nearest hundredth of a gram, then blended together in a polycarbonate plastic divider/mixer for three to four hours. This mixture was charged via a quartz funnel into the iridium crucible. The crucible was held at a temperature to cause pre-reaction and sintering of the powdered components. After adding and sintering the full charge, the crucible temperature was raised to the mixture's melting point of 1800-1900 °C where it was held for 24 to 48 hours to ensure complete mixing. Radial flow lines were clearly visible once the melting point of the charge was reached.

A standard Czochralski arrangement was used for all of the growth runs. A 3-inch diameter by 3-inch high iridium crucible, capped with a 3-inch iridium lid was used for the runs. The crucible was heated by an RF coil powdered by a 30 kW Lepel RF generator. The gaseous atmospheres for each run were either N_2 or forming gas of composition 99% N_2 - 1% H_2. The gases were fed through holes centered in the steel table and zirconia base plate. The gases exited from vertical slots cut in the zirconia support and flowed up through the zirconia grain insulation. The flow rates were generally in the range of 25 liters/min. The pulling rates and rotation rates were 0.1-1.0 mm/hr and 40-50 rpm. Melt drop compensation was employed during runs where a boule was pulled.

Bootstrapping seeds from previous growth runs has been most successful in generating new single crystal growth from subsequent melts. The first seeds were cut from spontaneously crystallized melts in 1.5 inch by 1.5 inch crucibles. These were used to grow 1.0 cm diameter boules which yielded larger seeds. The use of a 3-inch by 3-inch crucible and cooling the boule in contact with the melt has also produced cm sections of single crystal material from the frozen melt. Seeds were suspended from an alumina rod with an iridium wire hook. This has sometimes caused mechanical stability problems during growth. The larger single crystal pieces yielded larger seeds. The best growth direction appears to be parallel to the weak (100) cleavage plane. X-ray orientation of the seeds has also been used to determine more accurately the preferred growth orientation.. Some attempts at alternate orientations based on growth morphology and cleavage have been attempted.

A rapid polycrystalline survey of the potential for shifting the Ce^{3+} emission spectrum by compositional variations was developed. This was accomplished by a partial substitution of similar rare earths for the Gd in the Gd_2SiO_5 structure. Ten percent substitution of La, Y, or Lu was chosen as sufficient to cause an observable effect. This was achieved by using the pure component oxides and mixing formulation of the type $Gd_{1-x}Y_xSiO_5$ with x = 0.1. Samples weighed approximately 0.5-1.0 g each and were pressed into aspirin-sized tablets in a stainless steel die. The pellets were sintered on the bottom of an iridium crucible to produce the orthoscilicate structure. They were individually crushed and spectrally analyzed for their Ce^{3+} emission peaks by means of a demountable 1-inch CRT tube. Preliminary evaluations for luminescence were done under UV light. If a sample exhibited a visible response, then a small single or polycrystalline sample was prepared for cathode ray excitation. Upon exposure to the electron beam, the sample luminesces due to its Ce^{3+} content and this emitted radiation is scanned by an accurate monochromator. Its intensity versus wavelength profile from 300 nm to 700 nm is recorded on a strip chart. This allows evaluation of the relative strength and peak wavelength position of the Ce^{3+} emission spectrum.

3. Results and Discussion

A thorough examination of the available relevant phase diagrams for the R_2O_3-SiO_2 systems (where R is a rare earth) was performed at the start of our work. All the phase diagrams exhibited similar feature especially around the one-to-one mole ratio line. The existence regions for the R_2SiO_5 compositions are bounded by solidus and liquidus curves and two eutectics around the 1:1 line. The temperature ranges of these existence regions are also very closely centered around 1900°C. This would indicate that R_2SiO_5 single crystal material should be isostructural as R is varied from La to Lu. This allows mixed composition single crystal boules to be pulled stoichiometrically from the melt and various crystal fields about the R^{3+}. The first compound for single crystal growth was Gd_2SiO_5. The particular phase diagram is given in a standard reference [10]. Small single crystals were pulled and identified by powder X-ray diffraction. The pattern for Gd_2SiO_5 corresponded directly to the reported [11] phase of Y_2SiO_5 with small increases in lattice constant. A complete crystal structure of the Y_2SiO_5 phase has been reported [12]. We assume the rare earths correspond directly to this structure with appropriate allowance for size differential.

The crystals grown in the early series of experiments were doped by adding CeO_2 at about the 0.1% level of the component powder mixture. Conversion from Ce^{+4} to Ce^{+3} was detected to be incomplete in the early growth runs. This was revealed by a deeper yellowish coloration in portions of the melt and crystals. There was, however, always sufficient conversion to cause a very bright luminescence under ultraviolet illumination. More complete conversion appeared to be achieved in later runs by going from a flowing atmosphere of forming gas to only a flowing N_2 gas cover. This was accompanied by the growth of transparent water white boules which fluoresced brightly. The distribution coefficient (k) for Ce^{3+} in the Gd_2SiO_5 has not been measured directly yet but is assumed to be about 0.6. In most hosts the large size of the Ce^{3+} generally leads to a low distribution coefficient in Y or Lu analogs. The values are sometimes as low as 0.1-0.2. In the La_2SiO_5 analog the k values are close to 1. However, with the low dopant levels of Ce^{3+} (0.1%) which are needed, there are not large differences in the phase behavior or crystal growth. We fully expected that Ce^{3+} could be introduced in low amount (0.1%) in all crystals such as La, Gd, Y, or Lu compounds and so far this has been confirmed.

In one of our Gd_2SiO_5 growth runs the boule was cooled in contact with the melt causing the melt to crystallize out in very large pieces. Because of this, when growth of Y_2SiO_5 was to be attempted it was decided that slow cooling of the melt in contact with a Gd_2SiO_5 seed would cause the melt to crystallize out in a similar fashion yielding some single crystal Y_2SiO_5 suitable for seeds. As expected, large regions of the melt were high-quality single crystal. This leads one to believe that these systems would be good candidates for growth by the Bridgman technique due to the ease with which directional crystallization takes place. Figure 1 shows a crystallized melt of Gd_2SiO_5 while Figure 2 shows a seed crystal cut from this melt.

Both the melt crystallization and the growth of progressively larger boules of Gd_2SiO_5 have yielded regions of high-quality single crystal suitable for cutting into larger seeds. This in turn has helped produce boules with increasingly larger and longer sections of quality single crystal.

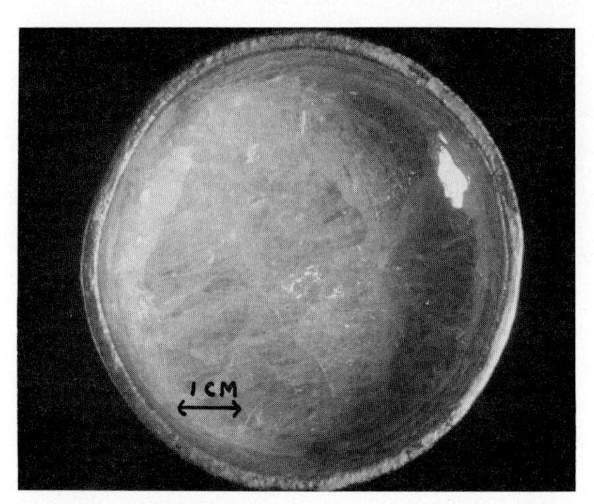

Fig. 1 Crystallized melt of Gd_2SiO_5 cooled from 1900 °C in a 2 x 2 inch iridium crucible

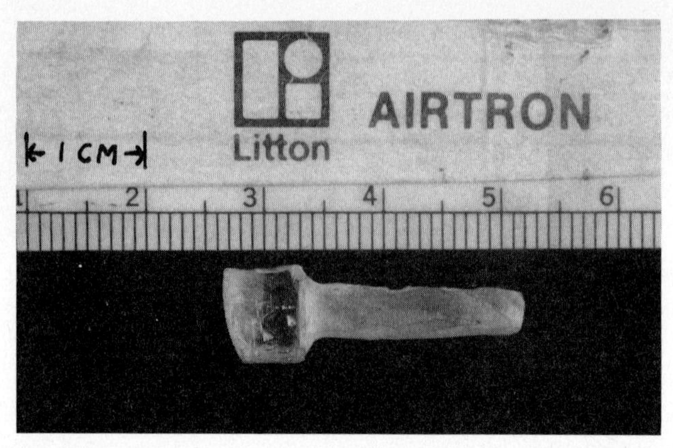

Fig. 2 Single crystal seed of Gd_2SiO_5 obtained from frozen melt

Seed rods have been cut from this material and growth was attempted with the seed rod held in a sapphire holder. Unfortunately the normal tight fit between seed rod and holder and the high temperatures encountered during growth along with the now obvious large differences in thermal expansion coefficient caused both seed rod and holder to shatter. Some modifications of either the fit tolerance or holder material will be necessary to use this technique successfully.

The last two experimental runs of $Gd_2SiO_5:Ce^{+3}$ have yielded excellent boules with large sections of single crystal of high quality. The first of these two runs yielded a boule 85 mm long by 30 mm diameter grown in a fashion similar to that of Airtron's Nd:YAG. Growth was terminated in this run due to the fear that the boule interface had undercut and might precipitate thermal runaway. This proved to be unfounded as the boule was firmly embedded in the frozen crucible mass. A section from this boule is shown in Figure 3. The crystal is water white transparent and fluoresces strongly in the blue. The word "Airtron" is being looked at perpendicular to the unpolished cleavage planes through about 20 mm of crystal. Looking closely at the letters one can see the birefringent effect of this crystal.

The last boule was grown from the same seed as the previous run. Similar rotation rates and thermal profiles were used but the boule was pulled at a slightly slower rate and it was kept to a smaller diameter. The result was a boule of an average diameter of 25 mm along a length of 157 mm. This boule is shown in Fig. 4. The growth was stopped at this length due to cracking at the upper end of the boule as it came out of the chimney. This cracking should be preventable by using an extended chimney with a ceramic lid with a smaller opening. The boule was pulled up 5 mm just

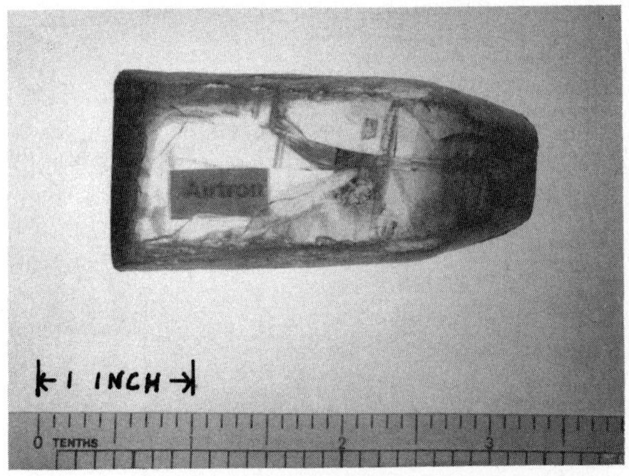

Fig. 3 Single crystal boule of Gd_2SiO_5 pulled from melt and cut

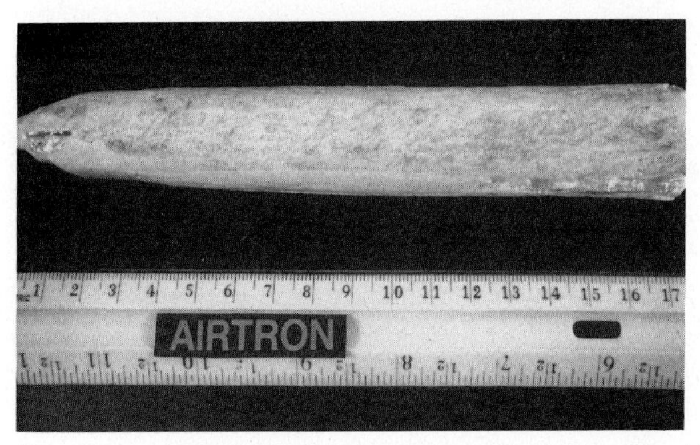

Fig. 4 Large boule of Ga_2SiO_5 grown from melt with surface layer intact

prior to cooling but was also firmly embedded into the frozen melt. This made removal, as was the case in the previous run, very difficult and is likely the cause of a good deal of the cracking observed. In the future boules will be pulled completely clear of the melt or necked in sufficiently to allow easy removal without damage to the crystalline quality. This last boule also displayed water white transparency with a bright blue fluorescence under ultraviolet illumination.

All of our single crystals were doped with 0.1% CeO_2 to investigate applications to tunable lasers. It was found that the peak fluorescent emission could be changed by 10-50 nm when various mixed crystals of Gd-Y, La-Gd, La-Lu, or La-Y were prepared. Most of the preliminary work was done with the Ce^{3+}:Gd_2SiO_5 crystals. Figure 5 shows an excitation spectrum

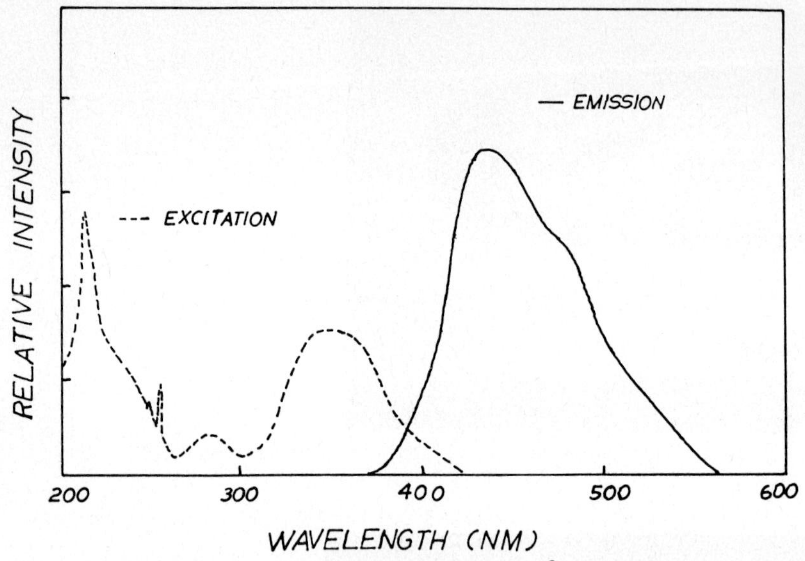

Fig. 5 Excitation and emission spectra of $Ce^{3+}:Gd_2SiO_5$

obtained as a function of wavelength where the monitoring wavelength was 423 nm. An absorption spectrum run on a Cary 17 spectrophotometer showed similar features. The sharp lines at 220 and 260 nm are from Gd transitions while the broad bands at 230, 280 and 350 nm are from Ce^{3+} transitions. The emission spectrum was obtained by monitoring the fluorescent output after excitation by 330 nm radiation. A number of samples were checked by cathodoluminescence with 15 kv electrons. These spectra were identical to our 330 nm excitation. The presence of two distinct emission peaks is probably related to the two site symmetries for Ce^{3+} in the Gd^{3+} positions. Gain measurement, determination of excited state absorption, and possible laser action in $Ce^{3+}:R_2SiO_5$ compounds are under investigation.

4. Conclusions

Single crystals of 1:1 rare earth silicates are congruently melting and can be grown by the Czochralski method. Samples of Gd_2SiO_5 doped with Ce^{3+} were investigated as possible tunable lasers involving 4f-5d transitions. Suitable absorption and emission bands are present but gain measurements and possible excited state absorption were not completed yet. These silicates should be appropriate hosts for other rare earth ions.

Acknowledgements

The authors thank the Office of Naval Research for their support of this work under contract N00014-85-C-0823. Excitation spectra were provided by Prof. B. Dunn of U.C.L.A.

References

1. G. Guisson and C. Michel, Mat. Res. Bull. 3, 193 (1968).
2. L.A. Harris and C.B. Fuich, Amer. Mineral. 50, 1493 (1965).
3. I.A. Bondar, L.N. Koroleva, and N.A. Toropov, Rost, Kristallov. Akad. Nauk SSSR Ins. Krist 6, (1965).
4. I.A. Bondar, L.N. Koroleva and D.P. Sipovskii, Rus. J. Inorg. Chem. 13, 1635 (1968).
5. B.M. Wanklyn, F.R. Wondre, G.B. Ansell, W. Davison, J. Mat. Sci. 9, 2007 (1974).
6. Kh. S. Bagdasarov, A.A. Kaminskii, A.M. Kevorkov, A.M. Prohkorov, S.E. Sarkison, T.A. Tevosyon, Soviet Physics-Doklady 18, (1973).
7. A.A. Kaminskii, et al., Inorganic Materials (USSR) 12, 1238 (1976).
8. Y.I. Smolin and S.P. Tkachev, Soviet Physics Crystal. 14, 14 (1969).
9. K. Takagi and T. Fukazawa, Appl. Phys. Letters 42, 43 (1983).
10. Phase Diagrams for Ceramists, 1969 Supplement, American Ceramic Society, Figure 2367.
11. Ito and Johnson, Am. Mineral. 53, 1940 (1968).
12. B.A. Maksimov, V.V. Ilyukkin, Yu A. Kharitonov, and N.V. Belov, Soviet Phys. Crystall. 14, 806 (1971).

Part IV

Chromium Tunable Lasers

The Chromium-Doped Rare-Earth Gallium Garnet Crystals as the Active Material for Solid-State Lasers

I.A. Shcherbakov

General Physics Institute of the USSR Academy of Sciences,
38 Vavilov Street, SU-117942 Moscow, USSR

By choosing the right composition of the rare-earth garnet crystals, it is possible to grow good optical quality large sized crystals with the rare-earth and transition ion concentrations varying over an extremely wide range. Using the processes of excitation transfer one can achieve considerable improvement of the energy performance characteristics of traditional lasers. A number of new schemes may also be possible /1/. Among such crystals there are, for example, the gallium garnet crystals doped with chromium and neodymium. Fig. 1 demonstrates the time evolution of the metastable state of neodymium for the YAG:Cr^{3+}, Nd^{3+} and the gadolinium scandium gallium garnet (GSGG:Cr^{3+}, Nd^{3+}) crystals. The normalization was carried out for the population of neodymium excited in the self-absorption bands. It can be seen that in the case of the GSGG crystal the maximum value of population is about a factor of 4 higher /2/. It is caused by the effective Cr $\rightarrow$ Nd energy transfer in GSGG crystal. The constant of the transfer here is about one order of magnitude higher than the YAG crystal /3/.

Due to the effect of sensitization, the efficiency of lasers based on GSGG:Cr^{3+}, Nd^{3+} crystals reaches 6-7% (Fig. 2) /4/.

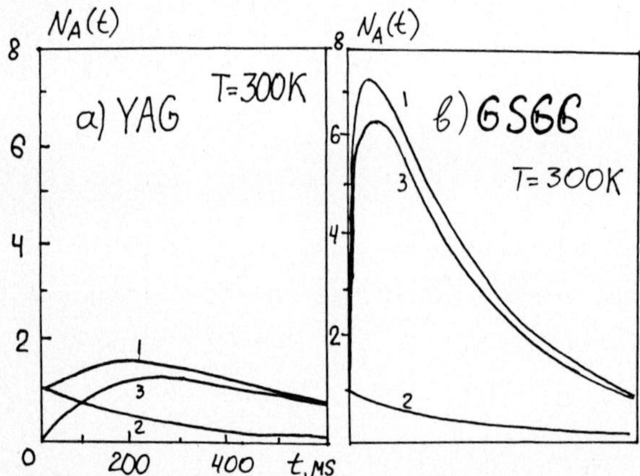

Fig.1. The time dependences of Nd^{3+} upper lasing level population at T=300 K in the crystals: **a**) YAG: Cr^{3+}, Nd^{3+}; $1.1\times10^{20}cm^{-3}$ Cr^{3+}; $0.9\times10^{20}cm^{-3}$ Nd^{3+}; **b**) GSGG: Cr^{3+}, Nd^{3+}; $3\times10^{20}cm^{-3}$ Cr^{3+}; $2\times10^{20}cm^{-3}$ Nd^{3+}. (*1*) The total excited state population of Nd^{3+}. (*2*) The excited state population of Nd^{3+} ions excited in the Nd^{3+} absorption bands. (*3*) The excited state population of Nd^{3+} ions excited in the Cr^{3+} absorption bands

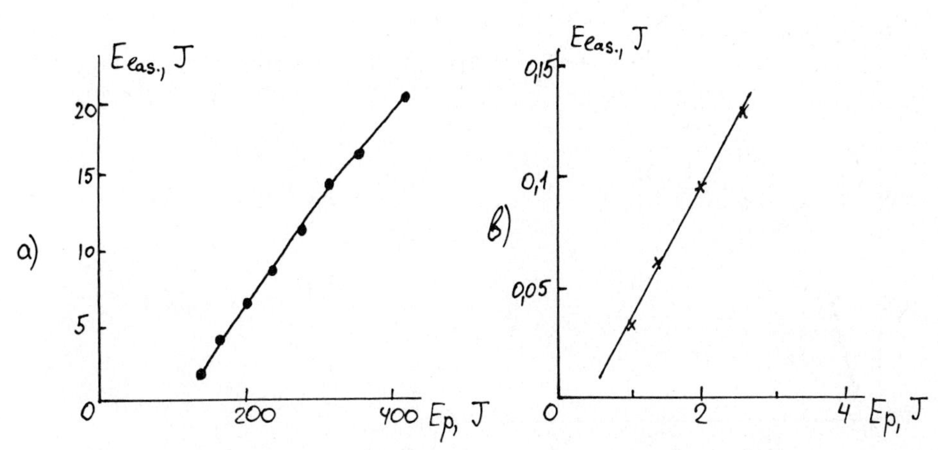

Fig.2. a) Lasing energy versus pump energy for GSGG:Cr^{3+}, Nd^{3+} transfer elements in the free running regime. The pumping pulse duration is 1.5 μsec. b) The pumping pulse duration is 70 μsec

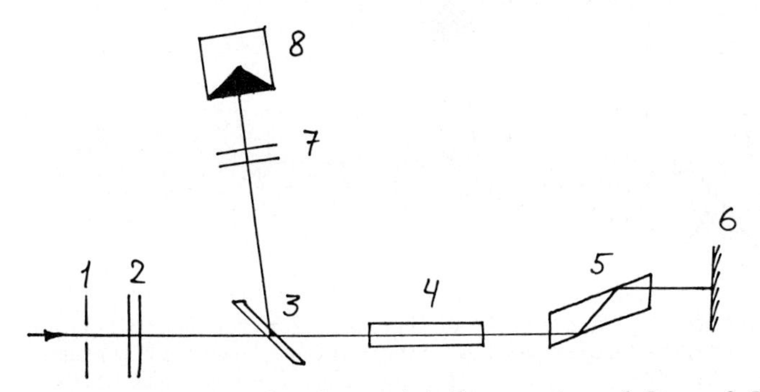

Fig. 3. The optical scheme. (*1*) The aperture Ø 5 mm; 2.7 - the neutral light filters; (*3*) the interference polarizer; (*4*) the active element of the amplifier based on the GSGG:Cr^{3+}, Nd^{3+} crystal; (*5*) the Fresnel rhomb; (*6*) the 100% mirror; (*8*) the photodiode and the calorimeter

The combination of high pump energy storage on the neodymium upper lasing level with the optimal effective lasing transition cross-section $\sigma_{eff} = (1.7 \pm 0.3) \times 10^{-19} cm^2$ makes it possible to create amplifiers and good efficiency and gain. Fig.3 shows the scheme for measuring the gain coefficient /5/. Fig.4 shows the results of measurements for different pump systems and active elements of different size. The obtained results, particularly the measured gain coefficient $N\sigma = 0.7\ cm^{-1}$, confirm that highly efficient small-sized laser amplifiers based on GSGG:Cr^{3+}, Nd^{3+} crystal can be created. The total gain of the double pass amplifier, defined as the ratio of the signal amplitude at the end of the "excited" (W_p = 155 J) to that at the end of the "cold" amplifier by the input radiation energy $E_{inp} = 1 - 10\ \mu J$, is 1200 ± 200. Fig. 5 shows the two-

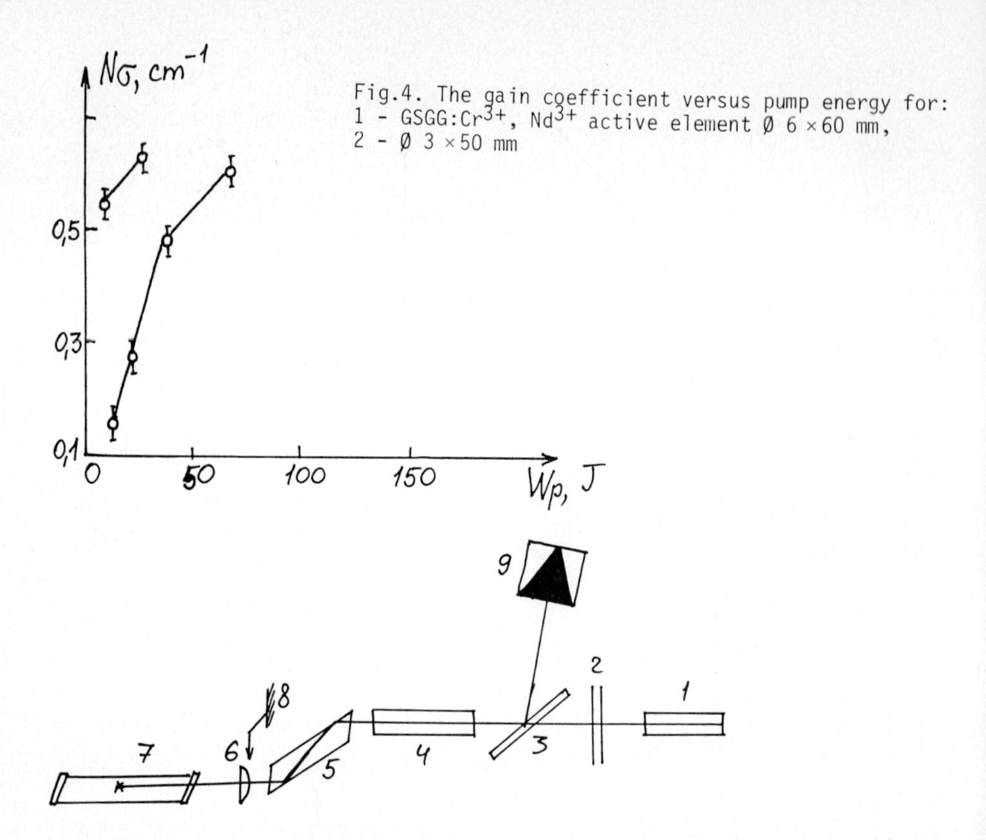

Fig.4. The gain coefficient versus pump energy for: 1 - GSGG:Cr^{3+}, Nd^{3+} active element Ø 6 × 60 mm, 2 - Ø 3 × 50 mm

Fig.5. The optical scheme. 1 - The master oscillator; 2 - the neutral light filters; 3 - the interference polarizer; 4 - the active element of the amplifier; 5 - the Fresnel rhomb; 6 - the focusing lens; 7 - the cell with $TiCl_4$; 8 - the dielectric 100% mirror; 9 - the calorimeter

pass amplifier with a phase conjugation scheme. The results of the energy measurements are given in Fig. 6 /6/. Note that the replacement of the ordinary mirror by the SBS mirror leads to a decrease of the output energy by more than 25%. However, the beam divergence becomes practically diffraction limited. The measurement shows that the diffraction cone of the beam contains 90% of all the output energy. The role of the SBS mirror was played by a 20 cm length cell containing $TiCl_4$; the focal length of the lens was 20 cm.

The GSGG:Cr^{3+}, Nd^{3+} laser operating in the periodic regime /7/ is now discussed. Development of solid-state lasers working in the Q-switching regime with high average lasing power has enabled solution of the principle problem of energy distribution inhomogeneity along the cross-section of the output beam, caused by the inhomogeneity of the crystal, nonuniform pumping, and depolarization of the radiation in the element due to thermally induced birefringence. The above-mentioned effects, to a greater or smaller degree, occur in every active element and lead not only to the degradation of the laser beam quality but also to the

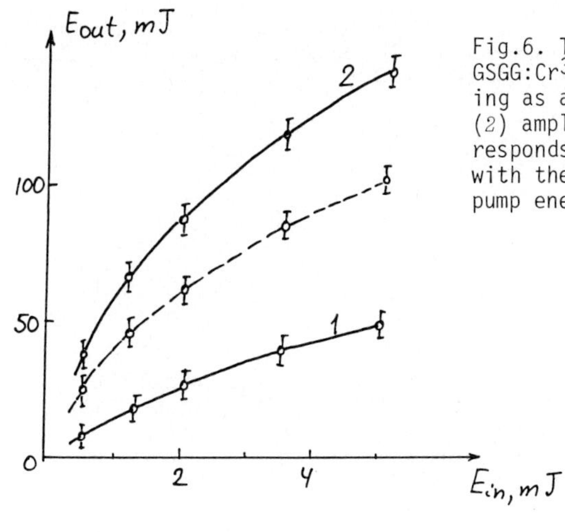

Fig.6. The gain characteristics of the $GSGG:Cr^{3+}$, Nd^{3+} active element operating as a one pass (*1*) and two pass (*2*) amplifier. The dashed line corresponds to the two pass amplifier with the SBS mirror. The amplifier pump energy is 38 J

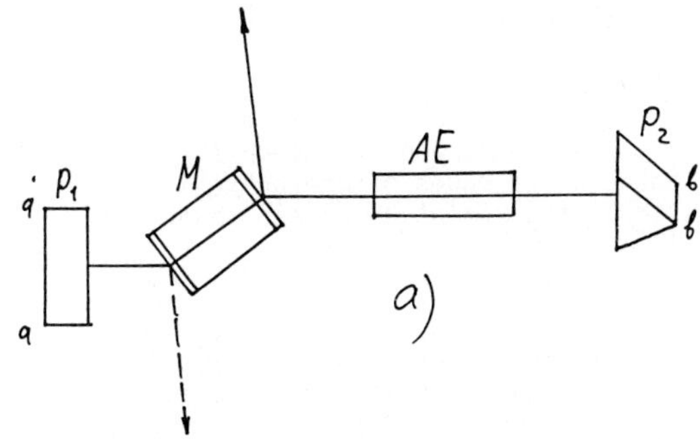

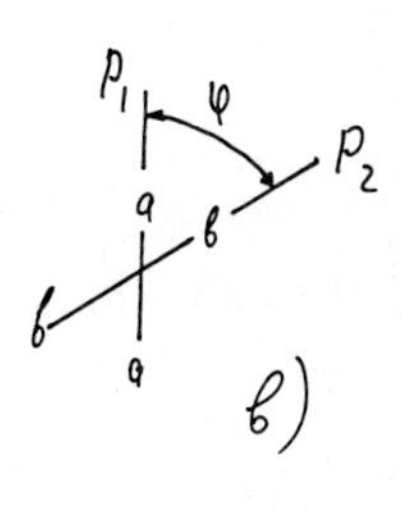

Fig.7. a) The optical scheme of the laser. AE - the active element; M - the electropic shutter; P_1, P_2 - 90° prism; 1,2 - directions of the output beams. **b**) The mutual location of the P_1 and P_2 prism edges (line of sight is in the direction of the optic axis of resonator)

decrease of laser energy performance characteristics. They are also quite considerable in the $GSGG:Cr^{3+}$, Nd^{3+} crystal /1/.

Within the limits of the scheme shown in Fig. 7, which has turned out to be quite effective in the case of the $GSGG:Cr^{3+}$, Nd^{3+} laser, we have managed to overcome these distortion difficulties to a great extent.

The problem of depolarization of radiation in the active element was solved by the well-known method of polarization coupling of the radiation out of the cavity /8/. In our case this consisted of obtaining linearly polarized output radiation by the reflection from the polarizer of the modulator. Instead of $\lambda/4$ plates and 100% mirrors, 90-degree prisms were used (Fig. 8). The edge of prism III was made parallel to the direction of the linear polarization passing through the modulator polarizers (Fig. 7). By this location of the prism, the polarization of the radiation coming through the modulator and reflected

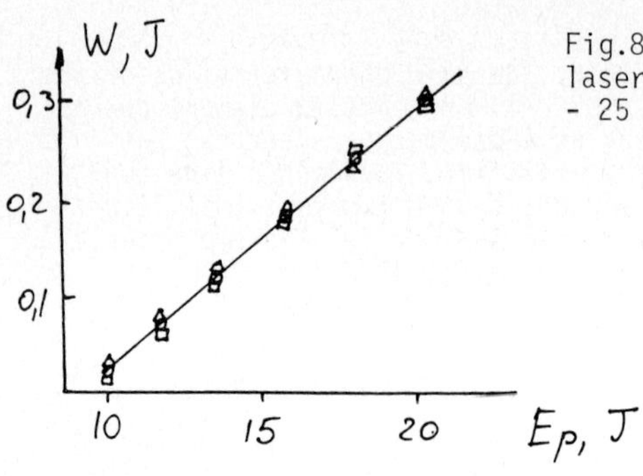

Fig.8. The energy characteristics of the laser. o - the repetition rate 12.5 Hz, - 25 Hz, - 50 Hz

by the prism remained linear. The edge of the prism p2 was turned through the angle φ with the direction of the linear polarization of the radiation passing through the modulator. Therefore, the radiation reflected by the prism p2 became elliptically polarized. The degree of depolarization of the radiation by the prism depends on the refractive index of the prism p and on the angle φ. It is essential that in the described scheme the radiation depolarized by the active element is not lost but represents the linearly polarized output laser radiation.

The use of the prism resonator also solves the problem of inhomogeneous energy distribution at the output beam cross-section caused by the imperfection of the active element and the nonuniform pumping /9/. Indeed, the angle φ between the edges of the prism directly defines the number of passes of the radiation in the resonator for which the beam trajectory repeats itself: $m = \text{int}\,\frac{2\pi}{\varphi}$. The beam describes the cylindrical surface around the resonator axis, thus permitting averaging of all the intensity inhomogeneities, and results in considerably better laser beam homogeneity along its cross-section.

The active elements made of the GSGG crystal had the dimensions 5.5 x 65 mm, with chromium and neodymium concentrations of 2 x 10^{20}cm^{-3} and 2 x 10^{20}cm^{-3}, respectively. They were put in the silver-plated quartz reflector Ø 30 x 60 mm.

The obtained results are given in Fig. 8. In spite of the considerable losses (~14% per pass), a 10 nsec giant pulse with energy of 0.3 J has been realized. The repetition rate of the pulses was between 12.5 to 50 Hz. No deterioration of the energy characteristics has been observed. At the pump energy of 20 J the efficiency of 1.5% and the slope efficiency of 3% have been reached. No giant pulse energy drop with respect to the free-running oscillation energy has been found. The implementation of the prism resonator has led to the principle improvement of the energy distribution homogeneity at the beam cross-section. Thus, using the ordinary "100% mirrors" the ratio of the energy at the "hot points" of the cross-section to the average along the cross-section energy reached one power value. In the case of the prism resonator no "hot points" were observed.

It is noteworthy that these results were obtained on different samples having varying optical quality and, particularly, different depolarization properties in the cold condition. The conditions under which the experiments were carried out differed only in values of the angle φ which, depending on the particular crystal and pump level, changed from 5 to 45 degrees.

The total laser beam divergence of our laser was 8-10 angle minutes at the pulse repetition rate of 12.5 Hz; and 20 and 30 angle minutes at the rates of 25 and 50 Hz, respectively. In all the cases, by compensation of spherical aberration with an intracavity lens divergences not greater than 5 angle minutes were obtained.

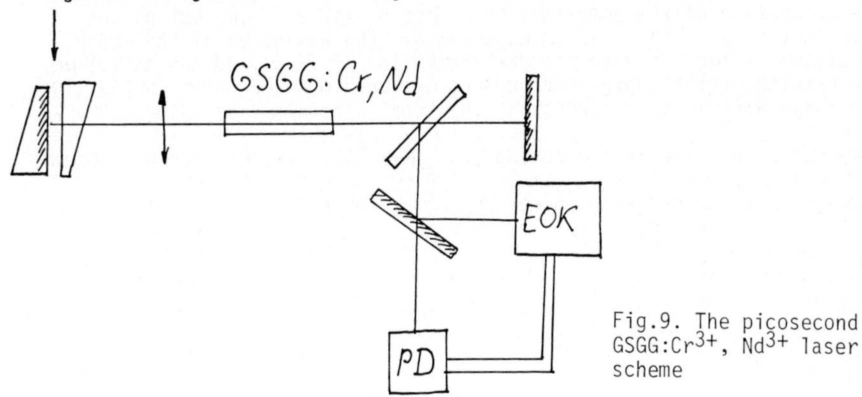

Fig.9. The picosecond GSGG:Cr^{3+}, Nd^{3+} laser scheme

The GSGG:Cr^{3+}, Nd^{3+} laser was also operated in the mode-locking regime /10/. The scheme of the experiment is given in Fig. 9. A solution of dye in nitrobenzene with a 14 nsec decay time was used to obtain passive axial mode locking. The output laser oscillation parameters were the following: pulse train energy E_{output} = 7 mJ with the pump energy E_p = 12.5 J; pulse train duration T_{pt} = 80 - 100 nsec, @ the number of pulses in the pulse train 16 - 20. Digital processing of the pulse duration gave the value of τ_{pulse} = 7.5 $\pm$ 0.7 nsec. Use of a dye with the decay time of 2.7 nsec made it possible to shorten the pulse duration up to 4.2 $\pm$ 0.4 nsec.

Saturated absorption at the oscillation wavelength was studied /11/. Reference /12/ reported self-Q-switching of the laser based on the GSGG:Cr^{3+}, Nd^{3+} crystal. The effect of self-induced anti-reflectivity occurred when the crystal was doped with a whole number of impurities /12/. The laser active element was made of the GSGG:Cr^{3+}, Nd^{3+} crystal containing photochromic centers and had the dimensions 4 x 50 mm and an initial transmittance on the oscillation wavelength of 16%. The pump duration was 80 sec. The pulse energy obtained was 65 mJ with pump energy of 6 J; the pulse repetition rate was 25 Hz and the oscillation pulse duration 20 nsec. It is known that neodymium ions can oscillate not only on the main ${}^4F_{3/2} \rightarrow {}^4I_{11/2}$ (λ_L = 1.06 μm) transition but also on the transition connecting the fine structures of the ${}^4F_{3/2}$ and ${}^4I_{13/2}$ multiplets (λ_L = 1.32 μm). However, the transition cross-section in the latter case is smaller. The implementation of resonators with selective elements helped us to depress the lasing on the $^4F_{3/2} \rightarrow {}^4I_{11/2}$ transition and to obtain oscillation on the $^4F_{3/2} \rightarrow {}^4I_{13/2}$ transition. Its efficiency was not very high. This can be explained by the high level of the parasitic stimulated emission of radiation on the main transition. We attempted to get rid of this channel of loss by using the absorption of the complexes that absorb near 1.06 μm, but do not bring any loss at the 1.32 μm wavelength. The resonator of the laser consisted of mirrors with the reflectivities of 99.8% and 73% at 1.3 μm. The results of the energy performance measurements are shown in Fig. 10. The slope efficiency exceeded 2%; the total efficiency by using a 9 J pump was 1.3%. The lasing on the $^4F_{3/2} \rightarrow {}^4I_{11/2}$ transition was absent throughout the whole pump energy range (up to 15 J).

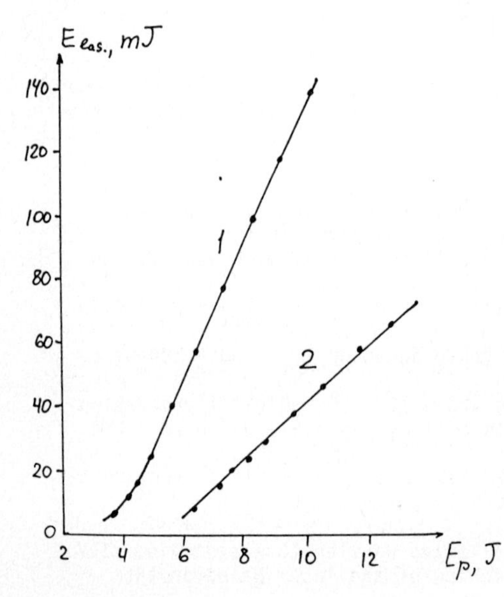

Fig.10. The energy characteristics of the GSGG:Cr^{3+}, Nd^{3+} laser, operating on the $^4F_{3/2}$ $^4I_{13/2}$ transition with $\lambda = 1.32$ μ. (*1*) The active element with the photochromic centers providing 85% absorption on $\lambda = 1.06$ μ. (*2*) The active element without the photochromic centers. The 85% absorption on $\lambda = 1.06$ μ was provided by the LiF crystal with F_2^--F centers

We must note that oscillation on the transition $^4F_{3/2} \rightarrow {}^4I_{13/2}$ in the same crystal under the identical conditions but without the photochromic centers was not demonstrated. The oscillation took place only when an additional selector was put into the cavity. Fig. 10 gives such energy characteristics for the case when a LiF crystal with F_2^- - F centers, having 15% transmittance at the 1.06 μm wavelength and transparent at λ = 1.32 μm, was chosen as such a selector. It is seen that the efficiency of this laser is considerably lower.

The data from Fig. 10 demonstrate not only the possibility of obtaining effective oscillation on the $^4F_{3/2} \rightarrow {}^4I_{11/2}$ transition, but what is more essential.the advantage of using photochromic complexes for depressing parasitic stimulated transitions over other commonly used selectors. The reduced losses, caused by photochromic complexes, can be easily created in GSGG:Cr^{3+}, Nd^{3+} crystal and their value can be varied over a wide range.

The remarkable peculiarity of the chromium-doped gallium garnets is the absence of sensitivity to γ-radiation /13, 14/. Unlike other well-known laser crystals, γ-radiation at room temperature does not influence the energy performance characteristics of the rare-earth garnets doped with chromium (Fig. 11). The photochromic centers are also not sensitive to γ radiation doses up to 2 x 10^7rad at prolonged high-temperature treatment /14/.

Thus, the rare-earth gallium garnets open wide possibilities toward developing a new promising class of solid-state lasers.

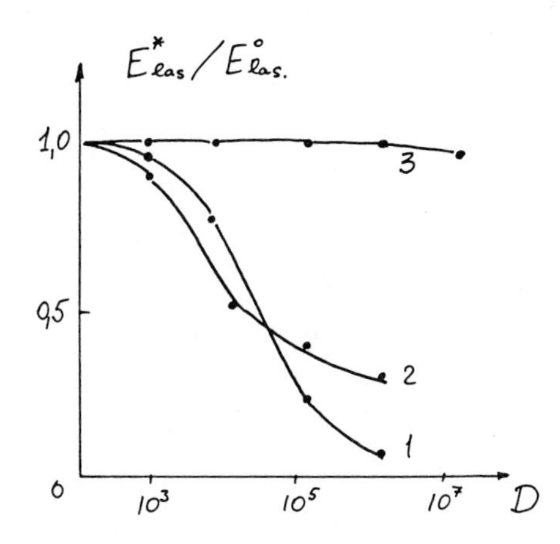

Fig.11. The relative change of the output laser energy versus the γ-radiation dose of the active elements. *1* - YAG, *2* - KGV, *3* - GSGG

REFERENCES

1. E. V. Zharikov, V. V. Osiko, A. M. Prokhorov, I. A. Scherbakov: Izv. AN SSSR, ser.fiz. 48, No. 7, 1330 (1984).
2. V. G. Ostroumov, Yu. S. Privis, V. A. Smirnov, I. A. Scherbakov: JOSA B 83, 81-94 (1986).
3. E. V. Zharikov, N. N. Il'ichev, V. V. Laptev, A. A. Malyutin, V. G. Ostroumov, P. P. Pashinin, A. S. Pimenov, B. A. Smirnov, I. A. Scherbakov: Kvantovoya Electron. 10, No. 1, 140-144; preprint FIAN, M., No. 14, 15 (1982).
4. A. V. Dobrovol'sky, A. S. Dorkin, M. B. Zhitkova, I. I. Kuratev, V. A. Lipatov, G. I. Narkhova, M. Yu. Nikitin, V. A. Pashkov, N. S. Ustimenko, E. M. Shvoi, A. V. Shestakov: Izv. AN SSSR, ser.fiz. 48, No. 7, 1349 (1984).
5. E. V. Zharikov, V. V. Lipatov, S. Yu. Natarov, V. V. Osiko, P. P. Pashinin, A. M. Prokhorov, E. I. Shklovsky, I. A. Scherbakov: Kvantovaya Electron. 12, No. 11, 2198 (1985).
6. T. T. Basiev, E. V. Zharikov, S. B. Mirov, S. Yu. Natarov, V. V. Osiko, P. P. Pashinin, A. M. Prokhorov, E. I. Shklovsky, I. A. Scherbakov: Kvantovaya Electron. 13, No. 2, 412 (1986).

7. V. N. Grigoriev, G. N. Egorov, E. V. Zharikov, V. A. Mikhailov, S. K. Pak, Yu. A. Pinsky, E. I. Shklovsky, I. A. Scherbakov: Preprint FIAN, M., No. 52 (1986).
8. V. G. Mikhalevich, G. P. Shipulo: Kvantovaya Electron. 1, No. 2, 455 (1974).
9. V. G. Mikhalevich, G. P. Shipulo: Kvantovaya Electron. 1, No. 2, 688 (1974).
10. A. V. Babushkin, N. S. Vorob'ev, E. V. Zharikov, V. V. Laptev, V. V. Osiko, A. M. Prokhorov, Yu. N. Serdyuchenko, M. Ya. Schelev, I. A. Scherbakov: Kvantovaya Electron., No. 2 (1986).
11. E. V. Zharikov, A. M. Zabaznov, V. V. Osiko, A. M. Prokhorov, Z. S. Saidov, V. A. Smirnov, A. P. Shkadarevich, I. A. Scherbakov: Preprint IOFAN, M., No. 238 (1985).
12. J. A. Caird, W. F. Krupke, M. D. Shinn, L. K. Smith, R. E. Wilder: In CLEO-85, Digest of Techn. Papers, (Baltimore, Maryland, 1985) p. 232.
13. M. Kh. Ashurov, E. V. Zharikov, V. V. Laptev, I. N. Nasyrov, V. V. Osiko, A. M. Prokhorov, P. K. Khabibullaev, I. A. Scherbakov: Izv. AN SSSR, ser.fiz. 48, No. 7, 1343 (1984).
14. E. V. Zharikov, I. I. Kuratev, V. V. Laptev, S. P. Nasel'sky, A. I. Ryabov, G. N. Toropkin, A. V. Shestakov, I. A. Scherbakov: Izv. AN SSSR, ser.fiz. 48, No. 7, 1352 (1984).

Single-mode Operation of Cr-Doped GSGG and $KZnF_3$

P. Fuhrberg, W. Luhs, B. Struve, and G. Litfin

Spindler & Hoyer, Postfach 33 53,
D-3400 Göttingen, Fed. Rep. of Germany

1. Introduction

Chromium doped garnets and perovskites are highly promising materials for tunable high-power lasers, as well as ion laser-pumped cw single-mode lasers.

So far chromium doped GSGG ($Gd_3(Sc,Ga)_2Ga_3O_{12}$) and $KZnF_3$ have been studied in cw multimode operation using an ion laser as pumpsource by STRUVE et al. /1/ and BRAUCH et al./2/, respectively. The authors obtained highest output power and lowest threshold when pumping with the red lines of a krypton laser (647 nm and 676 nm).

This contribution focusses on cw single-mode operation of GSGG and $KZnF_3$. In addition, recent results of flashlamp pumped chromium doped and chromium sensitized garnets in comparison to alexandrite and Nd:YAG are briefly given.

2. CW Single-Mode Operation

To investigate the spectral behaviour of these new laser materials, our laser experiments were performed by using two different types of resonators:

1. A three mirror astigmatically compensated set-up, whereby the Brewster's angle orientation of the laser crystal defines the direction of polarization (Fig. 1).

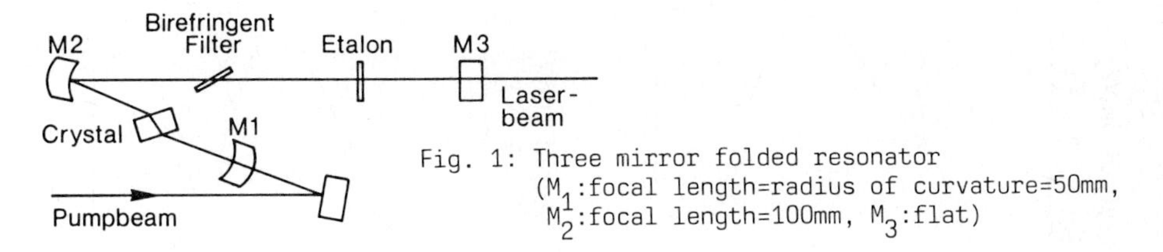

Fig. 1: Three mirror folded resonator (M_1:focal length=radius of curvature=50mm, M_2:focal length=100mm, M_3:flat)

2. A four mirror travelling wave ring cavity (Fig. 2). Uni-directional operation was accomplished by means of a Faraday rotator with FR5 glass. The backrotation was performed by a non-coplanar mirror geometry.

The output power versus pump power of GSGG and $KZnF_3$ measured in the standing and travelling wave resonator respectively are given in Fig. 3 . Tuning curves of both laser materials are shown in Fig. 4 and Fig. 5 . A single-plate birefringent filter was used as tuning element.

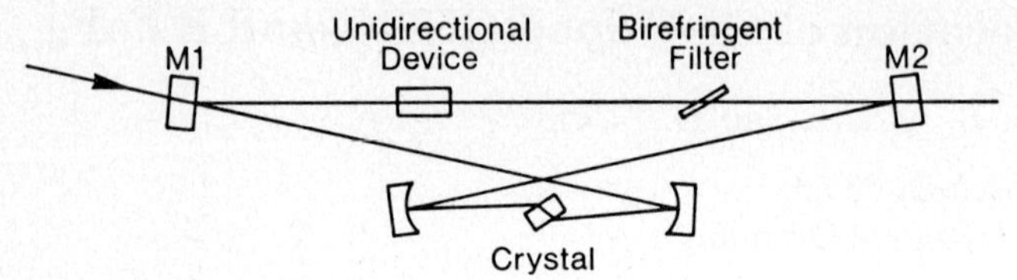

Fig. 2: Four mirror ring resonator
(M_1,M_2:flat, M_3,M_4:radius of curvature=100mm)

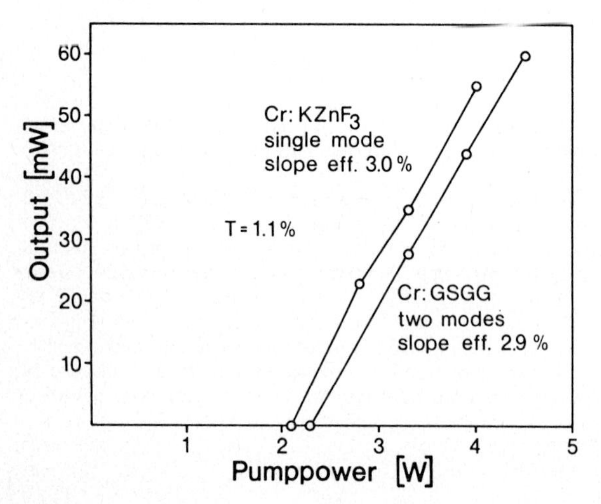

Fig. 3: Output power versus pump power for $KZnF_3$ and GSGG
(pump wavelength 647/676nm)

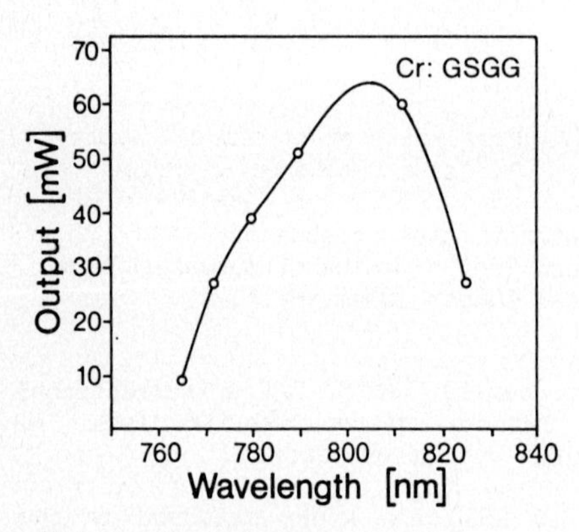

Fig. 4: Output power versus laser wavelength for GSGG (two holeburning modes, 4.5 W input power)

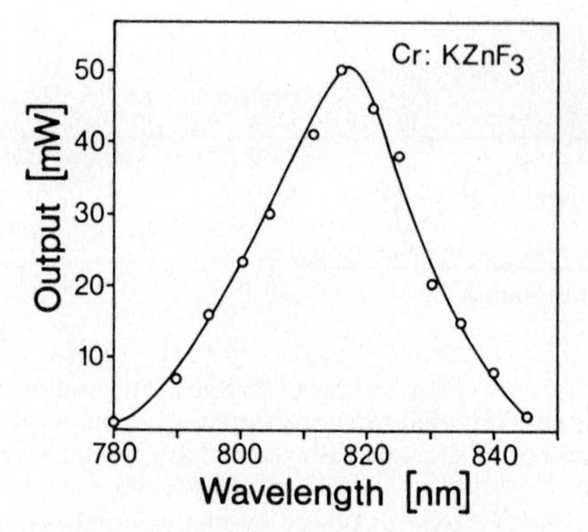

Fig. 5: Output power versus laser wavelength for $KZnF_3$ (single mode, 4.5 W input power)

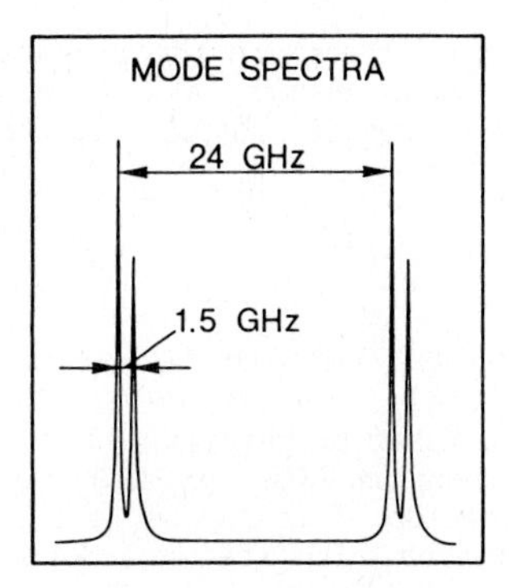

Fig. 6: Mode spectra of GSGG in the folded cavity

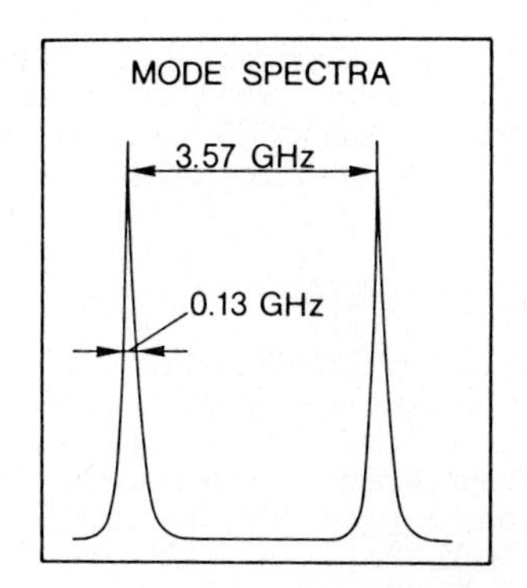

Fig. 7: Mode spectra of the ring cavity (the linewidth of 130 MHz is due to the resolution of the spectrum analyzer)

In the folded cavity two hole-burning modes with a mode spacing of 1.5 GHz (Fig. 6) were observed, when using a single etalon. Under acceptance of considerable decrease of output power single-mode operation can be achieved inserting a second etalon.

With $KZnF_3$ stable single-mode operation in the travelling wave resonator could be obtained even without etalon (Fig. 7). In the case of GSGG we could not realize stable unidirectional oscillation in the ring cavity although both materials exhibit similar spectroscopic properties . This unexpected difference in laser behaviour can be explained by differences in their optical- and thermo-physical properties, mainly the thermally induced effects on the index of refraction. These effects can appear as thermally induced birefringence as well as thermal lensing. Thermally induced birefringence is characterized by the photoelastic tensor of the material. In GSGG the elasto-optic tensor is anisotropic $((p_{11}-p_{12})/2p_{44}$=0.42 /3/, =0.23 /4/) . Therefore thermal birefringence will occur producing an inhomogeneous polarization distribution inside the beam profile. This behaviour partially cancels the optical diode effect of the unidirectional device preventing unidirectional operation.

Furthermore thermally induced birefringence causes considerable intracavity losses due to reflection of the depolarized laser beam at surfaces under Brewster's angle . For a single surface we measured losses of about 0.6% for GSGG and 0.1% for $KZnF_3$. In $KZnF_3$, however, no thermal birefringence was observed, hinting towards photoelastic isotropy of this material $((p_{11}-p_{12})/2p_{44}$ nearly unity).

In addition to thermally induced birefringence thermal lensing effects in GSGG were also observed. Due to thermal lensing the stability range of the resonator reacted sensitively on pump power variations. Thermal lensing is essentially described by the value of dn/dT. We determined the variation of the refractive indices with temperature to be $13.10^{-6}\ K^{-1}$ for GSGG respectively $-5.8.10^{-6} K^{-1}$ for $KZnF_3$. We also measured the coefficients of thermal expansion to be $7.9.10^{-6}\ K^{-1}$ for GSGG and $27.10^{-6}\ K^{-1}$ for $KZnF_3$.

To sum up, we obtained stable single-mode operation of chromium doped $KZnF_3$ with negligible additional losses over multimode operation. A homogeneous broadened laser transition and favorable optical properties are the reasons for this behaviour . Mainly the photoelastic isotropy was beneficial, whereas the anisotropy of GSGG caused considerable depolarization losses due to thermally induced birefringence.

3. Flashlamp Pumping

The goal of our investigations was to compare Cr:Nd:GSGG with Nd:YAG and the tunable material alexandrite with Cr:$Gd_3Sc_2Al_3O_{12}$ (GSAG). This material (GSAG) is similar to GSGG with respect to spectroscopic and laser properties /5/, but less susceptible to solarization. Therefore we focussed on GSAG. The GSAG rod (6 x 75 mm, n_{cr} = 0.5%) was manufactured by Union Carbide. This rod as well as a Cr:Nd:GSGG rod (6 x 75 mm, Airtron), was compared with the well-known materials Nd:YAG (6 x 75 mm) and alexandrite (5 x 75 mm, n_{cr} = 0.17%) in identical laser resonator configurations.

The experiments with free-running lasers were carried out in a silver plated single-ellipse pump cavity using a 5 x 75 mm i.d. flashlamp. A dielectric UV-filter was inserted between rod and lamp. In the experiment with GSAG where the heat load and the solarization drastically affects the laser efficiency we additionally used a cerium-doped quartz lamp (4 x 63 mm i.d.) to block the UV-radiation below 400 nm. The resonator consisted of a 1.5 m radius mirror and a 5 m radius output coupler. The transmission of the output coupler was 60% for the neodymium-doped and 6% for the chromium-doped laser. Fig.8 shows the output energy as a function of the pump energy in single-shot operation.

The codoped material Cr:Nd:GSGG showed under flashlamp excitation a slope efficiency of more than 6% thus exceeding by far the value of Nd:YAG . The tunable material Cr:GSAG has made substantial progress when considering first laser data /6/.

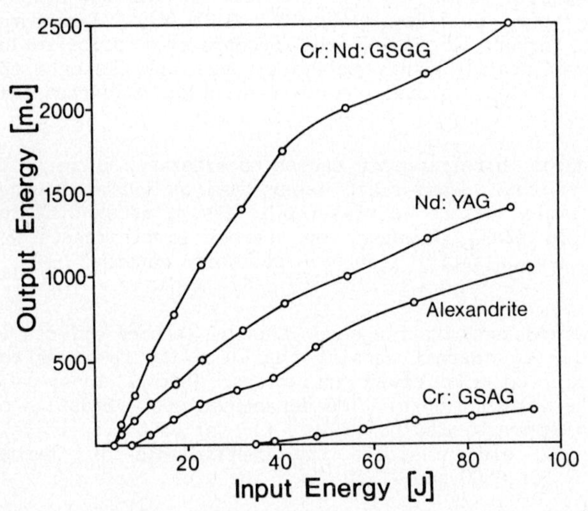

Fig. 8: Output energy versus input energy for Cr:Nd:GSGG, Nd:YAG, alexandrite, and Cr:GSAG

4. Conclusion

Our investigations prove that the discussed tunable cw single-mode solid-state lasers with an output range in the near infrared are simple in design and allow stable single frequency operation. This is especially true when using chromium doped $KZnF_3$ with its favourable thermo-physical properties.

Despite the fact that the GSAG performance seems to be mediocre, one has to recognize that little work has been done on this relatively new member of solid-state laser materials. Longterm we expect significant improvements on laser performance when crystal growth technology has progressed.

Acknowledgements: We thank G. Huber (University of Hamburg) and U. Dürr (University of Stuttgart) for crystal samples. Support for the work on flashlamp pumped systems by BMFT is kindly acknowledged.

References

1. B. Struve, G. Huber, V.V. Laptev, I.A. Shcherbakov, E.V. Zharikov: Appl. Phys. B30, 117 (1983)
   B. Struve, G. Huber: J. Appl. Phys. 57(1), 45 (1985)

2. U. Brauch, U. Dürr: Opt. Commun. 49, 61 (1984); Opt. Lett. 9, 441 (1984)

3. E.V. Zharikov, V.F. Kitaeva, V.V. Osiko, I.R. Rustamov, N.N. Sobolev: Sov. Phys. Sol. St. 26(5), 922 (1984)

4. W.F. Krupke, M.D. Shinn, J.E. Marion, J.A. Caird, S.E. Stokowski: J. Opt. Soc. Am. B 3(1), 102 (1986)

5. J. Drube, B. Struve, G. Huber: Opt. Commun. 50, 45 (1984)

6. G. Huber, presented at the Topical Meeting on Tunable Solid-State Lasers, May 16-17, Arlington, Virginia, Th A3-1 (1985)

Flashlamp-pumped Cr^{3+}:GSAG and Cr^{3+}:GSGG : Slope Efficiency, Resonator Design, Color Centers and Tunability

J. Drube[1], *G. Huber*[1], *and D. Mateika*[2]

[1]Inst. f. Angewandte Physik, Universität Hamburg, Jungiusstraße 11, D-2000 Hamburg 36, Fed. Rep. of Germany
[2]Philips GmbH Forschungslaboratorium Hamburg, Voigt-Kölln-Str. 30, D-2000 Hamburg 54, Fed. Rep. of Germany

$Cr^{3+}:Gd_3Sc_2Al_3O_{12}$ (GSAG) and $Cr^{3+}:Gd_3Sc_2Ga_3O_{12}$ (GSGG) are low field garnets which perform flashlamp-pumped tunable laser action at room temperature. We obtained 200 mJ output in the free running mode and a tuning range from 750 nm to 800 nm from 2 inch long Cr^{3+}:GSAG rods.

1. Introduction

Besides the well known Ga-garnets [1] (e.g. GSGG), the family of garnets with a low crystal field at the octahedral lattice site includes also Al-garnets as $Gd_3Sc_2Al_3O_{12}$ and $Y_3Sc_2Al_3O_{12}$ [2,3]. In spite of a decreased lattice constant in comparison to the analogous Ga-host, the Sc sublattice in GSAG offers a sufficiently large octahedral site which ensures a high population of the upper $^4T_2$-Cr^{3+}-laser level.

Flashlamp-pumplight induced color center absorption, which overlaps with the Cr^{3+} absorption bands, decreases the pumping efficiency. The comparison of these color centers in GSAG and GSGG shows a reduced influence in the Al-garnet. Both hosts yield improved laser performance under pumping conditions with blocked UV- and blue pumplight. Without any blocking color centers inhibit lasing in GSGG, whereas laser action still occurs in GSAG. From crystal growth considerations we presume that Al-garnets have advantages over the Ga-garnets concerning the susceptibility to color center formation. Furthermore, the heat conductivity of GSAG is expected to be slightly higher.

2. Crystal growth

From $Gd_3Sc_{1.6}Ga_{3.4}O_{12}$ and $Gd_3Sc_2Al_3O_{12}$ melt compositions Cr-doped GSGG and GSAG single crystals of 110 mm length and 25 mm diameter were grown in [111]-direction using the Czochralski technique. The Ga-garnet was grown with a flat interface, whereas the Al-garnet has a convex interface and exhibits a core region in the center due to the formation of (221)-facets. Recently, the congruent melting composition of the Ga-garnet was determined to be $Gd_{2.96}Sc_{1.90}Ga_{3.14}O_{12}$ [4]. The composition of the GSAG crystal grown from the nominal $Gd_3Sc_2Al_3O_{12}$ melt was determined by x-ray fluorescence analysis and is given in table 1. The lattice parameter increases slightly from 12.397 Å to 12.400 Å over the total length of the crystal. High-purity Gd_2O_3(5N), Al_2O_3(5N), Ga_2O_3(5N), Sc_2O_3(4N), and Cr_2O_3(6N) were used. High optical quality crystals could be obtained without codoping for interface stabilization or impurity compensation.

Table 1. Cr^{3+}:GSAG crystal composition in formula units

element	melt composition	crystal composition top	crystal composition bottom
Gd	3.000	2.92	2.93
Sc	1.988	1.92	1.94
Al	3.000	3.15	3.12
Cr	0.012	0.01	0.01

3. Purity requirements

Divalent cations like Mg^{2+} or Ca^{2+} as impurities in raw materials lead to additional absorption bands beside the Cr^{3+} spectrum (Fig.1). A high Ca concentration was added to the melt to increase the accuracy for the separation of the Cr^{3+}-absorption and the impurity-induced absorption. However, the actual Ca content in this crystal is much less due to a small Ca distribution coefficient. Although this impurity absorption appears only in combination with chromium doping, the nature of the associated optical active center is far from being clear. In our point of view there is no strong evidence for a simple compensation mechanism via Cr^{4+}.

Besides using high-purity raw materials there are other possibilities to avoid this additional impurity absorption. Annealing

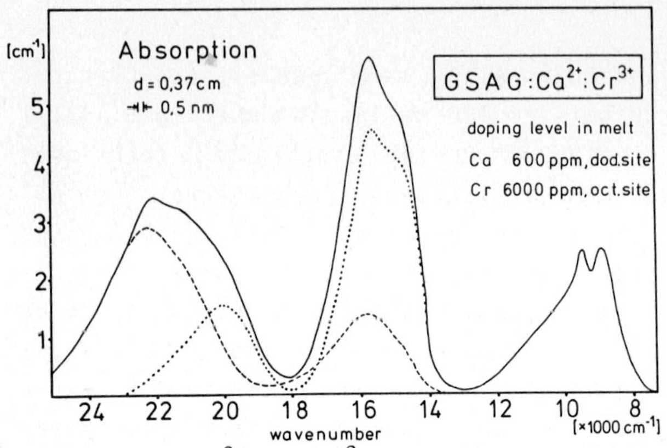

Fig. 1. Ca^{2+} or Mg^{2+} induced impurity absorption in Cr-doped GSGG and GSAG. Solid line: measured spectrum; dashed line: Cr^{3+}-absorption bands; dotted line: impurity absorption in the visible region. No absorption is present in the region from 1.3 μm to 2.5 μm.

(reducing atmosphere, e.g. N_2+10% H_2, 1400 C) of samples showing this absorption leads to impurity absorption-free crystals. In the same way, crystal growth from contaminated raw materials in reducing atmosphere (e.g. N_2+5% H_2) gives "clean" crystals. However, Ga-garnets need an oxidizing atmosphere and even for GSAG the most suitable atmosphere is slightly oxidizing (e.g. N_2+1% O_2) to avoid iridium inclusions. Codoping of Si^{4+}, Zr^{4+}, or Ce^{4+} compensates divalent ions as Ca^{2+} or Mg^{2+}. An overcompensation does not cause new problems if an oxidizing growth atmosphere is used.

4. Color centers

UV-light induced transient and stationary color center absorptions strongly influence the efficiency of Cr:GSAG and Cr:GSGG lasers. Flashlamp-excitation induces broad stationary color centers which overlap with the $^4T_1$- and to a smaller extent with the $^4T_2$-Cr^{3+} absorption (Fig.2, Fig.3). They show a qualitatively similar spectral appearance in both hosts, but are much less pronounced in Cr:GSAG. In very strongly contaminated crystals these detrimental absorptions run out into the IR-region and even inhibit lasing.

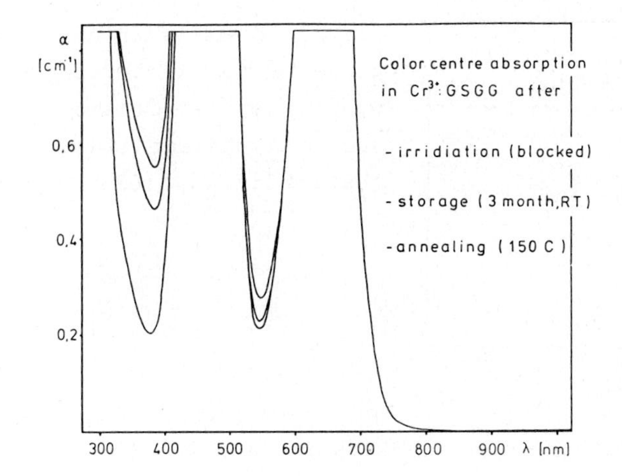

Fig.2. Stationary color center absorption after flashlamp irridiation (upper curve) and pure Cr^{3+} absorption after annealing(lowest curve)

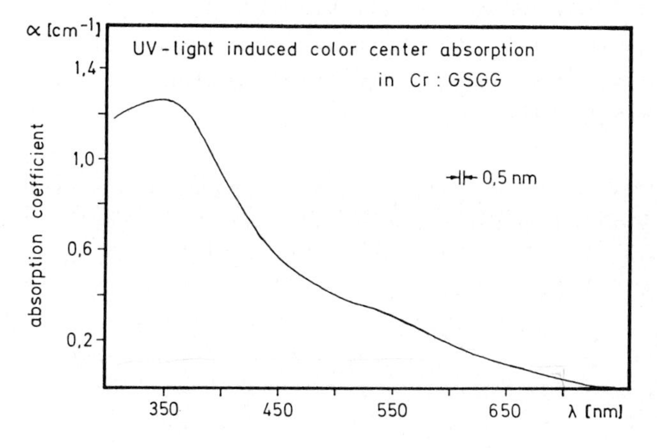

Fig.3. Separated stationary color center absorption

Transient color centers could only be observed in Cr:GSGG. The degree of accumulation and recovery of this transient absorption depends on the flashlamp repetition rate (Fig.4). The color center decay time is in the range of 10 ms to seconds and depends on raw material and growth procedure.

The smaller susceptibility to color center formation in Cr:GSAG is not really understood. However, the most striking difference in crystal growth is the chemical instability of Ga_2O_3 in comparison to Al_2O_3. At higher temperatures Ga is capable to form a suboxide Ga_2O, which may lead to an increased oxygen vacancy concentration.

A K_2CrO_4 solution in the cooling water was used to block the UV- and blue pumplight. For xenon flashlamp pumping the laser

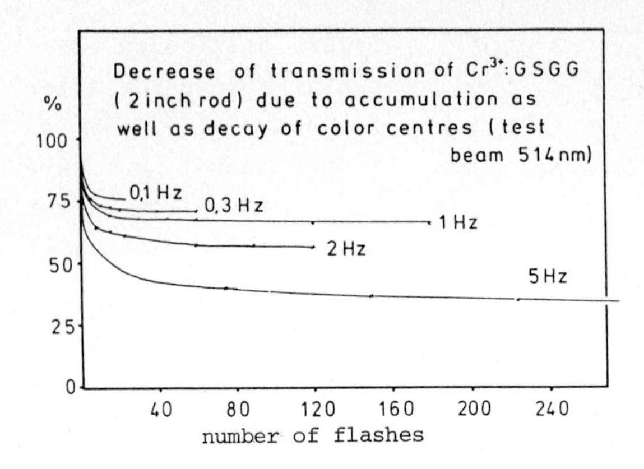

Fig.4. Transmission dependence for different pumping repetition rates due to transient color centers in Cr^{3+}:GSGG

efficiency and the threshold depend strongly upon the amount of blocking. A suitable blocking wavelength turned out to be 420 nm. For Cr:GSAG, the previous history (annealing procedure, unblocked operation) mainly determines the achievable output. Cr:GSGG exhibits a uniformly decreasing laser output which has recovered after an operation break. Unblocked operation causes the Cr:GSGG laser to fade out immediately. Cr:GSAG could be grown with such a minimum amount of interfering color center absorption so that unblocked laser operation was possible. The transient color centers in the Ga-garnet inhibit unblocked operation and are also responsible for the recovery mechanism in this host.

5. Laser experiments

Laser experiments were carried out on 2 inches long Cr:GSAG and Cr:GSGG rods from boules free from impurity absorption and with a minimum amount of color centers. For optimum results, the competing color center absorption was decreased in both hosts by blocking the pumplight below 420 nm.

The effect of the repetition rate-dependent transient color center absorption strength on Cr:GSGG laser action can directly be seen in Fig.5: a faster rep-rate decreases the pumping efficiency and lowers the output.

Cr:GSAG rods under identical pumping conditions do not show such a behavior. Due to the absence of transient color centers the energy per output pulse is stable. In addition, the output is higher due to a smaller amount of stationary color centers (Fig.6). A direct comparison of both hosts which are doped with $5\times10^{19}\,cm^{-3}$

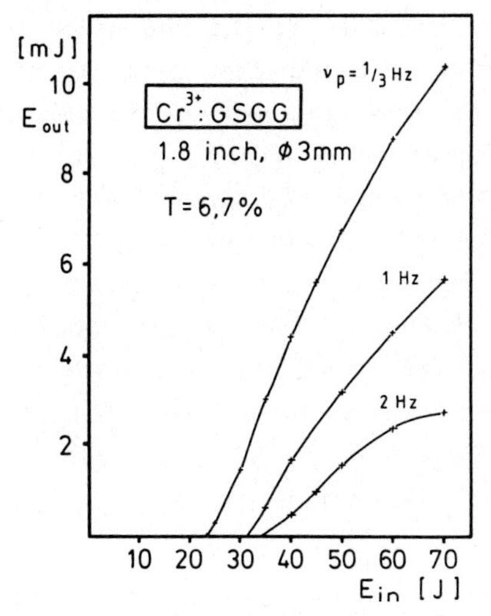

Fig. 5. Repetition rate-dependent Cr^{3+}:GSGG laser output

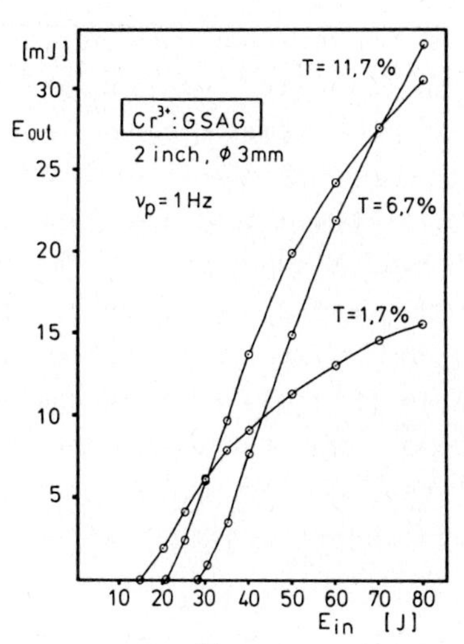

Fig. 6. Cr^{3+}:GSAG laser output for different outcoupling transmissions

Cr^{3+} can be made for the curves with T=6.7% output coupling and 1 Hz rep-rate. At 70 joules the output is five times larger for Cr^{3+}:GSAG.

Up to now, 206 mJ output (T=6.7%) and 0.24% slope efficiency (T=11.7%) are obtained from 5 mm diameter rods in single-shot operation (Fig.7). Mirror curvatures below 1 m were necessary to

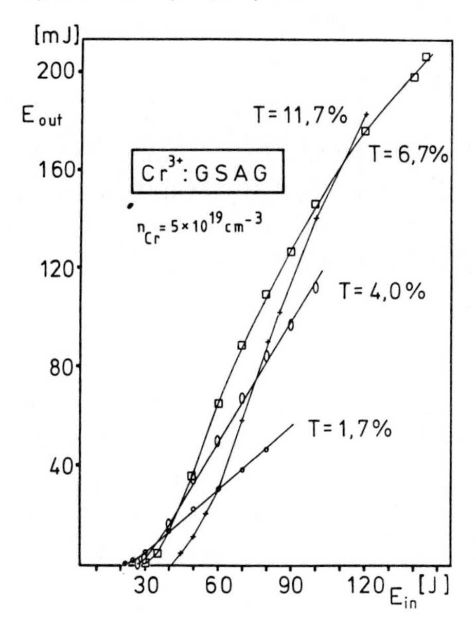

Fig. 7. Xenon flashlamp-pumped Cr^{3+}:GSAG laser rod performance in a single-elliptical pump chamber

suppress losses due to thermal lensing during the 100 μs pump pulse. A resonator configuration with 5 m mirrors leads to an output 3 to 4 times lower. The modes are centered at 780 nm in the free-running operation and are about 10 nm wide.

Tunability was achieved with a birefringent filter and a Brewster prism. A one-stage birefringent filter is not capable to cover the whole gain curve. Best tuning results were obtained with a prism and high-reflecting mirrors in the range from 750 to 800 nm (Fig.8). From an optimized set up we anticipate an expansion of the tuning range to both wavelength sides. This is underlined by the tuning range from 735 nm to 820 nm which we obtained under longitudinal krypton laser pumping of Cr^{3+}:GSAG.

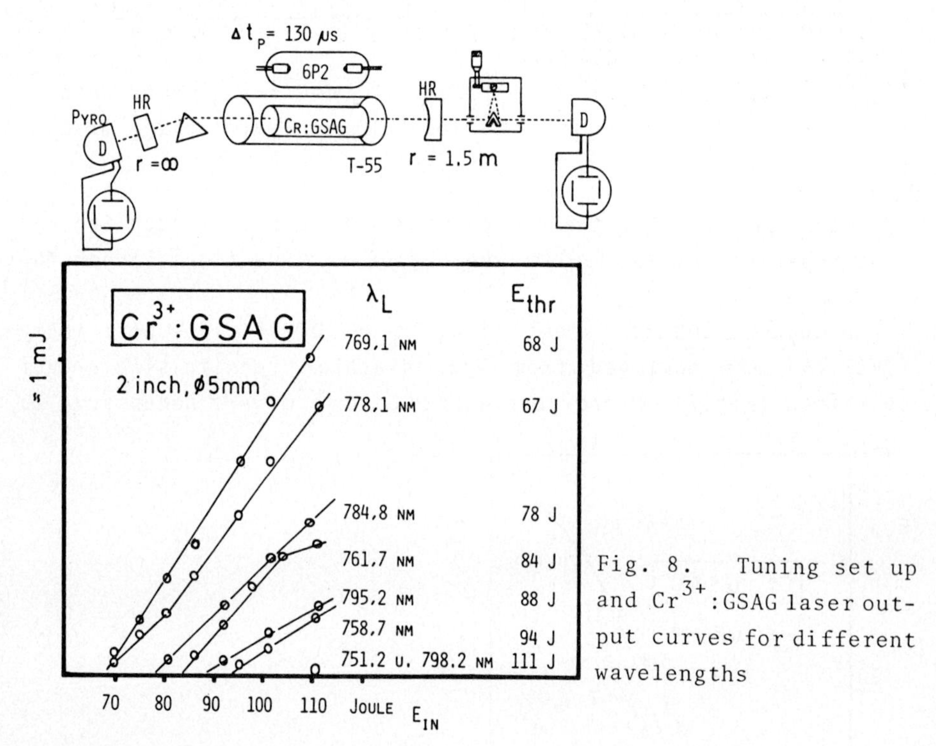

Fig. 8. Tuning set up and Cr^{3+}:GSAG laser output curves for different wavelengths

6. Conclusion

Due to broad absorption bands in the visible and long lifetimes Cr^{3+}:GSAG and Cr^{3+}:GSGG are promising solid-state laser materials for flashlamp pumping. Large boules with good optical quality and Cr^{3+} distribution coefficients near unity can be grown by the

Czochralski technique. Color centers which are created under optical pumping decrease the efficiency. Efforts to achieve a reduction of color centers seem to be more promising for Cr:GSAG. Up to now, the Cr^{3+} laser ion profits mainly by its $^4T_2$-absorption band. Garnet crystal material which allows for a blocking wavelength shifted to the blue would improve drastically the laser performance. Higher output is anticipated from a further optimization of the resonator geometry and from 3 to 4 inches long rods which are more favorable for these low-gain laser materials.

References

1. B.Struve and G.Huber, J.Appl.Phys. 57(1), 45 (1985)
2. J.Drube, B.Struve and G.Huber, Opt. Com. 50(1), 45 (1984)
3. J.Drube, G.Huber and D.Mateika, CLEO'85, paper PD3, Baltimore, USA (1985)
4. V.J.Fratello, C.D.Brandle and A.J.Valentino, CLEO'86, paper FA5, San Francisco, USA (1986)

Laser Action in Flashlamp-Pumped Chromium: GSG-Garnet

M.J.P. Payne and H.W. Evans

Royal Signals and Radar Establishment, St. Andrew's Road, Great Malvern, Worcs, WR14 3PS, UK

This paper is concerned with two quite different aspects of recent work on materials for broadly tunable solid-state lasers. We shall firstly describe a programme of work aimed at the improvement of the performance in long pulse mode of flashlamp pumped lasers based on crystals of chromium-doped gadolinium scandium gallium garnet (Cr:GSGG). The second part of the paper details some constraints on the properties of potential laser media if they are to be considered suitable for efficient and broad-band operation.

1. Introduction

Although cw-pumped tunable lasers employing Cr:GSGG crystals are available commercially [1], flashlamp-driven operation with the purpose of generating high power output pulses has so far yielded disappointing performance [2-4], considerably worse than that expected from the basic spectroscopic properties of the material [e.g. 5, 6].

Various effects may be considered as possibilities in limiting laser action in Cr:GSGG. They include the optical distortions and losses arising from crystal inhomogeneities which may be either grown into the material (eg inclusions; grown-in strains) or thermally induced (eg lensing). Unwanted absorption of both pump and laser radiation may occur. At the lasing wavelengths, absorption due to Cr^{3+} ions in both ground and excited states acts to increase threshold and restrict the tuning range. Excited state absorption of pump radiation may lead directly to degradation of pumping efficiency. A further important loss mechanism is the formation of colour centres which absorb both pump and laser radiation. These effects are sensitive to the spectrum of pump radiation. The purpose of the present work was to test the extent to which manipulation of the pump spectrum can be employed to improve the long pulse lasing performance of Cr^{3+}:GSGG laser rods. No special effort has been made to maximise energy extraction.

2. Colour Centre Absorption Spectra

The crystals used were grown by the Czochralski method using 5N grade starting materials.

Initial experiments were performed on an undoped sample to determine the absorption spectrum of the colour centres. Measurements were made using a Cary 14 spectrometer. The results are shown in Fig. 1, with a change of vertical scale at 700nm and an error about .01 cm^{-1} in the relative values of absorption coefficient. The crystal as grown showed an absorption spectrum represented by curve A. Absorption decreased through the visible to a minimum near 1200nm. The sample was exposed to unfiltered flashlight,

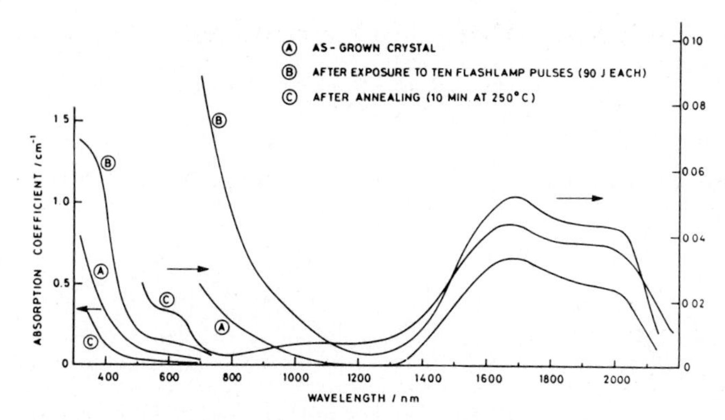

Fig.1. Absorption spectra of undoped GSGG

in the laser pump chamber to be described, resulting in a marked increase of absorption in the visible (curve B). In particular, the loss at the free lasing wavelength, 800nm, increased greatly. This process proved to be readily reversible and the absorption near 800nm was found to be essentially completely removed by annealing the sample for 10 minutes at 250°C (curve C). Quantitatively similar changes in absorption coefficient are observed for chromium doped samples, at least in the regions outside the chief absorption bands.

The origin of the colour centres is unknown but they are clearly not directly associated with the presence of chromium. The most likely defect centres are oxygen vacancies; our crystals, in common with others, have been shown to be oxygen-deficient. The cause of the infrared absorption band near 1700nm is also not known, but its constancy and presence in all samples indicates a likelihood of its association with an impurity.

3. Laser Performance

Experience with other materials indicates that crystal coloration is commonly generated by ultra-violet or blue radiation. Since the absorption generated by only a few flashlamp shots is sufficient to degrade the performance of a Cr:GSGG laser very severely, a series of tests of laser performance was made with various portions of the short wavelength pump spectrum filtered out by means of long wave pass filter glasses.

A laser rod with diameter 6.35 mm, length 76 mm and chromium density $2.3\ 10^{19}\ cm^{-3}$ was pumped in a flooded cavity using an elliptical reflector of axes 31.4mm and 28.2mm manufactured by CVI. The flashlamp arc length was 75mm and the bore 4mm, with a cerium-doped silica envelope and a fill of 470 torr xenon. The optical pump pulse length was 100μs FWHM. The circulating coolant fluid carried a UV absorber in solution. A 3mm-thick glass filter (Schott OG or GG series) between lamp and rod provided attenuation of short wavelength visible light as well as providing some, inefficient, fluorescence conversion. The lamp envelope, coolant and glass together provided very strong attenuation of the ultra-violet.

Some results gained using a resonator of length 24cm, output mirror transmission 7% and a total reflector of radius of curvature 2m are shown in Fig. 2. The laser output vs flashlamp capacitor-stored energy is shown for

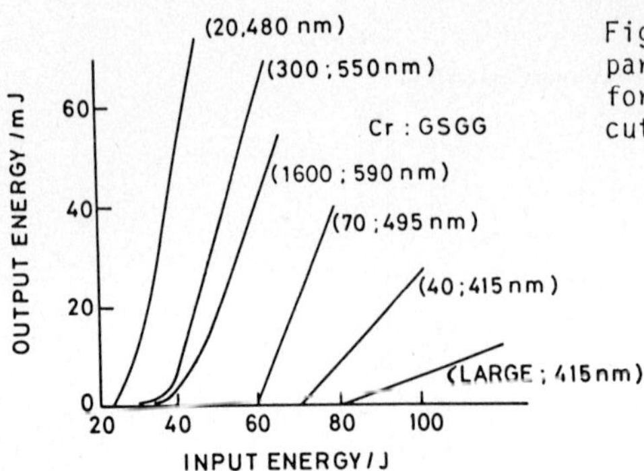

Fig 2. Laser performance; in parentheses - number of shots for 20 to 10 mJ fall in output; cut-off wavelength.

a series of spectral cut-off wavelengths. Ignoring for the present the curve furthest to the left, the progression in improved performance for cut-offs of 415nm, 495nm and 550nm is apparent in both increased laser output and reduction in degradation rate. For longer cut-off wavelength the laser efficiency falls because too much pump light is lost without further major reduction of the laser loss. However, the life of the laser as measured by the number of shots for output to fall from 20mJ to 10mJ continues to increase rapidly, indicating a sharp reduction in the rate of colour centre generation. For 610nm cut-off the life reaches 3000 shots.

These data were found using a freshly annealed, and hence initially colour centre free, laser rod but following a laser alignment procedure of roughly 100 shots during which output degradation occurred. They do not therefore represent the potential initial performance for each cut-off. The initial performance was found using a resonator essentially free of the necessity of alignment (the previous 2m radius total reflector was replaced by one of 50cm radius). The resulting input/output graph is shown for 480nm cut-off as the left-hand curve in Fig. 2. A large improvement in efficiency is evident although with rapid degradation. Slope efficiency was 0.57% with 24J threshold.

The flashlight-induced loss at the lasing wavelength may be estimated as a function of cut-off from the above results and measurements of the pump light and rod absorption spectra. Relative values of the loss induced per shot have been derived and are shown in Fig. 3, where the information is displayed in two ways, both normalised to the value at 415nm. There is a factor of about 500 in the loss generation rate between cut-offs of 400nm and 600nm. In addition to the lasing losses, the generation of colour centres produces loss of pump light, but we have detected no serious performance loss from this cause.

The best performance recorded in Fig. 2 was very short-lived. After only one hundred shots the threshold was nearly doubled and slope efficiency halved. In an attempt to improve this situation and gain the benefits of good utilisation of the pump energy while maintaining the slow degradation rate shown to be achieved by restricting the pump spectrum to long wavelengths only, we have employed a fluorescence conversion technique. This has been implemented using the pump chamber already described but without the use of a filter glass. Instead, a flow tube around the laser rod carried a laser dye in the 1mm thick annular space.

Figure 4 shows the spectral scheme used. The absorbance scale is representative of the attenuations employed between lamp and rod. A coolant

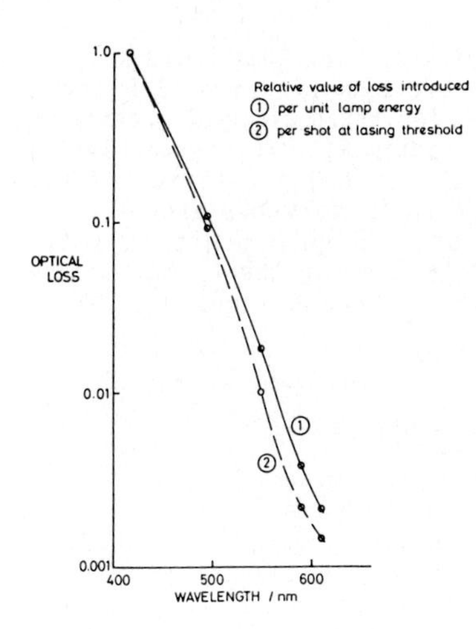

Fig.3. Relative optical loss vs. pump spectrum cut-off.

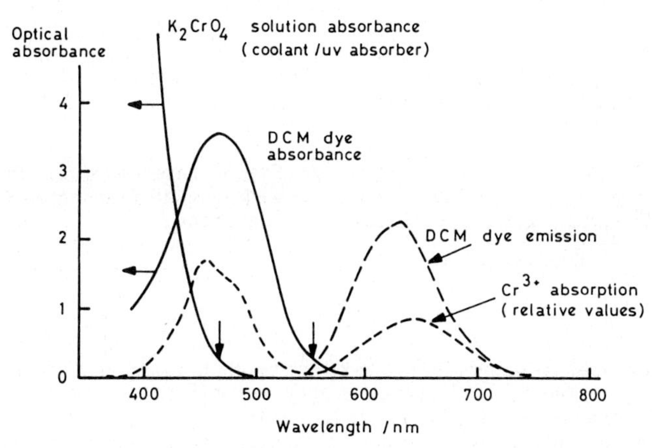

Fig.4. Fluorescence conversion spectra

solution of potassium chromate in water filled the pump chamber and filtered the UV and violet components of pump radiation. DCM dye in methanol was used in the annulus around the rod. It had the double role of protecting the rod from blue and green components of the pump light and of converting this radiation to beneficial wavelengths greater than 600nm. The attenuation through the dye had peak absorbance about 3.5. The corresponding cut-off wavelength, where transmission through to the rod was 50%, was at about 550nm as shown by the right-hand vertical arrow. The left hand vertical arrow indicates the UV absorber cut-off at 470nm. Thus the spectrum from 470 to 550nm was available for conversion to the dye emission band. This band matches the absorption spectrum of chromium quite well. A narrower emission line is desirable for optimum optical coupling but this would in practice inflict the use of an undesirably narrow absorption band. DCM seems about optimum for this application.

The effectiveness of this form of fluorescence conversion was tested by comparing the lasing performance with that achieved by the previous pumping arrangement using a glass filter having a similar cut-off wavelength, 550nm. Slope efficiency of 0.42% was attained, against 0.30% with the glass filter.

Threshold was 30J in both cases. The number of shots for 20mJ to 10mJ laser output degradation was somewhat worse with the dye solution (250 shots against 360 shots) probably because of the smaller attenuation of green pump components. The concentrations used in this experiment were, respectively, 2.4 g/l K_2CrO_4 in water and 0.24 g/l DCM in methanol. A factor of two reduction in both caused no essential change in performance except for a worsening of lasing 'life' to only 100 shots. Figure 5 displays the degradation of laser output over a 900 shot test using the former set of concentrations. The main effect is a change in threshold energy. Slope efficiency is unchanged.

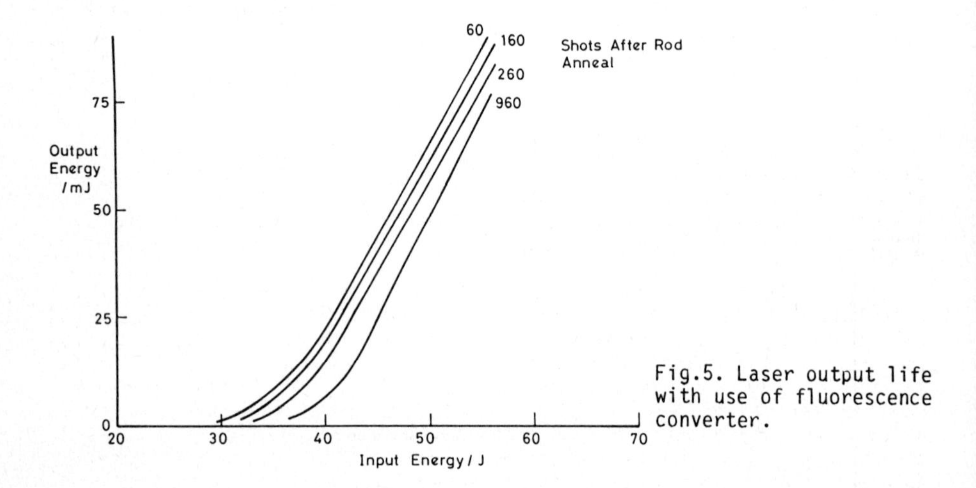

Fig.5. Laser output life with use of fluorescence converter.

Realistic values of useful laser life are clearly much in excess of the 20mJ to 10mJ life. For instance, 75mJ output energy was maintained over the 900 shots by a 5% increase in input energy. The input required for constancy of output energy increases by about 4mJ/shot. There appears to be a good possibility of extracting much higher output energy within the same degradation rate. The ten times lower rate of loss of output which can be achieved with longer wavelength filtering of the pump spectrum may then yield laser life useful for at least some applications. Furthermore, a slow spontaneous recovery of laser performance has been observed, presumably due to some annealing of the colour centres at room temperature. In the case of Fig. 5, for instance, laser output regained the 260 shot line after a two day rest.

Analysis of the lasing threshold and slope may be expected to reveal which factors represent the chief limits on performance and whether any major unknown detrimental mechanisms are operating. The peak stored energy has been calculated for several conditions of pumping from measurements of the fluorescence intensity emitted from the end of the laser rod. Values of stored energy efficiency range from 5.4% for 415nm cut-off wavelength to 2.0% (610nm) and 3.2% with fluorescence conversion. (These values were not noticeably dependent on numbers of shots, so the pumping efficiency is scarcely affected by colour centre formation.) The stored energy efficiency allows calculation of the gain at threshold, taking the emission cross section near 800nm to be 6.0 10^{-21} cm^2 [5].

GSGG : 2.3 10^{19} Cr^{3+}/cm^3
Rod cross-section = 0.317 cm^2
Semi-confocal resonator; 91.3% R output reflector
DCM dye pump conversion + direct flashlight
Threshold energy = 30J
Differential efficiency = 0.43%
Lasing area = 7 mm^2
Lasing wavelength = 792 nm

THRESHOLD
At 30J input
Measured peak stored energy = 0.93J
Hence, gain coeff, g = 0.070
Cr^{3+} absorption loss, $\alpha\ell$ = 0.019
Output loss, ℓn (1/R) = 0.095
Round trip gain
$2g - 2\alpha\ell - \ell n(1/R)$ = $0.007 \simeq 0$

EFFICIENCY
At 50J input
Measured stored energy efficiency = 3.2%
Output coupling efficiency = 67%
Lasing area/rod area = 22%
Differential efficiency = 0.47%

Fig. 6. Laser performance analysis.

The area in the rod cross-section in which lasing occurred was well defined and approximately circular. It was measured using a Reticon photodiode array close to the laser output mirror. The lasing wavelength was measured with accuracy about 2nm using a Jarrell-Ash Monospec 27 spectrograph with a 150 g/mm grating and a Reticon array replacing the output slit.

An example of the analysis of laser action is shown in Fig.6. The calculation of the round-trip gain at threshold includes the rod single pass gain coefficient, g, calculated from the stored energy, the absorption due to trivalent chromium in the laser terminal levels and the output coupling loss. The combination of these yields a calculated round-trip gain coefficient of 0.007. This value is much less than any of the individual gain or loss components and is essentially zero as expected for lasing threshold. The difference from zero may be accounted for in various ways. An error of 2nm in the wavelength measurement would suffice, since the Cr^{3+} absorption varies rapidly with wavelength. Alternatively, the discrepancy might lie in the presence of a small transient loss due to colour centres or excited state absorption, neither of which seems likely, however, from observations we have made of such transients.

The calculation of slope efficiency also shows self-consistency. The slope is influenced by three main factors. These are the pumping efficiency, the output coupling defined by the ratio of output coupling loss to total loss, and the rod-filling factor. The product of these, 0.47%, is close to the 0.43% slope found from lasing experiments.

The measured rod pumping efficiency conforms well to estimates made from the lamp radiative efficiency, optical density of the laser rod, useful pump spectral widths and so on. We conclude, therefore, that no major unknown effects limit the performance of the laser under the operating conditions used in this work.

4. Conclusions to Flashlamp Pumped Cr:GSGG Study

Improved performance of a Cr:GSGG flashlamp-pumped laser has been achieved by the elimination of most colour centre formation due to high-energy pump photons. Overall best performance in terms of laser output and rate of degradation was achieved with a spectral cut-off in the region 550 to 600nm. In this case, compared with the number of photons which would be absorbed by the rod if the UV alone were filtered, only about one third remained to generate laser action. There are thus possibilities for further improvements if a more effective fluorescence conversion scheme can be utilised. Added benefits may be gained by the optimisation of both the chromium concentration and the rod-filling factor.

5. Limits on the Properties of Vibronic Media for Efficient Laser Action

A large number of possible combinations of active ion and host may be considered for investigation as vibronic solid-state laser media. Guidance on the selection of potentially suitable materials using simple measurements on small samples is of great use prior to undertaking the difficult process of producing good quality large samples. Rules for selection may be exemplifed by the choice of site symmetry to minimise non-radiative rates and, for Cr^{3+} ions, by the application of the Tanabe-Sugano diagram in the selection of host material with low ligand fields. The purpose of the present paper is to present some additional constraints on the properties of materials for use in broadly tunable and efficient lasers and amplifiers.

The approach taken is the simple, and approximate, derivation of the range of parameter values of the configurational co-ordinate (CC) diagram for the material which are consistent with various features required for effective lasing. The representation of the energy-level structure of fluorescent media by CC diagrams has several well-known defects but it is nonetheless widely used and usually offers a reasonably good description. Our results lead to some bounds on useful values of some parameters such as the Stokes shift and localised phonon frequency.

A very basic model is employed, consisting of a two-band CC diagram representing the linear coupling case. The two bands are taken to be parabolic with equal curvature. Any additional feature such as Jahn-Teller splitting or the presence of other bands is ignored. The results can be described in terms of three normalised parameters:-

$$e = E/\hbar\omega \quad ; \quad T' = kT/\hbar\omega \quad ; \quad S,$$

where E is the photon energy at the fluorescence peak, S is the Huang-Rhys factor, T is the absolute temperature and $\hbar\omega$ is the phonon energy.

Upon this energy-level structure we have imposed the following specific conditions, which are required for an efficient, broad band laser material.

i. Tunability - fluorescence linewidth greater than 10% of wavelength of peak emission. Two linewidth values have been evaluated, corresponding to intensities of 50% and 90% of the peak fluorescence intensity. The 90% value covers the desirable case where gain is fairly uniform over the operating bandwidth. The broad absorption line width needed for efficient pumping by flashlamp is coupled to the tunability criterion.

ii. Four-level operation - population of terminal levels of radiative transitions across the required bandwidth less than 0.03% of lower state total population.

iii. Non-radiative transition rate less than radiative rate. Three non-radiative processes have been included:-

a. Pump excitation to a vibronic level above the cross-over of the CC curves leads directly to non-radiative decay with no fluorescence. This was confirmed for a range of materials by Bartram and Stoneham [7].

b. Thermal excitation to the cross-over within the vibronic levels of th excited state.

c. The mechanism whereby vibronic levels near the excited state minimum are coupled to the ground state by appropriate vibrational modes is strongly related to b. The low temperature limit of this process is used in the present analysis and numerical values can be readily derived from the work of Struck and Fonger [8] with the exception of an unknown frequency constant. This constant is generally of the same order as, or perhaps identical to, the phonon frequency. It occurs in the present work in combination with the radiative rate. Values of the product ranging from 10^7 to 10^9 make little difference to our results and correspond to radiative lifetimes between 1 and 100 microseconds with a frequency constant of 10^{13}/sec.

Other strong non-radiative processes which are not readily evaluated in terms of CC diagrams are not included here. Such processes are, for example, coupling to impurity ion sinks or, as in glasses, the effect of static deviations from site symmetry. Likewise, other important practical requirements on laser media, such as limits on useful values of emission cross-section, are not considered in the present context.

We shall give our results here without further explanation of the working. In Fig. 7 the region in the (S/T', e/T') plane which conforms to the imposed conditions is shown for a number of values of S. Figure 8 shows the boundaries for the particular case S = 20. Each region is limited by an upper sloping line governed by the linewidth requirement. The high and low temperature segments have different slopes. The left-hand, vertical boundary indicates the terminal state population limit. The lowest boundary is generated by the various non-radiative decay mechanisms, with the Bartram-Stoneham limit giving a major constraint. The limit imposed by the 'tunnelling' process is relevant only for S less than 20.

Some comments may be made concerning Fig. 7.

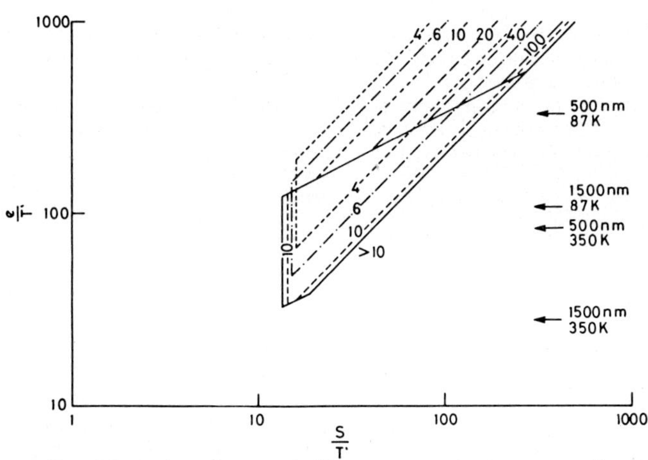

Fig.7. Allowed values of fluorescence energy vs Huang-Rhys factor(S), both normalised to temperature (e/T', S/T'), for various values of S.

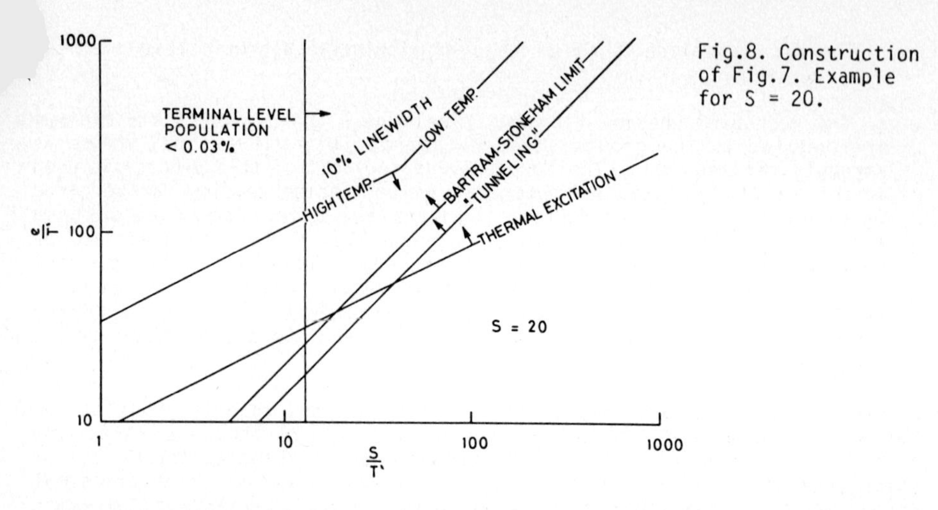

Fig.8. Construction of Fig.7. Example for S = 20.

i. The terminal state population exceeds the desired limit for materials with S/T' less than about 15. At 300K this indicates need of a Stokes shift greater than 6000 cm^{-1}.

ii. The region excluded by thermal activation of non-radiative decay is small. This shows that materials with parameters otherwise suitable for efficient laser action (and described by the two-band model) will not be very temperature sensitive.

iii. Figure 7 shows results when linewidth is taken as the fluorescence FWHM. For the 90% linewidth mentioned previously the upper boundary is lower by a factor of about 2.5, reducing the allowed region very greatly.

iv. Currently-employed vibronic laser materials have S about 10 or 20. The diagram indicates that e/S about 4 or 5 is then optimum, giving points centrally in the allowed region. This implies that the wavelength ratio between peaks of fluorescence and absorption should be about 1.5.

v. Table 1 shows e/T' values for visible and NIR media at two possible laser operating temperatures (temperatures above the likely ambients to allow for heating effects).

Table 1

Temperature [K]	87	87	350	350
Wavelength [nm]	1500	500	1500	500
e/T'	110	530	27	82

e/T' = 27 lies virtually completely outside the permitted region, indicating that an efficient room-temperature laser emitting near 1500nm cannot be constructed from materials conforming to a two-band CC model.

500nm, 350K operation (e/T' = 82) is possible across a wide band of parameters (S=4 and 14, S/T' 42). Good operation at 87K, 1500nm is expected for materials with S=4 and 14, S/T' 55. For each case appropriate limits on Stokes shift and phonon frequency can be derived easily.

vi. The experimental values of fluorescence peaks and Stokes shifts of individual materials may be plotted on Fig. 7, for various working temperatures. The approximate areas occupied by certain classes of materials are shown in Fig. 9. (Some of these classes are known to be not well described by our model, however). Ti^{3+}-doped crystals, represented here only by Ti:sapphire and Ti:YALO, lie in the optimum region. Cr^{3+}-doped crystals lie well to the left of the permitted area. This represents the large overlap of their absorption and emission bands. A similar remark is true for the several dye families encompassed in Fig. 9.

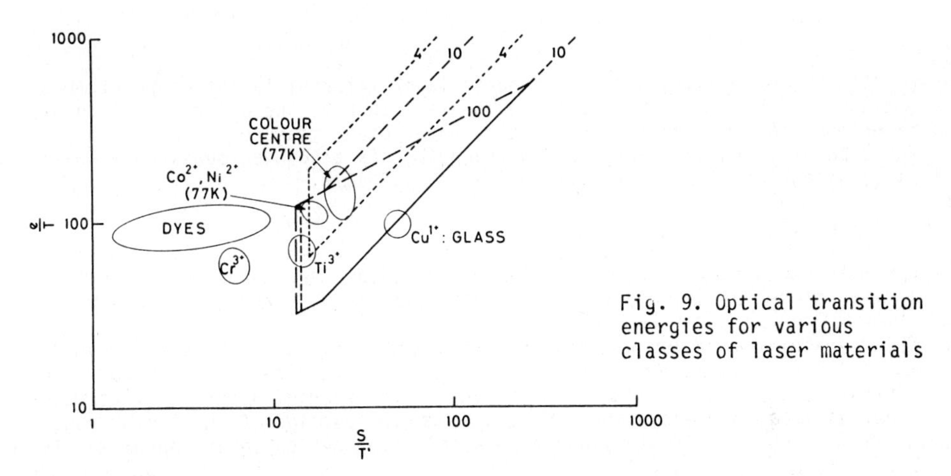

Fig. 9. Optical transition energies for various classes of laser materials

Acknowledgements

We are grateful to our colleagues for their contributions. B Cockayne and J Plant grew the GSGG crystals. Laser rods were fabricated by R Vivian and coated by N A Lowde. Discussions with S J Till added greatly to the CC diagram work.

References

1. Spindler and Hoyer data sheet.
2. E V Zharikov et al: Sov. J. Quantum Electron. 13, 1274 (1983).
3. M J P Payne and H W Evans: OSA Topical Meeting on Tunable Solid-State Lasers, Arlington, VA (1985).
4. R C Simpson and K L Schepler: ibid.
5. B Struve and G Huber: J Appl Phys 57, 45 (1984).
6. E V Zharikov et al: Sov. J. Quantum Electron 12, 1124 (1982).
7. R H Bartram and A M Stoneham: Solid St. Comm. 17, 1593 (1985).
8. C W Struck and W H Fonger: J. Luminescence 10, 1 (1975).

Cr:YSAG – A Tunable Near-Infrared Laser Material

N.P. Barnes [1]*, *D.K. Remelius* [1], *D.J. Gettemy* [1], *and M.R. Kokta* [2]

[1]Los Alamos National Laboratory, Los Alamos, NM 87545, USA
[2]Union Carbide Corporation, 750 S. 32nd Street, Washougal, WA 98671, USA

Cr:YSAG has been evaluated as a tunable laser material in the near-infrared region of the spectrum. Basic parameters such as the upper laser level lifetime and the fluorescent spectra have been measured. Gain, as a function of flashlamp energy and wavelength, has also been measured.as well as the flashlamp-induced loss at 0.532 and 1.064 μm.

1. Introduction

Cr has been widely recognized as a good candidate for tunable solid-state lasers since the discovery of the tunable Cr:BeAl$_2$O$_4$ laser [1,2]. While this system has proven to be both robust and versatile, there are several reasons for searching for other materials into which Cr can be incorporated. One of the primary reasons for searching for an alternate material is to achieve operation at wavelengths outside of the tuning range of Cr:BeAl$_2$O$_4$. Cr:BeAl$_2$O$_4$ has a peak emission wavelength at approximately 0.75 μm. The wavelength of peak emission is dependent on the temperature and, to some extent, concentration. However, the effective stimulated emission cross-section decreases to half of its peak value at approximately 0.73 and 0.79 μm. If a tunable laser is needed outside of this range, another laser material could be sought. The peak emission wavelength of Cr can be varied by selecting a laser material with a different crystal field parameter. Materials with a strong crystal field parameter, such as Cr:Al$_2$O$_3$, will demonstrate short wavelength, narrow transitions, from the ^2E to the $^4A_2$ level or will tend to have short wavelength, broadband transitions, from the $^4T_2$ to the $^4A_2$ level. On the other hand, materials with a weak crystal field parameter, such as Cr:LLGG, will tend to have long wavelength, broadband transitions, from the $^4T_2$ to the $^4A_2$ level. By selecting materials with different crystal field parameters, the peak emission wavelength can be varied.

Another important consideration when selecting a material to be a host for Cr is associated with the potential sites. It is desired to have a single site for the Cr and a site which will accommodate an atom the size of Cr. Toward this end, Sc bearing components have been investigated [3-7] because of the relatively good match between the ionic radii of Cr and Sc.

While several Sc bearing garnets have been investigated, Cr:YSAG has several properties which make it of interest. Primary among these

---

*Present address: Langley Research Center, Hampton, VA 23665-5225

properties is its similarity to YAG. Although not presently confirmed, it is suspected that this material will have thermal properties similar to the good thermal properties of YAG. Since the energy gap between the valance and conduction band of a material is dependent on the composition, it is suspected that YSAG may have somewhat better ultraviolet transmission than GSAG or GSGG. Better ultraviolet transmission may be useful for other dopant atoms as well as for Cr based systems. With these prospects in mind, Cr:YSAG was studied as a potential laser system.

2. Growth

Yttrium Scandium Aluminum Garnet, YSAG, belongs to the group of garnets in which the octahedral structural position is substituted by Sc ions. The substitution of scandium into aluminum garnets was studied by Kokta [8] in 1973. The first yttrium scandium aluminum garnet crystals were grown in Bell Laboratories by Brandle and Vanderleeden [9]. An interest in scandium-substituted garnets was renewed when their use as an effective host for vibronic lasers was recognized [10].

The crystal growth of YSAG is similar in many respects to the growth of neodymium YAG, Yttrium Aluminum Garnet. For growth of YSAG chromium doped, relatively slow pull rates, 0.015"/hour, were used to ensure high optical quality. The dopant distribution throughout the crystal is uniform due to distribution coefficient approaching unity. All chemicals used in crystal growth (yttrium oxide, scandium oxide, aluminum oxide, and chromium oxide) were of 99.999+% purity. The starting materials were subjected to heating at 1100°C for 12 hours, in order to free them of moisture and carbon dioxide, prior to usage.

The crystal growth was carried out in a 3" diameter, 3" height iridium crucible heated by coupling RF energy via a coupling copper coil. The radio frequency power was provided by a 50 kW RF generator operating at 10 kHz frequency. Thermal insulation, used to protect the coil from the heat of the crucible as well as to retain the heat in the crucible, was made from stabilized zirconium oxide tubing. The molten charge was maintained under a protective atmosphere of nitrogen containing 800 ppm of volume of oxygen. The melting temperature of YSAG was found to be 1860°C $\pm$20°C as determined by an optical pyrometer uncorrected for emissivity.

All crystals were grown in <111> orientation, with diameter of 1.5" and length of 5". The first of the crystals was seeded on gadolinium scandium aluminum seed. Severe cracking occurred due to lattice parameter mismatch. The first boule was, however, used to fabricate satisfactory seed for a subsequent growth run.

The grown crystals showed typical garnet faceting characteristics of the <111> orientation. The interface shape was convex and resulted in a strained core area. The laser rod was drilled from an unstrained portion of the crystal.

3. Spectroscopy

Transmission spectra have been recorded for Cr:YSAG showing absorption peaks near 0.45 and 0.65 μm. More detailed information appears in Fig. 1. Like all garnets, Cr:YSAG is isotropic. Consequently, only a single transmission curve is included. Data was taken on a Beckman Acta MIV spectrophotometer. A Cr:YSAG sample with a nominal 0.01 atomic fraction

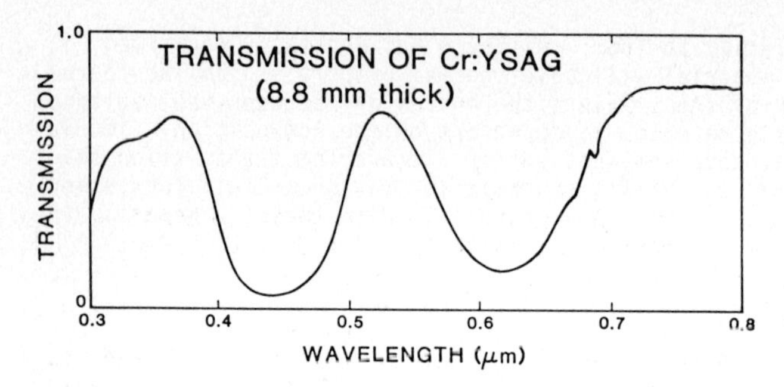

Fig.1. Transmission of Cr:YSAG

and a thickness of 8.8 mm was used to obtain the curve. The data has not been corrected for Fresnel reflections losses at the faces of the crystal. As the refractive index is about 1.87, peak transmission of the material with no doping should be about 0.83.

Fluorescent spectra for Cr:YSAG were recorded both at -189°C and at 35°C. To obtain the spectra, the Cr:YSAG sample was placed in a vacuum dewar. A Kr cw laser, operating primarily on the 0.647 μm line, was used as a pump. Fluorescent output from the sample was focused into a 1.0 m Spex monochromator. Output of the monochromator was incident on a photomultiplier tube with a S-1 response. Output from the detector was, in turn, sent to a lock-in amplifier which provided the input for a strip chart recorder. Spectral response of the monochromator and detector combination was normalized using the output of a tungsten lamp.

At reduced temperatures, two prominent lines appear at 0.687 and 0.688 μm as shown in Fig. 2. These spectral features are believed to be associated with the 2E to $^4A_2$ transitions. Vibronic transitions from the $^4T_2$ to the $^4A_2$ levels are believed to represent the broadband emission features between about 0.69 μm and 0.79 μm. Very little fluorescence is observed at wavelengths longer than 0.79 μm at the reduced temperatures.

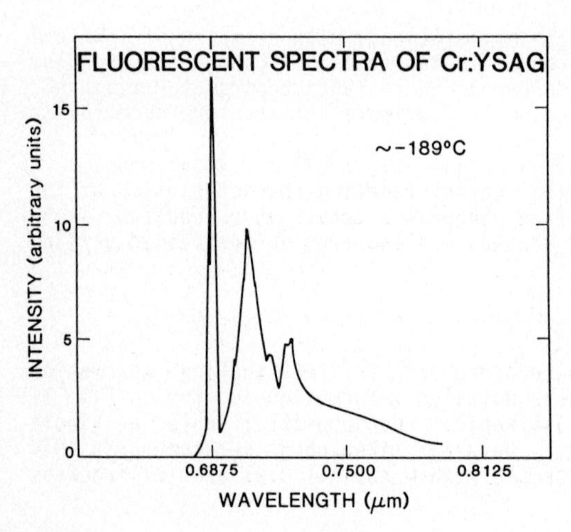

Fig.2. Fluorescent spectra of Cr:YSAG, -189°C

At room temperature, the narrow lines associated with the ^2E to $^4A_2$ transition have all but disappeared into the broad fluorescent features associated with the vibronic spectrum. The fluorescence can be observed at wavelengths as long as 0.87 μm. Due to the λ^5 dependence of the gain and the fluorescence spectrum, the gain peak should occur at considerably longer wavelengths than the fluorescent peak. It is estimated from the curves that the gain peak should be at approximately 0.73 μm. Three lines appear in the wings of the fluorescent spectrum. These lines do not appear in the fluorescent spectra taken at reduced temperatures nor do they correspond to normal Kr transitions. It appears that they may be associated with impurity levels which are not populated at reduced temperatures.

The lifetime of the upper laser level was measured at the two temperatures at which the fluorescent spectrum was taken. At reduced temperatures, the lifetime of Cr:YSAG was measured to be 1710 μsec. At room temperature, the lifetime had been reduced to 202 μsec. A comparison of other upper laser level lifetimes for Cr doped garnets grown by Union Carbide and recorded at Los Alamos include Cr:GSAG at 160 μsec, Cr:GSGG at 120 μsec, and Cr:LLGG at 71 μsec. Of the Cr-based garnets, Cr:YSAG has the most desirable lifetime from the viewpoint of optical pumping.

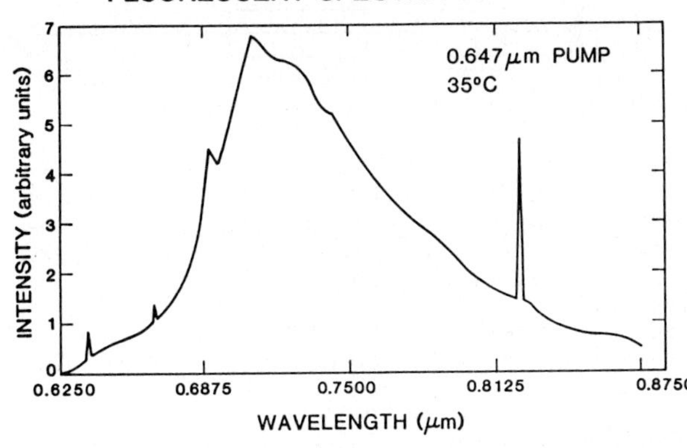

Fig.3. Fluorescent spectra of Cr:YSAG, 35°C

4. Gain Measurements

Gain of Cr:YSAG was measured by using a $Ti:Al_2O_3$ laser as a probe. A schematic diagram of the optical arrangement similar to the one used to measure the gain appears in Fig. 4. A frequency-doubled Nd:YAG laser system excites the $Ti:Al_2O_3$ laser. Tuning of the $Ti:Al_2O_3$ laser is accomplished by using a dispersive prism in the laser resonator and tilting the highly reflecting mirror to select the desired wavelength. A sample of the $Ti:Al_2O_3$ probe beam is taken before and after traversing the Cr:YSAG laser rod. After suitable spatial averaging, the two samples of the probe beam are detected using Si diodes. Output of the detector is recorded on two transient digitizers housed in a CAMAC crate. An IBM PC interrogates

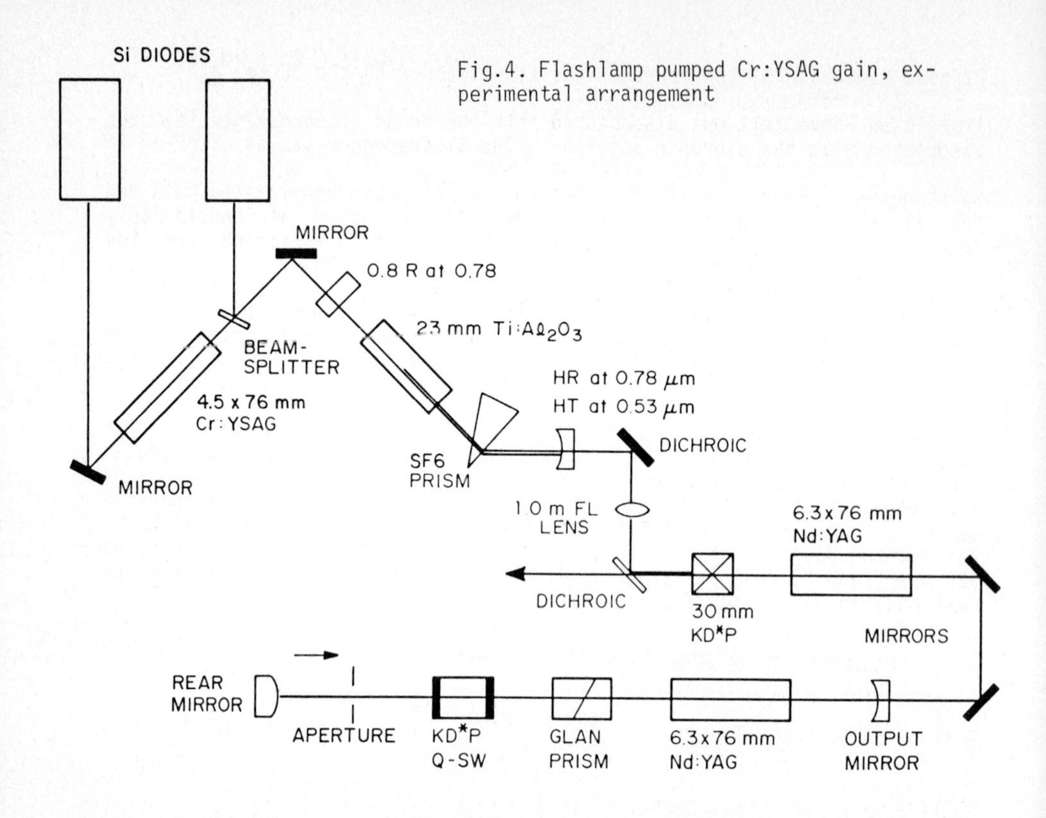

Fig.4. Flashlamp pumped Cr:YSAG gain, experimental arrangement

the transient digitizers, curve fits the recorded waveforms to determine the peak, and performs averaging. Typically, 10 to 50 measurements of the gain are recorded. From the data, the average and root mean square are determined. Gain is measured by flashlamp exciting the Cr:YSAG laser on every other shot of the probe laser. In this manner, gain is defined as the ratio of the measured transmission with and without flashlamp excitation. By using the method, long-term changes in the probe laser energy or the transmission of the Cr:YSAG laser rod are ratioed out.

An existing laser cavity and pulse forming network, normally used for Nd:YAG, was used to excite the Cr:YSAG. Major and minor diameters of the elliptical cavity were 65.0 and 63.6 mm, respectively. A silver reflective coating was used on the ellipse. Flow tubes of pyrex glass were used around both the flashlamp and the laser rod and the entire cavity was flooded. A 6.0 mm by 76.0 mm Kr filled flashlamp was used to excite the Cr:YSAG laser rod. A simmer current of about 35 mA was normally employed to keep the flashlamp alive. The pulse-forming network consisted of a 50 μF capacitor and a 21 μH inductor. Unfortunately, the Cr:YSAG laser rod was only 4.5 mm in diameter and 76 mm in total length. Excited length of the laser rod was only about 70 mm. Due primarily to the significantly smaller diameter of the laser rod with respect to the flashlamp, the coupling efficiency of this arrangement was only about 0.36. Much improved performance may be expected with a pumping arrangement design to optimize the efficiency of the Cr:YSAG laser.

The gain coefficient of Cr:YSAG exhibits a linear dependence on the electrical energy. Gain was measured at 0.741 μm near the wavelength of

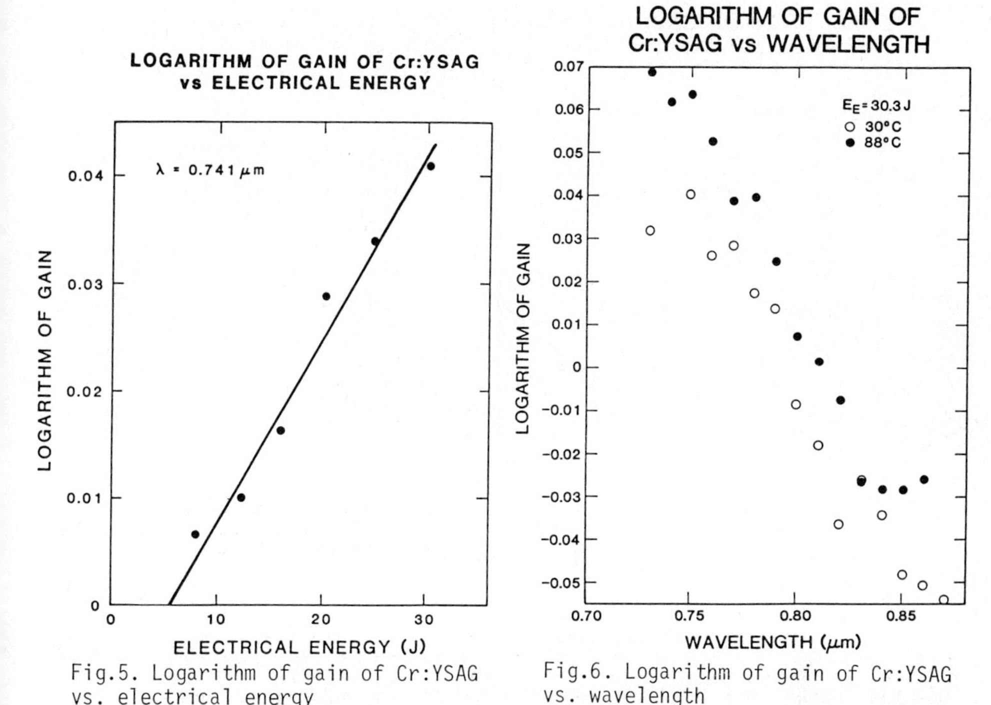

Fig.5. Logarithm of gain of Cr:YSAG vs. electrical energy

Fig.6. Logarithm of gain of Cr:YSAG vs. wavelength

maximum gain. Results displayed in the Figure 5 show the logarithm of the gain, or the gain coefficient, plotted as a function of the electrical energy stored in the capacitor. These gain measurements were recorded at room temperature. Even with the relatively inefficient pumping arrangement, a gain of 0.04 was achieved with a 30J input. The logarithm of the gain is expected to be directly proportional to the electrical energy, to first order approximation. As such, the experimental gain data was fit to a linear curve, as shown by the straight line in Figure 5. Although the data points can be well represented by a straight line, the line does not go through the origin. Similar results have been observed with Cr:GSGG and Cr:GSAG [4,6].

Gain of Cr:YSAG has been measured as a function of wavelength over much of the fluorescent spectrum. These results appear for two different temperatures in Figure 6. At room temperature, the gain appears to maximize at a wavelength of about 0.75 μm. Maximum gain with 30J of electrical energy is above 0.04. Gain then declines on either side and goes to zero somewhere between about 0.79 and 0.80 μm. Beyond 0.80 μm, the gain actually becomes negative, that is, a flashlamp-induced loss exists. Flashlamp-induced loss has also been observed in Cr:GSGG and Cr:GSAG [4,6].

At higher temperatures, about 88^{0}C, the maximum gain increases to about 0.07. Gain now persists at longer wavelengths, beyond 0.81 μm. It also appears that the flashlamp-induced loss is not as severe at the longer wavelengths.

Transmission through the excited laser rod was also measured at 0.53 μm. With the data processing technique described above, the static transmission

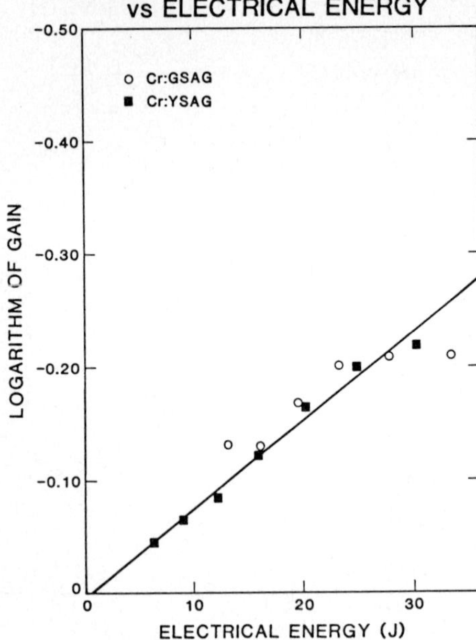

Fig.7. Logarithm of gain at 0.532 μm vs. electrical energy

of the laser rod at these two wavelengths is subtracted out of the measurements. Transmission at 0.53 μm as a function of the electrical energy appears to decrease in a similar manner to the decrease observed in Cr:GSAG. The logarithm of the gain, which is negative in this case, is plotted versus the electrical energy in Fig. 7. For reference, similar data for Cr:GSAG is also plotted. In this case, the negative logarithm of the gain increases linearly with the electrical energy. Moreover, the relationship is more nearly direct, that is the linear curve fitted to the data nearly goes through the origin. This can be contrasted to Cr:GSGG where the logarithm of the gain increases approximately quadratically with electrical energy. It may be noted that the loss at 0.53 μm is nearly an order of magnitude larger than the gain at 0.74 μm. Thus, a small offset could be somewhat masked by the larger losses.

Transmission was measured as a function of delay between the flashlamp excitation and the probe. Probes, both at 0.53 and 1.06 μm, were used for these measurements. Results of these measurements appear in Figures 8 and 9. Time equal to zero corresponds to the initiation of the flashlamp discharge. In both cases, the losses rise approximately sinusoidally. A sinusoidal increase may be expected of a nearly critically damped flashlamp current pulse and can be well approximated by a half sinusoid. After the cessation of the flashlamp discharge, the loss tends to decay away exponentially with a time constant characteristic of the lifetime of the upper laser level. However, the decay does not bring the flashlamp-induced loss back to the nominal zero level. After about 1.0 ms, the loss at 0.53 μm has decayed to about one fifth of its peak value. The decay rate at this point is much slower, having a time constant on the order of 1.2 ms. At 1.06 μm, the peak flashlamp-induced loss is significantly less than the flashlamp-induced loss at 0.53 μm. After about 1.0 ms, the loss at this wavelength has decayed to about one fourth of its peak value and appears to

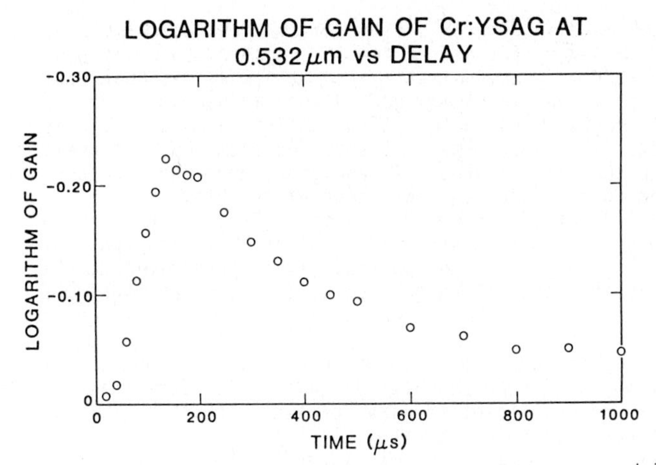

Fig.8. Logarithm of gain of Cr:YSAG at 0.532 μm vs. delay

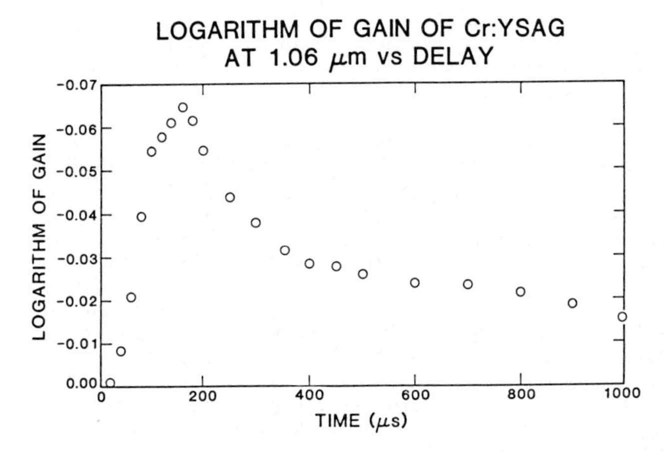

Fig.9. Logarithm of gain of Cr:YSAG at 1.06 μm vs. delay

be decaying away with a much slower time constant. The time constant for this decay is on the order of 0.7 ms.

5. Discussion

The experimentally observed data is consistent with two flashlamp-induced loss mechanisms. One of these loss mechanisms could be associated with the population of the upper laser level, that is excited state absorption.

Since the ^2E level in this material should be heavily populated at room temperature, a multitude of states exist to which excited state absorption could occur. For example, excited state absorption could exist at 0.53 μm between the ^2E level and the $^2T_1$ level. Similarly, excited state absorption at 0.74 μm could occur between the ^2E level and the $^2A_1$ level.

Excited state absorption would decay with a time constant characteristic of the upper laser level. However, not all of the loss appears to decay away in 5 upper laser level lifetimes, approximately 1.0 ms. This residual loss may be associated with a second flashlamp-induced loss mechanism such as color centers. Color centers could be self-annealing and essentially dissipate before the next flashlamp pulse. Absorption from color centers tends to display a broad spectral nature. In addition, more than one type of color center could be formed under the influence of flashlamp radiation. A complex nonexponential decay of the residual loss, observed in Cr:GSGG for example, tends to support this contention. Given multiple types of color centers and their broad absorption spectra, it is possible that the residual loss observed in the spectral range from 0.53 to 1.06 μm might be caused by this effect.

Although color center absorption may be present, it would not necessarily be proportional to the electrical energy. If, for example, only a limited number of color centers were present in the crystal, absorption by the color centers would saturate. In this case, a fairly low electrical energy input to the flashlamp may saturate the color center absorption mechanism. Thus, at low electrical energies color center loss would compete effectively with the gain resulting from the increased upper laser level population. After saturating the color centers, further increases in the electrical energy would produce a gain. A mechanism such as this could explain the apparent threshold in the plot of the logarithm of the gain versus electrical energy.

1. J. C. Walling, O. G. Peterson, H. P. Jenssen, R. C. Morris, and E. W. O'Dell: "Tunable Alexandrite Lasers," IEEE J. Quant. Elect. QE-16, 120, (1980).

2. J. C. Walling, O. G. Peterson, H. P. Jenssen, and R. C. Morris: "Tunable CW Alexandrite Laser," IEEE J. Quant. Elect. QE-16, 120, (1980).

3. B. Struve and G. Huber: "Laser Performance of Cr^{3+}: GD(Sc,Ga) Garnet," J. Appl. Phys. 57, 45-48, (1985).

4. N. P. Barnes, J. V. Meier, and D. K. Remelius: "Flashlamp-Pumped Cr:GSAG Laser," Conference on Lasers and Electro-Optics, Baltimore, Maryland, May 1985.

5. Flashlamp-pump single ellipse data was reported by J. Drube and G. Huber in a postdeadline paper at the Conference on Lasers and Electro-Optics, Baltimore, Maryland, May, 1985.

6. J. V. Meier, N. P. Barnes and Dennis K. Remelius: "Flashlamp-Pumped Cr^{3+}: GSAG Laser," Accepted by IEEE J. Quant. Elect., (1986).

7. E. V. Zharikov et. al.: "Tunable Laser Utilizing An Electronic-Vibrational Transition in Gadolinium Scandium Gallium Garnet Crystal," Sov. J. Quant. Elect. 13, 1274-1275, (1983).

8. M. R. Kokta, J. Solid State Chem., 8, 39-42, (1973).

9. C. D. Brandle, J. C. Vanderleeden, IEEE J. Quant. Elect. QE-10, No. 2, 67-71, (1974).

10. D. Pruss, G. Huber, A. Belmowski, V. V. Lapteo, I. A. Shcherbakov, and Y. V. Zharikov, J. Appl. Phys. B28, 355-358, (1982).

A $ScBO_3$:Cr Laser

S.T. Lai[1], *B.H.T. Chai*[1], *M. Long*[1], *M.D. Shinn*[2], *J.A. Caird*[2], *J.E. Marion*[2], *and P.R. Staver*[2]

[1]Allied-Signal, Inc., P.O. Box 1021R, Morristown, NJ 07960, USA
[2]LLNL, Livermore, CA 94550, USA

1. INTRODUCTION

Since the first sucessful demonstration of room-temperature tunable laser operation in alexandrite [1], researchers have discovered several new room-temperature tunable lasers including emerald [2], GSGG:Cr [3], GSAG:Cr [4], $KZnF_3$:Cr [5], $SrAlF_5$:Cr [6] and Al_2O_3:Ti [7]. The majority of the newly discovered tunable lasers have Cr^{3+} as an active ion. This is partly due to the relatively high room-temperature fluorescence quantum efficiency of Cr^{3+}.

One of the objectives in tunable laser research is to extend the tunability range beyond that of the alexandrite laser. As is evident from the Tanabe-Sugano energy-level diagram for d^3 electron ions such as Cr^{3+}, the tunability range may be controlled by the crystal field strength. Since the transition of $^4T_2$ levels is spin-allowed, the $^4T_2$ transition time is on the order of few micro-seconds in a low symmetry site. Hence it is necessary to select a laser host which will have an inversion symmetry so that the fluorescence lifetime is sufficiently long for flash-lamp laser operation.

We report here a new laser, $ScBO_3$:Cr. The Cr^{3+} substitutes in the single 6-fold coordinated Sc site with ScO_6 forming a nearly perfect octahedron [8]. The compound melts congruently at 1610°C [9]. High optical quality single crystals of $ScBO_3$ have been obtained by slow cooling or by Czochralski pulling from the melt [8]. Laser action was achieved from crystals grown by either method. Preliminary results of the laser measurements were reported earlier. [10].

2. SPECTROSCOPY OF $ScBO_3$:Cr

2.1 Fluorescence

$ScBO_3$ samples for fluorescence measurement were made from single crystals doped with 0.3 or 0.4 at% Cr, and were oriented by X-ray diffraction. The room temperature fluorescence spectra for both π and σ polarization are shown in Fig. 1. The ratio of π to σ polarization intensities is about 1:0.8. The emission peak is at 815 nm which is significantly red-shifted as compared to alexandrite or GSGG. The fluorescence lifetime measured at room temperature is 115 μsec and the fluorescence decay is single valued and exponential (Fig. 2). The emission cross-sections have been calculated and are also shown in Fig. 1. The maximum emission cross-section is about 1.2×10^{-20} cm^2 and is located around 840 nm.

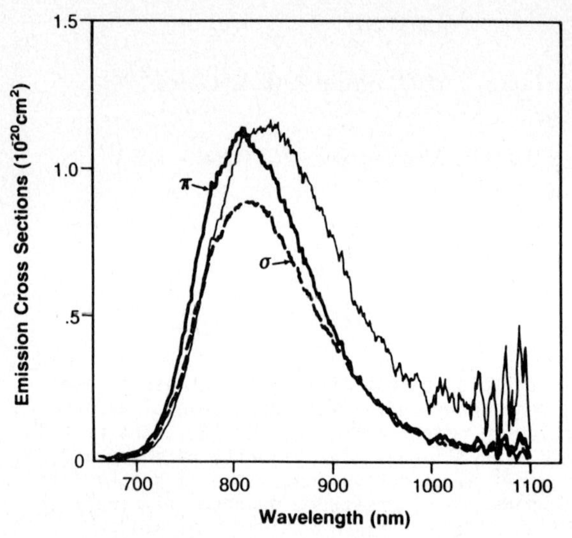

Fig. 1 Room-temperature polarized fluorescence spectra and the emission cross-section of $ScBO_3$:Cr.

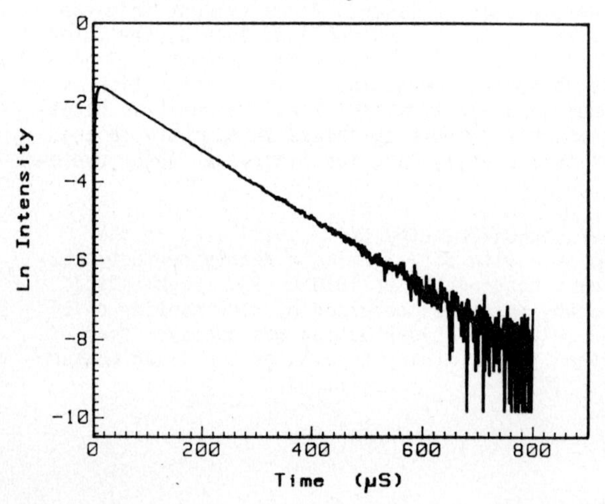

Fig. 2 Fluorescence decay of $ScBO_3$:Cr. The crystal is excited at 650 nm and monitored at 820 nm.

The fluorescence lifetime as a function of temperature was also measured and is shown in Fig. 3. In a semi-log plot, the rate change of fluorescence lifetime can be clearly divided into two regions with the crossover around 340°K (70°C). The fast lifetime decrease at higher temperature is due to multi-phonon-assisted non-radiative relaxation which reduces the laser efficiency. The slow lifetime decrease at lower temperature is most likely due to symmetry breaking by thermal phonons. The dynamic coupling of the lattice vibration at higher temperatures decreases the site symmetry but increases the oscillator strength of the $^4T_2$ to $^4A_2$ transition. Therefore, even though the lifetime decreases from 200 μsec at 8°K to 115

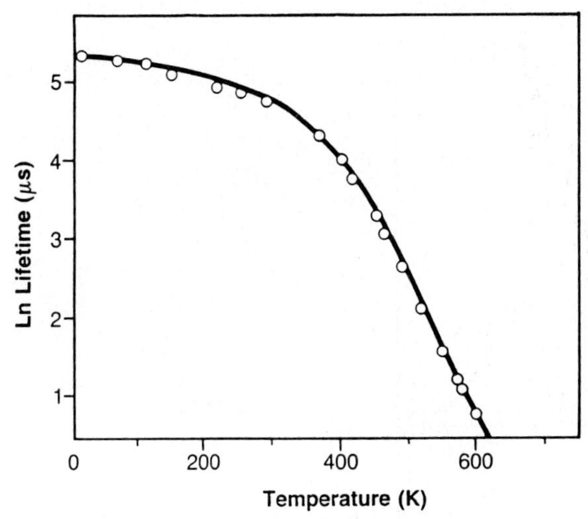

Fig. 3. Fluorescence lifetime of $ScBO_3$:Cr as a function of temperature.

μsec at room temperature, the quantum yield might still remain close to unity.

2.2 Absorption

The polarized absorption spectra of $ScBO_3$:Cr is shown in Fig. 4. It shows two absorption bands of Cr^{3+}, the $^4T_2$ and $^4T_1$ bands. Similar to the emission spectrum, both the absorption bands are also red-shifted. The dip near 673 nm in the $^4T_2$ absorption band is characteristic for low crystal

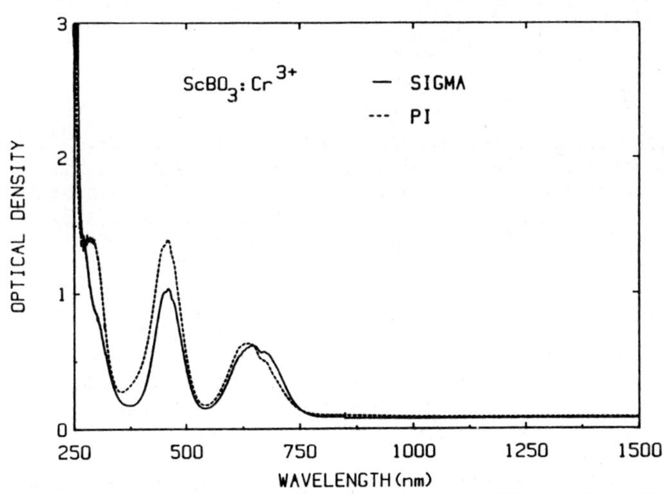

Fig. 4. Polarized absorption spectra of $ScBO_3$:Cr from 250 nm to 1500 nm. Notice the absence of anomalous absorption beyond 770 nm.

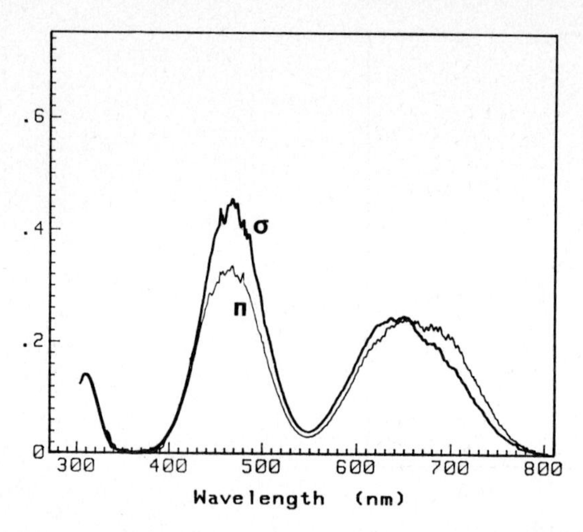

Fig. 5. Polarized excitation spectra of $ScBO_3$:Cr as monitored at 825 nm.

field Cr hosts, which is also seen in $KZnF_3$ [5] and $SrAlF_5$ [6]. The polarized excitation spectra were also measured and are shown in Fig. 5. The fluorescence intensity at 825 nm was monitored while the excitation wavelengths were scanned from 300 to 800 nm. Comparing Fig. 4 and Fig. 5, we can see perfect matching for the $^4T_2$ bands in both the π and σ polarization. For the $^4T_1$ band, the excitation spectra had slightly lower amplitudes than the absorption spectra. Nonradiative absorption is also noted at the higher energy shoulder of the $^4T_1$ band, in the 350 to 420 nm region. We attribute it to the impurity contamination in the starting materials. This absorption feature was much lower when a fresh melt was prepared with 99.99% Sc_2O_3. Discarding the impurity absorption, the fluorescence quantum efficiency is near unity for both $^4T_1$ and $^4T_2$ bands. There is also no anomalous absorption in the lasing region as seen in Fig. 4. This is essential for the low threshold, high output efficiency of the present laser.

3. TUNABLE LASER ACTION MEASUREMENT

Tunable laser operation was achieved in a 20 cm long nearly concentric cavity. An oriented $ScBO_3$ crystal (12 mm long, 4x4.5 mm^2 cross-section) was longitudinally pumped by a krypton ion laser at 647 nm. The pump beam was focused to a 60 μm beam waist in the crystal and was chopped at a 3% duty cycle to reduce local heating. The pump mirror had nearly 100% reflectivity in the lasing range and the output mirrors had transmissions of 0.53% and 1.8% at 842 and 848 nm, respectively. The free-running wavelength depended slightly on the output coupling. The laser output of the $ScBO_3$:Cr is shown in Fig. 6. At To = 0.53%, the slope efficiency was 19.5% and threshold was at 250 mW. At To = 1.8%, the slope efficiency increased to 29.2% whereas the threshold was at 400 mW. The estimated single pass loss was about 1.3%/cm. This included both scattering loss as well as possibly excited state absorption.

Tuning measurements were conducted in a high reflectivity cavity with an output coupler of 0.3% transmission. Tuning was achieved with a single element birefringent filter and a tuning range of 787-892 nm was obtained.

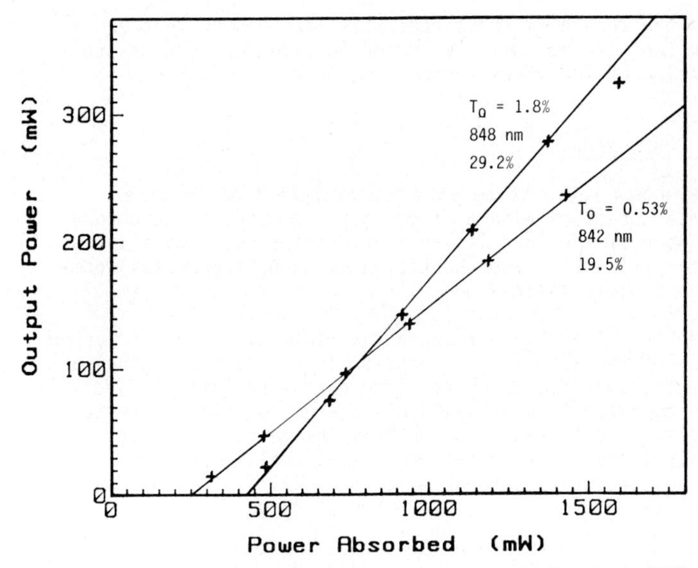

Fig. 6. Laser output efficiency of $ScBO_3$:Cr for two different values of output couplings. Cr concentration is 0.3 at.%.

Table 1. Comparison of Cr-doped room-temperature tunable lasers.

Host Crystal	Alexandrite	Emerald	GSGG	GSAG	$KZnF_3$	$ScBO_3$	$SrAlF_5$
Fluorescence Peak (nm)	697	746	750	755	765	810	840
Lifetime (μs)	260	60	114	150	80/270 (at 10°K)	115	93
Laser Tuning Range (nm)	701-818	720-842	742-842	740-820	785-865	787-892	852-947
Peak Emission Cross-Section ($x10^{-20}$ cm^2)	0.7	3.3	0.8	0.6	---	1.2	2.3
Kr laser pumped output efficiency	51%	51%	28%	18.5%	14%	29%	3.6%
Crystal Hardness	8.5	7.5	7.5	8	3	5	3
Remarks		Difficult to grow	Color center		Two lifetimes		Dead Cr site
Reference	[11]	[12]	[4]	[4]	[5]	This study	[6]

We expect that both the threshold and single pass loss can be further reduced, whereas the tuning range will be increased with improved crystal quality and elimination of the impurity content.

4. DISCUSSION AND CONCLUSION

We have demonstrated a new solid-state room-temperature tunable laser $ScBO_3$:Cr. The simple structure with a single high symmetry Sc octahedral site contribute to some of the good lasing characteristics. Table 1 lists some of the characteristics of a representative class of successful room-temperature Cr doped tunable lasers.

The lasing properties of $ScBO_3$:Cr compare favorably with other Cr-doped lasers. The absence of the anomalous absorption in the lasing region was verified by the spectroscopic measurement. The estimated laser loss of 1.3%/cm is also low compared to most newly discovered tunable solid-state lasers. Large single crystals can be grown from the melt. The laser efficiency can be further increased with better crystal quality and reduced impurities. The $ScBO_3$:Cr infrared tunable laser is likely to be a practical and efficient solid-state laser in the 810 - 870 nm range.

ACKNOWLEDGEMENT:

This work was supported by LLNL under DOE Contract W-7405-ENG-48.

REFERENCES:

1. J.C. Walling, H.P. Jenssen, R.C. Morris, E.W. O'Dell and O.G. Peterson, Opt. Lett., vol. 4, 182 (1979).
2. M.L. Shand and J.C. Walling, IEEE J. Quant. Electron. vol. QE18, 1829 (1982).
3. B. Struve and G. Huber, Appl. Phys. vol.B30, 117 (1983).
4. G. Huber and K. Peterman, in Tunable Solid State Lasers, Springer Ser. Opt. Sci., Vol. 47 (Springer, Berlin 1985)
5. U. Brand and U. Durr, Opt. Lett., vol. 9, 441 (1984).
6. H.P. Jenssen and S.T. Lai, J. Opt. Soc. Am. vol B3, 115 (1986).
7. P.F. Moulton, J. Opt. Soc. Am. vol. B3, 125 (1986).
8. B.H.T. Chai, M. Long, R.C. Morris and S.T. Lai, this issue, (1986).
9. E.M. Levin, J. Am. Ceramic Soc., vol. 50, 53 (1967).
10. S.T. Lai, B.H.T. Chai, M. Long and R.C. Morris, IEEE J. Quantum Electron., Oct. issue, 1986, to be published.
11. S.T. Lai and M.L. Shand, J. Appl. Phys. vol. 54, 5642 (1983).
12. S.T. Lai, Proceedings SPIE, vol. 622, 146 (1986).

Tunable Lasers with Transition Metal Doped Fluoride Crystals

U. Dürr [1,2] *and U. Brauch* [1]

[1]Physikalisches Institut, Universität Stuttgart, D-7000 Stuttart 80, Fed. Rep. of Germany
[2]LASAG AG, CH-3600 Thun, Switzerland

1. Introduction

Since the original work of JOHNSON et al [**1**] on tunable solid-state lasers on the basis of transition metal doped fluoride crystals much progress has been made in crystal growing technologies, laser technology and the understanding of the lasing properties of these vibronic laser materials. Also several new laser crystals-especially Cr(3+) doped systems - have been reported. A survey on actual cw-operated laser crystals is given in Fig. 1 together with the tuning range [**2,3,4**]. Information about oxide crystals may be found in reference [**2,4**]. The successful laser systems given in Fig. 1 only represent a very small selection of the large variety of candidates which not only may include simple cubic, fluoride crystals but also host crystals like Kryolites, Elpasolites or even fluoride garnets. These host crystals can be doped with transition metal ions from the 3d, 4d, 5d group. A large number of combinations has been investigated and evaluated with respect to their suitability as vibronic laser crystals. The main reasons for the relative low success rate are crystal growing difficulties,

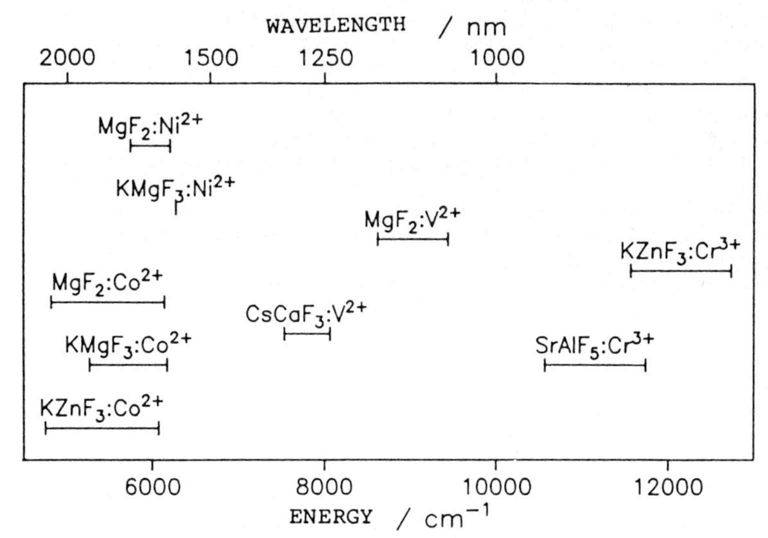

Fig. 1. Tuning range of cw operated transition metal doped fluorides.

optical quality of the systems and also the fact that not all transition metal ions show reasonable emission properties in dielectric host crystals when located on undistorted or distorted octahedral site symmetry. Even combinations with excellent fluorescence properties fail because of excited state absorption or reabsorption processes.
These problems are discussed in more detail in reference [5] from a theoretical point of view. Most crystal systems investigated by group are 3d ions in simple cubic fluorine perovskites. From our many experiments some general conclusions can be drawn which may be compared with the tendencies given by CAIRD [5].

V(2+) typically shows a high quantum efficiency of fluorescence. Unfortunately the spin-allowed transitions in the visible are weak and very often excited state absorption limits the lasing properties [1,2,6].

V(3+) does not emit sufficiently well

Ni(2+) in many host crystals shows a high quantum efficiency at least at 77K. To date similar to V(2+) excited-state absorption limits the tunable systems to MgF_2 : Ni(2+). It is worth pointing out that there are additional emissions in the visible due to transitions between higher Ni(2+) electronic states which should be investigated more carefully [1,2,7,8]

Cr(2+) no emission was observed

Cr(3+) is the most successful laser ion. In cubic perowskites the only problem lies in the fact that a doubly charged metal ion is substituted, in which case charge compensation has to take place in the host crystal [2,3].

Fe(2+) has very efficient radiationless transitions between the two lowest levels.

Mn(2+) is a very interesting ion not only because of the high quantum efficiency of fluorescence in the visible spectral region but also because this is the only electronic configuration where the spacing beween the lowest levels increases with decreasing crystal field. Unfortunately the transitions are also spin-forbidden. Therefore the lifetime of the excited state is extremely high. Also, efficient pumping seem to be difficult because the main absorption bands are in the near UV.

Co(2+) has been extensively studied in the past. Reasonable low-temperature quantum efficieny and broad tunability motivated the development of prototype laser systems for high-resolution spectroscopy [2,8,9].

First promising results on 4d laser ions in fluorides have also been reported [10]. In what follows only Cr(3+) doped perowskites of the $AMeF_3$ type are discused in more detail . The reader may find more information about other systems in the given references. The last chapter is devoted to discussions of the possibility to increase the efficiency of lamp-pumped solid-state lasers using an energy transfer from host lattice excitations to the laser ion. This so-called excitonic pumping scheme may be realized for example in transition metal doped magnetic transition metal fluorides.

2. Cr(3+) in $AMeF_3$

Different host single crystals were doped with Cr(3+) and spectroscopically investigated. The absorption bands are listed in table 1. Also given are the crystal field parameters calculated from the absorption and emission spectra. The corresponding emission data are given in table 2.

Table 1: Splitting of the lowest electronic states of Cr(3+) and calculated crystal field parameters

Cr^{3+}	$KZnF_3$	$RbZnF_3$	$RbCdF_3$	$CsCdF_3$
	($\tilde{\nu}$ / cm^{-1})			
$^4T_2$	14900 ± 1200	15400 ± 1300	14300 ± 1100	14100 ± 2000
^2E (#)	15750 ± 100	15720 ± 50	15700 ± 100	15700
$^2T_1$(#)	16390 ± 200		16260 ± 200	
$^4T_1(^4F)$	22000 ± 1700	22600 ± 2000	21400 ± 1700	21200
$^4T_1(^4P)$	34000 ± 3000		31800 ± 4000	
Dq / cm^{-1}	1490	1542	1436	≃ 1410
B / cm^{-1}	760	757	757	≃ 760
Dq / B	1,96	2,04	1,90	≃ 1,85

Table 2: Emission range and peak position of the vibronic sideband and 0-phonon line of the $^4T_2$ - $^4A_2$ transitions

Cr^{3+}	$KZnF_3$	$RbZnF_3$	$RbCdF_3$	$CsCdF_3$
$\tilde{\nu}(\sigma_{em}^{max})$ / cm^{-1}	12 350	12 050	11 750	11 500
$\frac{1}{2}\Delta\tilde{\nu}(\sigma_{em})$ / cm^{-1}	1 000	1 000	1 000	1 100
$\tilde{\nu}$(0-Ph) / cm^{-1}	14 088 13 763	≃ 14 300	≃ 13 300	≃ 13 060

The temperature dependence of the quantum efficiency in Fig. 2 clearly demonstrates that $KZnF_3$ is the laser material where the room-temperature threshold is lowest [12]. It is also clear that $CsCdF_3$ is not a suitable host for this laser ion. The dramatic low-temperature decrease of the efficiency is not yet clear because no structural phase transition is reported

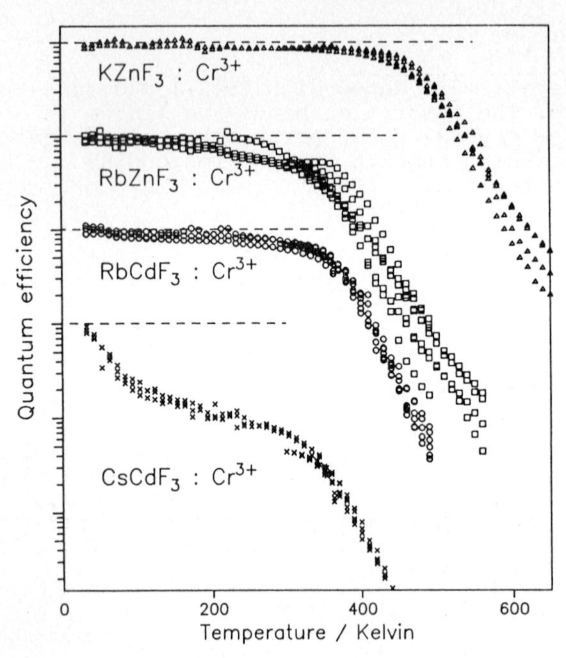

Fig. 2: T dependence of the quantum efficiency of Cr(3+) in different host crystals. The dashed horizontal lines mark quantum efficiency 1.

for this host crystal. Indeed $KZnF_3$: Cr(3+) is easily cw and single mode operated at room temperature [11]. Tuning range and gain can be seen in Fig. 3. The laser slope efficiency is only about 14% with an output coupling of 1.6%. One reason might be that only one of the two Cr(3+) defects always present in $KZnF_3$ [13] is participating in the laser process, while both centers are absorbing the pump radiation. As the two centers differ in their radiative lifetime (500µs, 200µs), dynamical laser properties can be used to decide this question.

For this purpose the temporal and spectral behaviour of the laser relaxation oscillations was investigated. Figure 4 shows the spiking when switching on the laser in the free-running modality at room temperature. Numerical simulation of the spiking with help of the photon- and inversion rate equations clearly demonstrates that the best fit of the experimental results is achieved when the rate equations are extended to two centers with radiative lifetimes given by the experiment (200µs, 500µs). The change in the spectral width during the relaxation oscillation in shown in Fig. 5a. The strong decrease from the first pulse (bandwith $\approx$ 100 cm^{-1}) to the cw laser bandwith of about 10 cm^{-1} is due to the different degree of inversion during the laser transient. Note the shift of the emission maximum in time. This behaviour is not observed for V(2+) in $CsCaF_2$ (Fig. 5b) where no charge compensation is necessary and therefore only one laser center is present. Once more

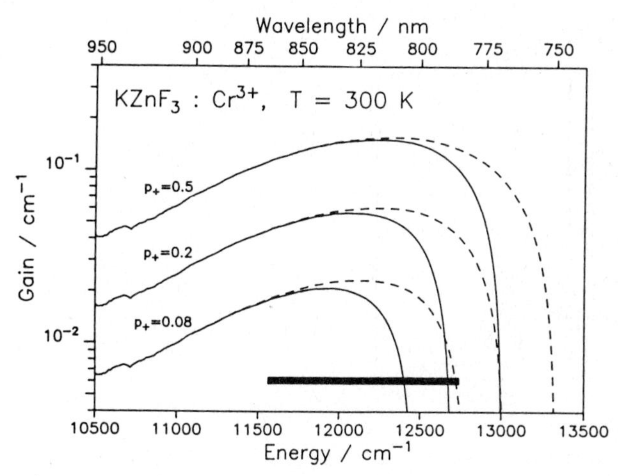

Fig. 3: Gain curves for different pump intensities (given in relative inversion) calculated from the emission spectrum [**14**]. The bar shows the tuning range for a pump intensity twice threshold.

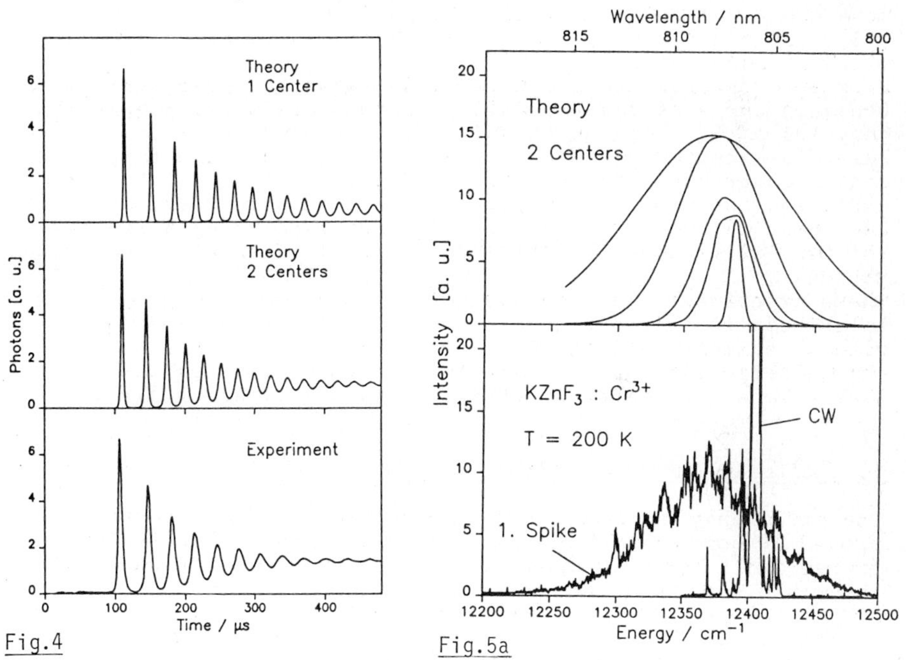

Fig. 4: Relaxation oscillation of the $KZnF_3$: Cr(3+) laser in comparison with the best theoretical fit with one or two different laser centers.

Fig. 5a: Spectral behaviour during relaxation oscillations. The 5 theoretical curves correspond to 110µs (1. spike) 174µs (3. spike), 300µs, 500µs and 5ms

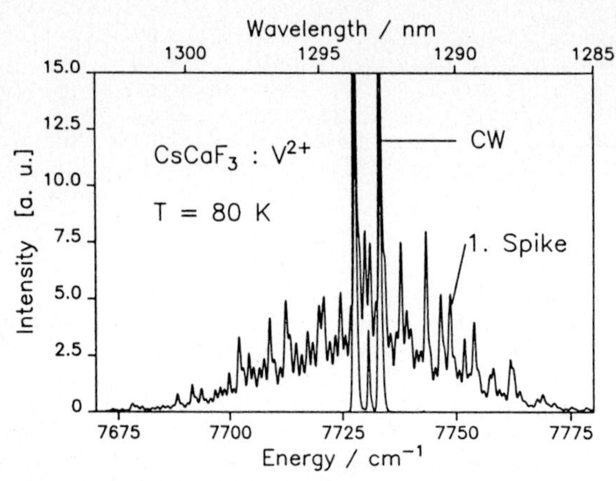

Fig. 5b: Spectral behaviour during relaxation oscillation for $CsCaF_3$: V(2+)

the behaviour can be numerically simulated with the extended rate equations and numerical methods described by BRUNNER et al [**15**]. 10^5 modes were taken and a Gaussian gain profile with a width of 2000 cm^{-1}.
From specroscopic measurements it is known that the maximum of the gain curve of the two Cr(3+) centers are shifted approximately 100 cm^{-1}. Therefore a shift of the emission maximum of the laser is expected, because at the beginning only the fast laser center contributes to the laser radiation while during time evolution the slow center with the shifted gain curve starts to take part in the laser process. The numerical simulation taking the two Cr(3+) centers into account nicely fits the observation.
These experiments, in combination with spectroscopic data, lead to the conclusion that both centers, the unperturbed center with O_n symmetry and the C_3- center with the neighbouring charge compensation, take part in the cw laser process. The low laser slope efficiency in $KZnF_3$: Cr(3+) is still unclear.

3. Excitonic pumping of laser centers

The efficiency of lamp-pumped solid-state lasers can be improved by codoping [**4**]. A natural consequence of this idea is to use excitations of host crystals with strong absorptions in the visible or near UV. These excitations should be delocalized and be trapped at the laser centers. This can be achieved with excitons. Crystals where this excitonic pumping mechanism can be used are, for example, magnetic insulators like $KMnF_3$. One reason to use Mn - compounds is the large energy gap between the ground state and the excited states. Therefore most transition metal ions lead to impurity states far below the excitation band, one condition for efficient energy transfer even at room temperature. The energy transfer to impurities has been investigated spectroscopically under various aspects [**6,16**]. Even laser operation was reported for the system MnF_2 : Ni(2+).

Detailed investigation with V(2+) in $KMnF_3$ or $RbMnF_3$ [6] shows that the excitonic transfer of the absorbed energy takes less than 10 ns with an efficiency of about 100%. The only problem may arise from excited state absorption during the laser process. Up to now it is not clear if this is the reason why the laser experiments have not been successful with these crystals.

4. Summary

Transition metal doped perowskites operate in the spectral range from 2.1 - 0,78μ. Spectroscopic data indicate that laser operation should be possible between 3μ and 0,5μ with the various systems. The lasers show efficient tunable cw single mode behaviour. Therefore these crystals seem suitable for tunable high-resolution laser systems. Only the future can decade whether these laser materials can be used in flashlamp-pumped high-power lasers.

Acknowledgements

The authors thank the BMFT for financial support

References

1 L.F. Johnson, H.J. Guggenheim, R.A. Thomas: Phys. Rev. 149, 179 (1966)

2 P.F. Moulton: In Laser Handbook ed. by M. Bass and M.L. Stick, p 203 ff, Elsevier Science Publishers BV (1985)

3 Tunable Solid State Lasers, Springer Series in Optical Sciences, Vol. 47 ed. by P. Hammerling, A.B. Budgor and A. Pinto 1985

4 Tunable Solid State Lasers, Technical Digest, Zigzag, Oregon (1986), this volume

5 J.A. Caird: talk WA5, this volume

6 W. Knierim, A. Honold, U. Brauch and U. Dürr: J. Opt. Soc. Am. B Vol 3 (1), 119 (1986)

7 L.F. Johnson, H.J. Guggenheim, D.Bahnck and A.M. Johnson: Opt. Lett, 8 (7), 371 (1983)

8 U. Dürr, U. Brauch, W. Knierim and W. Weigand in Proceedings of the International Conference on Lasers 83, ed by R. Powell, STS Press, Mc Lean VA p 142 ff

9 K. German, U. Dürr and W. Künzel: Opt. Lett 11, 12 (1986)

10 R.C. Powell: Talk 4A2, this volume

11 P. Fuhrberg, W. Luhs, B. Struve, G. Litfin: talk WB6, this volume

12 U. Brauch, U. Dürr: Opt Lett. 9, 441, (1984)

13 U. Brauch, U. Dürr: Opt. Comm. 49 (1), 61 (1984)

14 D.E. Mc Cumber: Phys. Rev. 134, A299 (1964)

15 W. Brunner, H. Paul: Opt. Quant. Elect. 14, 453 (1982)

16 L.F. Johnson, R.E. Dietz and H.J. Guggenheim;Phys. Rev. Lett. 17, 13, (1966)

Laser-Pumped Laser Measurements of Gain and Loss in $SrAlF_5$:Cr Crystals*

J.A. Caird[1], *P.R. Staver*[1], *M.D. Shinn*[1], *H.J. Guggenheim*[2], *and D. Bahnck*[2]

[1]University of California, Lawrence Livermore National Laboratory, P.O. Box 55 08, Livermore, CA 94550, USA

[2]AT & T Bell Laboratories, 600 Mountain Avenue, Murray Hill, NJ 07974, USA

1. Introduction

Strontium aluminum pentafluoride crystals doped with chromium, $SrAlF_5$:Cr have been shown to produce continuously tunable pi-polarized laser emission from 825 nm to 1010 nm with krypton ion laser pumping.[1-4] In the more recent experiments a slope efficiency of 15% was obtained for conversion of absorbed pump radiation at 647 nm to the output beam at 910 nm, with approximately 1% output coupling from the $SrAlF_5$:Cr laser resonator.[3] The observed efficiency was considerably less than the quantum defect-limited efficiency of 71%, and led to speculation that there was a high level of loss in the material. To check this hypothesis a series of experiments has been carried out,in which the magnitude of the linear loss coefficient and the gain per unit pump power have been determined as a function of wavelength.

2. Crystal Growth

The chemicals used in the synthesis of $SrAlF_5$:Cr crystals were: Fisher Scientific reagent grade $SrCl_2 \cdot 6H_2O$; Alpha reagent grade $Al(NO_3)_3 \cdot 9H_2O$; Alpha reagent grade $Cr(NO_3)_3 \cdot 9H_2O$. The fluoride compounds were synthesized by co-precipitating the respective salts from aqueous hydrofluoric acid. The compounds were then air dried at 100°C and finally sintered by slow heat treatment to the melting point in a flowing HF atmosphere.

High-quality single crystals were prepared by the horizontal Bridgeman method using a vitreous carbon boat contained in a platinum tube with a dynamic HF atmosphere. Typical growth rates were 2 to 4 mm per hour.

3. Laser Experiments

The configuration of equipment for the laser-pumped-laser experiments on $SrAlF_5$:Cr crystals is shown in Fig. 1. A crystal with 2 at.% Cr^{3+}, and thickness t = 1.93 mm was positioned at the mode waist of the nearly concentric resonator. The crystal's surfaces were polished flat and parallel, and were left uncoated. The c-axis of the uniaxial (tetragonal) crystal was nearly parallel to the polished surfaces, and was oriented in the plane of incidence on the birefringent tuner plate.

*Work performed under the auspices of the U.S. Department of Energy by Lawrence Livermore National Laboratory under Contract No. W-7405-ENG-48.

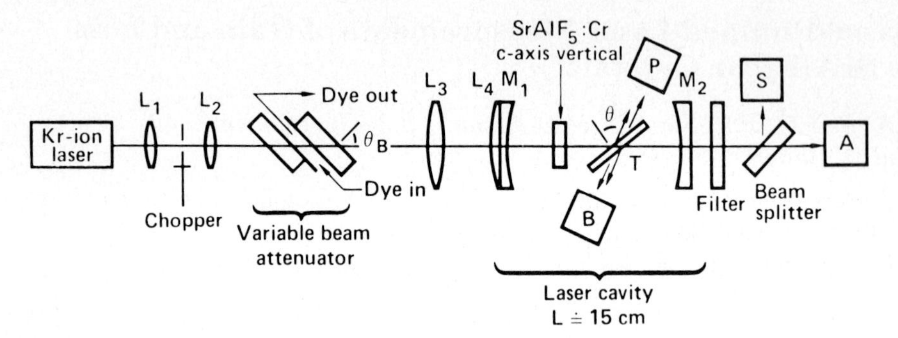

Figure 1. Experimental arrangement for laser-pumped laser diagnostic tests on $SrAlF_5$:Cr laser crystals. $L_1 = L_2 = 15$ cm focal length lenses, $L_3 = 20$ cm focal length lens, $L_4 = 10$ cm focal length lens, M_1 = 5 cm radius high-reflectivity mirror, M_2 = 10 cm radius high-reflectivity mirror, T = crystal quartz birefringent tuner, P = thermopile power monitor, B = beam dump, S = spectrometer/optical multichannel analyser, A = photodiode/signal averager, θ = output coupling angle.

The pump source was a linearly polarized krypton ion laser operating at 647 nm and was focused on the sample through one of the resonator mirrors which had high reflectivity (peak $>99.5\%$) at the output laser wavelengths, and high transmission ($\sim$ 80%) at 647 nm. The pump beam polarization was also oriented in the plane of incidence on the birefringent tuner plate, for convenience. The pump laser was mechanically chopped, yielding a 2% duty factor so that thermal loading of the sample was minimized.

In order to avoid changes in the pump beam spot size at the sample, the Kr^+ laser was operated at its maximum output power throughout the experiments. The pump power incident on the sample was varied by attenuation of the beam in a flowing solution containing dye of variable concentration. The dye cell was oriented at the Brewster angle to minimize its insertion loss. The pump spot size at the sample was determined by measuring the transmission of a number of circular pinholes centered on the pump beam in place of the sample and then fitting the results to a Gaussian profile. This method gave a pump spot radius of 88 microns.

Tuning of the output wavelength was accomplished using a single-plate crystalline-quartz birefringent filter oriented near the Brewster angle. Reflection of the resonator mode from the tuner plate was used to provide variable output coupling. The tuner plate was tilted away from the Brewster angle so as to vary the angle of incidence, θ, of the resonator mode on it, thereby varying the output coupling fraction in a systematic manner.

A series of experiments was performed to determine the achievable wavelength tuning range as a function of absorbed pump power, for various levels of output coupling (determined by the angle of resonator mode incidence on the tuner plate). The end points of the observed tuning ranges are plotted in Fig. 2 as a function of absorbed pump power, and parametric in tuner plate reflectivity per surface, $R(\theta)$. Two data sets were taken with the tuner plate at $R(\theta)=0.08\%$. The variability of the results obtained with this tuner plate orientation is thought to be typical of the effects of random resonator alignment errors.

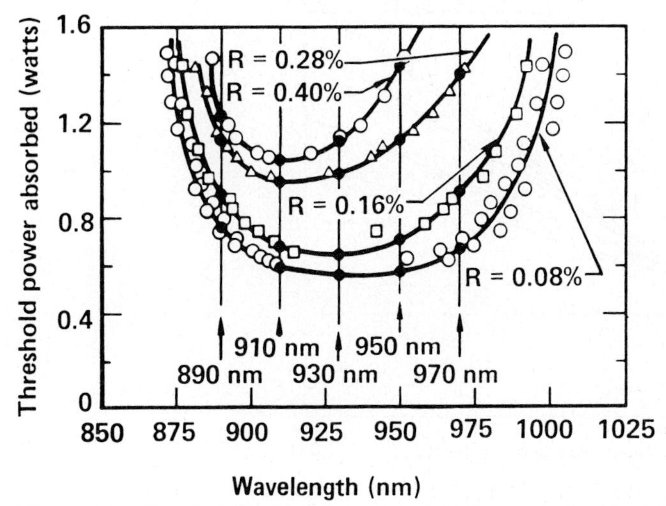

Figure 2. Wavelength tuning range limits vs. absorbed pump power for various tuner plate reflectivities (per surface). Open symbols represent data points while the hand drawn smooth curves represent the locus of threshold absorbed pump power as a function of output wavelength. Solid circles indicate points plotted in Fig. 3.

4. Analysis and Discussion

The analysis of the data utilized a modified Findlay-Clay[5] technique in which the threshold pump power for laser oscillation is plotted as a function of the output coupling fraction. At any given wavelength the total round-trip resonator output coupling fraction, C_{out}, is given by

$$C_{out} = 4\,R(\theta) + T_{M1} + T_{M2}\,, \tag{1}$$

where T_{M1} and T_{M2} are the wavelength-dependent transmissions of the resonator mirrors M_1 and M_2, respectively. In the small gain limit the threshold pump power, P_{th} is related to the total resonator output coupling fraction by

$$G_p P_{th} = C_{out} + L, \tag{2}$$

where L is the passive linear insertion loss of the laser active crystal, G_p is the round-trip gain produced in the crystal per unit of absorbed pump power.

The threshold pump powers from Fig. 2 are replotted in Fig. 3 versus output coupling fraction, parametric in output wavelength. In Fig. 3 the intercepts at the vertical axis represent the insertion loss, L, due to the laser crystal at each output wavelength. The apparent crystal insertion loss is plotted as a function of laser wavelength in Fig. 4. Also shown in Fig. 4 is a plot of the pi-polarized absorption spectrum of the laser crystal. The uncertainty in the baseline of the absorption spectrum is estimated to be ±0.17%, as indicated in the figure. It can be seen that the measured insertion losses agree very well with the more standard absorption measurement, indicating that all other contributions to the insertion loss are negligible by comparison. The apparent loss coefficient (L/2t) at 910 nm is about 2% cm^{-1}. While this level of loss would

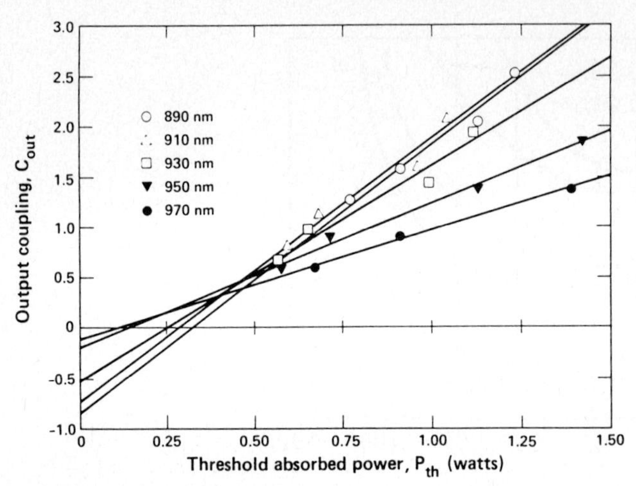

Figure 3. Pump power absorbed by the sample at threshold for laser oscillation vs. total output coupling fraction (in % per round trip) for wavelengths from 890 nm to 970 nm in 20 nm increments.

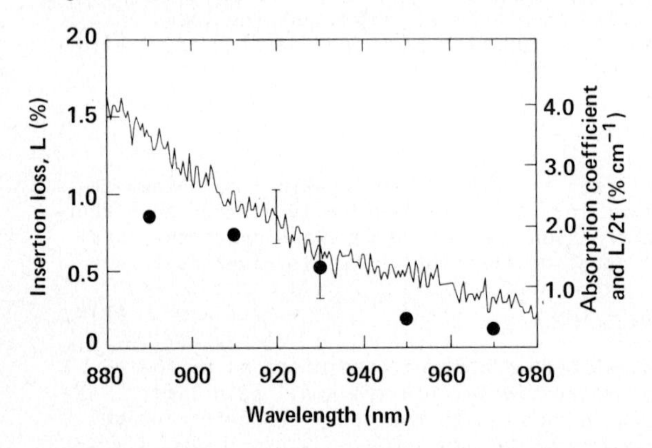

Figure 4. Laser resonator insertion loss vs. wavelength (from Fig. 3), and absorption coefficient measured with a Perkin-Elmer Lambda-9 spectrophotometer.

be problematic for operation of a low-gain laser with flashlamp pumping, it is not high enough to explain the low lasing efficiency achieved to date in the laser-pumped laser cavity. The level of losses required to explain the low lasing efficiency would have been about 10% cm^{-1}.[3]

The slopes of the lines in Fig. 3 determine the gains achieved in the $SrAlF_5$:Cr laser crystal per unit pump power, G_p, at each output wavelength. These values are plotted as a function of wavelength in Fig. 5. A decrease in gain with increasing wavelength is apparent, and was expected due to the declining cross-section for stimulated emission. However, the magnitude of the gain achieved is considerably lower than expected.

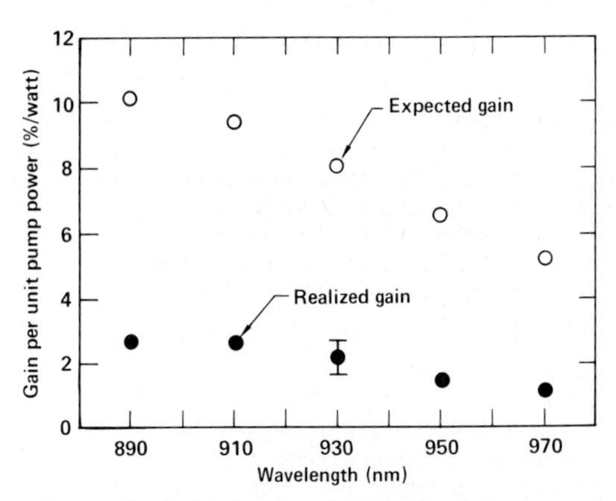

Figure 5. Gain in the $SrAlF_5$:Cr laser crystal per unit pump power absorbed vs. wavelength. Upper curve: gain expected; lower curve: gain realized.

In the absence of nonradiative losses and excited state absorption, the gain per unit absorbed pump power would be expected to be given by

$$G_p = \frac{4\lambda_p \tau_{f\ell} \sigma(\lambda_\ell)}{\pi \omega_p^2 hc} \quad , \tag{3}$$

where λ_p is the pump wavelength, $\tau_{f\ell}$ is the fluorescence lifetime of the excited state, $\sigma(\lambda_\ell)$ is the emission cross-section at wavelength λ_ℓ, and ω_p is the gaussian pump beam radius at the laser crystal. The emission cross-section was determined as a function of wavelength from measurements of the fluorescence spectrum and radiative lifetime. The gains calculated using (3) were found to exceed the gain deduced from the Findlay-Clay measurements by factors of 3 to 5, depending on wavelength, and are also plotted in Fig. 5.

The gain predicted by (3) assumes that the active region is pumped uniformly across the laser resonator mode. This is only true if the pump mode diameter is significantly greater than the resonator mode diameter. Since the gain declines off-axis, the average gain experienced by a resonator mode with a finite beam waist radius is less than the on-axis gain. The decrement in gain is expected to be small if the resonator mode diameter is significantly smaller than the pump mode diameter. Given a TEM_{00} resonator mode of radius ω_r and a gaussian pump mode of radius ω_p, it can be shown that the average gain experienced in the resonator is reduced from the gain at the center of the pump beam by a factor of[6]

$$\delta = 1 \, / \, \left(1 + \frac{\omega_r^2}{\omega_p^2} \right) . \tag{4}$$

Based on measurements of the resonator length, with mirror radii of 5 cm and 10 cm, the fundamental resonator mode radius was estimated to be less than 1/2 of the pump mode radius. The average gain experienced by a resonator mode of this size should have been greater than 80% of the on-axis gain produced by the pump. Thus, this factor is inadequate to explain the observed gain reduction.

Two mechanisms which could explain the further reduction of gain from the value predicted by (3) are excited state absorption, and absorption of pump radiation by non-radiative chromium sites. It should be noted that there are four inequivalent octahedral sites on which chromium can reside in this crystal,[7] some of which could be nonradiatively relaxed at room temperature. Evidence of the existence of at least one nonradiative site has been shown by comparison of the absorption and excitation spectra of a $SrAlF_5$:Cr crystal at cryogenic temperatures.[4]

5. Conclusion

Passive linear insertion losses for a $SrAlF_5$:Cr laser crystal were measured and found to be comparable to values obtained from a simple absorption spectrum. The level of the passive insertion loss (2% cm^{-1}) is substantially smaller than would be required to explain the low observed lasing efficiency in laser-pumped laser experiments (10% cm^{-1}).[3] The gain per unit of absorbed pump power was also determined and found to be significantly less than expected (neglecting non-radiative losses and excited state absorption). It is currently thought that the low slope efficiency achieved in the laser-pumped laser experiments and the reduced gain are due to the same factors. Excited state absorption at the output wavelength and absorption of pump energy by non-radiative chromium ions are potential explanations for both.

References

1. J. A. Caird, W. F. Krupke, M. D. Shinn, and P. R. Staver, presented as a post deadline paper at the Topical Meeting on Tunable Solid State Lasers, Arlington, VA, May 16-17, 1985.

2. S. Lai and H. P. Jenssen, presented at the Topical Meeting on Tunable Solid State Lasers, Arlington, VA, May 16-17, 1985, paper FA7-1.

3. J. A. Caird, W. F. Krupke, M. D. Shinn, and P. R. Staver, presented at the First International Laser Science Conference, Dallas, TX, November 18-22, 1985, Bull. Am. Phys. Soc., 30, 1857 (1985).

4. H. P. Jenssen and S. T. Lai, J. Opt. Soc. Am. B, 3, 115-188 (1986).

5. D. Findlay and R. A. Clay, Phys. Lett., 20, 277-278 (1966).

6. P. F. Moulton, IEEE J. Quantum Electron., QE-21, 1582 (1985).

7. S. C. Abrahams, J. Ravez, A. Simon and J. P. Chaminade, J. Appl. Phys., 52, 4740 (1981).

Part V

Alexandrite Lasers

Alexandrite Laser Amplifiers

J.A. Pete, J. Krasinski, T. Chin, and D.F. Heller

Allied Corporation, 7 Powder Horn Drive, Mt. Bethel, NJ 07060, USA

Recent performance of alexandrite single- and double-pass amplifiers is described. Amplification factors of 5x per pass are reported at pulse energies of several Joules/pulse and pulse powers exceeding 100 MW.

1. Introduction

Recent advances in the development of tunable alexandrite laser amplifiers, in single-pass and double-pass configurations, show promising trends toward higher average power and higher brightness performance. As the low emission cross-section of alexandrite ($\approx$5-7 x 10^{-21} cm^2 @ 25°C and 1 x 10^{-20} cm^2 at 95°C at band center) indicates, alexandrite is a low to moderate gain material[1,2]. However, under pumping conditions compatible with reasonable flashlamp lifetimes (>10^6 shots), single pass-amplification factors of four to seven have been achieved using 0.635 cm diameter by 10 cm long rods operated at a temperature of 95°C. Alexandrite's ability to store large amounts of energy ($\sim$6 J/cm^3 at 95°C) coupled with its intrinsically high damage threshold (>30 J/cm^2 for 30 ns duration pulses) and modest thermal lensing rate ($\sim$0.3 diopters/kW) make it an excellent material for use in high pulse energy, moderate repetition rate applications.

Since alexandrite's energy storage efficiency is high, efficient energy extraction is key to efficient amplifier operation. At 95°C alexandrite's saturation fluence is about 20 J/cm^2 at band center (755 nm). The implication is that relatively large fluences are required to extract stored energy efficiently. To date, reliable single-pass amplifier output energies of 3.0 Joules in 30 nsec pulses have been demonstrated. The 3.5 mm beam diameter used in these studies produced exit fluences in excess of 30 J/cm^2 and peak power densities > 1 GW/cm^2. Under these conditions, more than 20% of the energy stored in the amplifier rod was extracted. Improved energy extraction, for fluences comparable to the saturation fluence, was achieved using double-pass configurations.

2. Background

The impetus for the development of laser amplifiers is to enhance brightness and output power while maintaining predetermined beam characteristics (wavelength, bandwidth, divergence)[3]. The optical design of alexandrite oscillators and amplifiers does not follow the traditional design approaches used for higher cross-section, fixed frequency lasers. Designs for an alexandrite laser must take into account: 1) the temperature and wavelength-dependent cross-section, 2) the gain, 3) the energy storage, 4) the fluorescence lifetime, and 5) the material damage limitations. In addition to these factors, there are a host of operational constraints imposed on system design by auxilliary components such as flashlamps, power supplies, heat exchangers, etc.

The parametric dependence of the cross-section on temperature permits certain important design flexibilities that are not possible for fixed cross-section materials. Energy storage in alexandrite (at 95°C) can be nearly 20 times that of Nd:YAG. Although the saturation fluence is correspondingly high (20 J/cm^2), alexandrite's observed optical damage threshold is good enough to tolerate the high fluence levels required for efficient extraction. Both energy storage and saturation fluence continue to drop as temperature increases; at 250°C $E_{stor} \approx 3$ J/cm^3 while $E_{sat} \approx 7.5$ J/cm^2. Figure 1 shows laser gain as a function of wavelength for different temperatures.

Temperature-Dependent Gain in Alexandrite

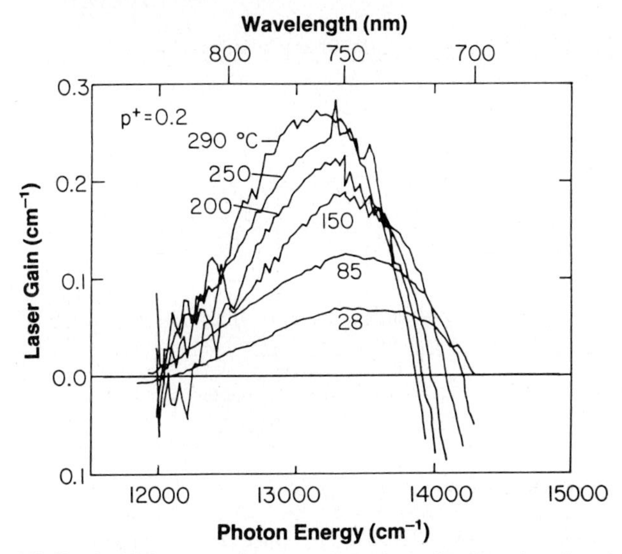

M.L. Shand and H.P. Jenssen, "Temperature Dependence of The Exited-State Absorption of Alexandrite".

Figure 1: Temperature dependency of laser gain for different wavelengths.

3. Energy Storage

Energy storage in all solid-state amplifiers is limited by parasitics, such as Amplified Stimulated Emission (ASE). Nd:YAG is severely limited by these processes: from a maximum energy storage density of 25 J/cm^3, for 100% inversion of 1% Nd^{3+}, to a realizable energy storage of only about 0.2 J/cm^3 in a typical rod geometry. In Ruby, besides being constrained by its three-level lasing dynamics, ASE limits useful energy storage to about 2.5 J/cm^3 [3]. Even under vigorous pumping, producing substantial inversion, energy storage in alexandrite is not strongly limited by ASE (see below). Figure 2 shows some important comparisons for common solid-state laser materials.

The useful laser energy stored in a gain medium during pumping is given by:

$$E_{store} = [(h\nu)\, n^*]V, \tag{1}$$

where n^* = excited state population density of Cr^{3+} ($^4T_2+^2E$) mirror sites
V = volume of material
$h\nu$ = energy of laser photon

In alexandrite only 78% of all Cr^{3+} ions are in mirror sites (the rest are in electric dipole-forbidden inversion sites), and only these contribute to gain. Therefore, the maximum energy that can be stored in a laser rod 0.635 cm in diameter and 10 cm in length, at a 0.12 atom % chromium concentration, is approximately 10 J/cm^3. This corresponds to a 100% inversion. In practice, parasitics limit useful inversion levels and thus energy storage. However, useful storage energy densities of 3 to 5 J/cm^3 (30-50% inversion levels) can be (and have been) attained (See Fig. 2).

Comparison of Some Common Solid State Laser Material Characteristics

	$E_{storage}$ (J/cm^3)							
MATERIAL	Max. inv.	ASE	Operational Max./Typ.	$\eta_{storage}$ (%)	E_{sat} (J/cm^2)	G (J^{-1})	f (m^{-1}/kW)	H (m^{-1}sec^{-1})
Nd:YAG (1 at. %)	25	0.2	0.2/0.2	1-2	0.5	0.02	0.6	0.33
RUBY (0.1 at. %)	2.5	2.5	2.5/1.6	0.3	11	0.0027	0.3	0.009
ALEXANDRITE (0.12 at. %)	10	6	3/2	1-2	20*	0.003	0.3	0.010

* 95°C, 750 nm

Figure 2: Characteristics of three solid-state lasers.

In the absence of saturation and appreciable intrinsic (ground or excited state) absorption in the amplifier, gain can be simply related to the stored energy and saturation fluence.

The unsaturated gain is given by:

$$G = \sigma_{eff}\, n^* L, \tag{2}$$

where L is the gain length. The effective cross-section, σ_{eff}, is defined by:

$$\sigma_{eff} = \sigma_{em(\lambda)}\,(1-\gamma) - \sigma_{2a} - \sigma_{a_\gamma}, \tag{3}$$

where $\gamma = \exp[(E-E^*)/kT]$ and σ_{em}, σ_{2a}, σ_a are the emission, excited state absorption, and ground-state absorption cross-sections, respectively.

Here E^* is the effective no-phonon energy level taken to be 14,700 cm^{-1} in alexandrite (at 296°K), k is Boltzmann's constant, and T is the temperature in degrees Kelvin.

For negligible ground and excited state absorption (95°C and 755 nm),

$$\sigma_{eff} \approx \sigma_{emission} \tag{4}$$

and thus

$$G \approx E_{store}/AE_{sat}. \tag{5}$$

Here the gain G is averaged over the extracted area, A, of the amplifier rod. Thus, a measurement of (small signal) gain for fixed operating conditions (temperature and wavelength, thus E_{sat}) can be used to determine stored energy. (Care should be exercised to do this only at low average power, since thermal gradients over the extracted area drive gain gradients in alexandrite).

The storage efficiency, is:

$$\eta_s = E_{store}/E_{in}, \tag{6}$$

where E_{in} is the input energy to the flashlamps.

η_s can be factored into constituent process efficiencies that aid design and system performance analysis[3]

$$\eta_s = \eta_e \cdot \eta_t \cdot \eta_c \cdot \eta_f. \tag{7}$$

These efficiencies represent electrical, transfer (pumped light coupled by the reflector), color, and fluorescence efficiencies, respectively. Typical values are $\eta_e = 0.7$; $\eta_t = 0.4$ for dual ellipse, 0.7 for single ellipse; $\eta_c = 0.15$; and $\eta_f = 0.6$. This produces overall storage efficiencies of 3-5%.

4. Energy Extraction

A principal goal of amplifier design is to extract stored energy efficiently. Careful design of both oscillators and amplifiers requires a balance to be maintained between energy storage and energy extraction, the latter being largely determined by gain and damage limits. The extraction efficiency can be defined by:

$$\eta_{ext} = \frac{E_{ext}}{E_{store}} = (A_G-1)\, E_{inj}/E_{store}, \tag{8}$$

where: E_{ext} is the energy extracted from the amplifier ($E_{out} - E_{in}$); E_{inj} is the injected energy and A_G is the amplification. In the small gain limit ($A_G-1 \approx G$) equations (5) and (8) give: $\eta_{ext} = E_{inj}/A\,E_{sat}$; the extraction efficiency is independent of stored energy.

In alexandrite, the dependence of the emission cross-section on temperature permits trade-off of energy storage for gain. In an optimal design the temperature is increased until the gain is sufficient to extract the stored energy.

Under well-controlled conditions, studies using (10 ns) single-mode focussed probe beams to damage alexandrite specimens (for both bulk and surface) have shown that local material damage thresholds can exceed power densities of 20 GW/cm^2 and fluences of 250 J/cm^2.[4] Under typical laboratory conditions, our studies have shown that alexandrite laser amplifiers can withstand fluences in excess of 30 J/cm^2 in a large aperture beam that extracts most of the rod cross-section[5]. Alexandrite's high damage threshold permits the effective manipulation of gain and energy storage in spite of the high saturation fluence required to extract the energy.

5. Single-Pass Amplifiers

Amplifier systems have been reliably operated at fluences well above the saturation fluence at 95°C, where amplification factors greater than five have been obtained. In operation (at 10 Hz PRF at 95°C), an alexandrite laser system composed of an oscillator and a single-pass amplifier (each containing 0.635 cm diameter x 10 cm long rods) has delivered over 100 Megawatts of peak power in a 7x diffraction limited beam. This same system produced over 20 Megawatt pulses in a near (1.2x) diffraction-limited beam.

Figure 3 depicts a typical operating configuration for an alexandrite single stage, single pass amplifier. The oscillator determines the beam parameters (divergence, linewidth, pulse duration, wavelength, etc.). In our studies it was composed of the high reflector, pump chamber, (usually double ellipse, flooded or unflooded), aperture, Brewster angle Q-switch, three-element birefringent tuner (for wavelength control), and output coupler. An external pump chamber and rod served as an amplifier stage. The setup also used an extra-cavity (3.5 mm) aperture to prevent amplifier rod damage from overfilling its aperture. The oscillator and amplifier rod temperatures were maintained at 95°C. In both cases a 10 cm length was optically pumped. The electronics consisted of two separate (Allied-Signal Corporation-designed) power supplies whose lamp current delivery to the pump chambers was synchronized using a delay generator (California Avionic, Inc.). The lamp current pulses were critically damped at 100 to 180 μsec between 10% power points. Long pulsed and Q-switched oscillator pulses having incident energies of 200 mJ to 1 J seeded the single-stage amplifier.

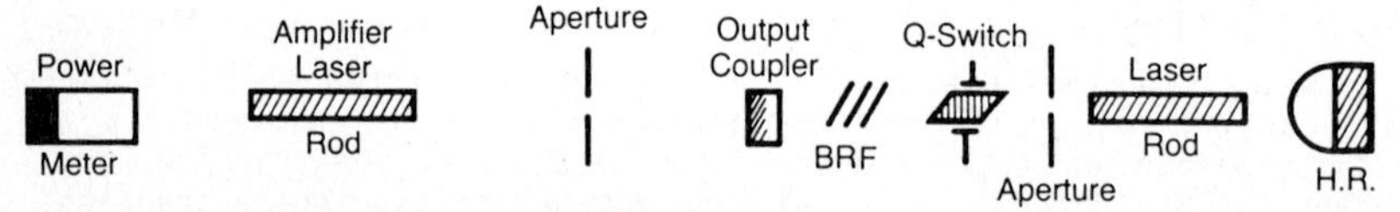

Fig. 1. Typical SINGLE PASS Oscillator/Amplifier Configuration. Aperture is used to prevent Rod "Overfilling".

Figure 3: Typical single-pass oscillator/amplifier configuration.

For 95°C operating temperature, where the saturation fluence (E_{sat}) is ≈20 J/cm^2, amplification factors of 4x to 5x have been attained and storage efficiencies that exceed 3% have been measured. For alexandrite the maximum stored energy (limited by ASE) is approximately 6 J/cm^3. Under our typical conditions between 2 and 3 J/cm^3 is stored, with values in the 5 to 6 Joules/cm^3 range have been measured. Storage efficiencies are >3%. Better storage efficiency can be realized, particularly by improving the coupling of pump light energy into the absorption band of alexandrite. Methods are now under study to accomplish this.

Extraction and storage can also be controlled by selection of rod Cr^{3+} concentration and diameter, since this affects excited state population density for a given pump power. Amplifier rods between 0.4 cm and 0.635 cm in diameter were investigated. The smaller diameter rod has larger gain, but also experiences a larger thermal gradient and thus more thermal lensing[6]. At a given power loading, the thermal lens of a 4 mm rod is about 60% greater than for a 0.635 cm rod. Thermal lensing can strongly influence resonator design and its aspheric components can drive uncorrectable beam divergence. For lower gain systems, such as alexandrite, a figure of merit for high average power operation is given by the ratio of gain to thermal lens strength:

$$H = G \cdot f^{-1} \tag{9}$$

$$= \text{Gain/Thermal Lens Strength} .$$

For a given repetition rate and lamp spectrum, H is an intrinsic property of the gain medium, i.e., it is independent of pumping efficiency, pump length, Cr^{3+} concentration, and most other operating conditions. It is a parameter of central significance in the design of high average power laser systems that are "extraction" (i.e., gain) as opposed to "energy storage"- driven. Figure 2 compares the H parameter for several common solid-state laser materials.

Single pass amplifications as a function of amplifier input powers are shown in Fig. 4 for 5 mm and 6.35 mm diameter rods. New data using a single ellipse flooded pump chamber has yielded amplification of better than 6x, with extraction efficiencies of about 20%. No gain saturation effects have been observed for single-pass, single-stage amplifiers injected with energies of 0.5 to 0.75 Joules.

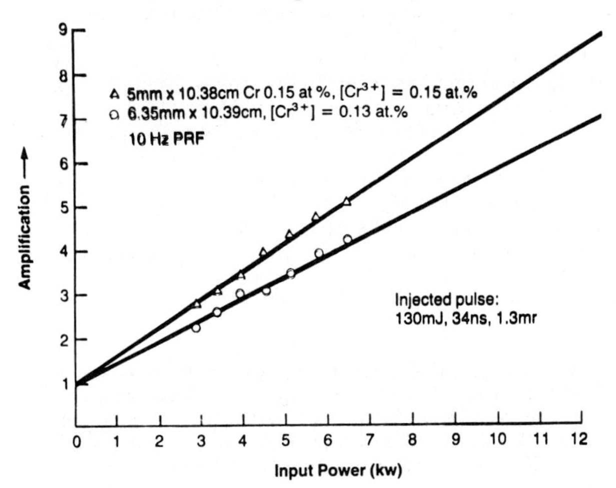

Figure 4: Alexandrite single-pass amplification as a function of lamp input power.

6. Double Pass Amplifiers

More efficient extraction, at low fluence, can be obtained by using a double-pass amplifier configuration. One such design employs a permanent magnet Faraday rotator (described below), and a polarization beam splitter. This permits both forward and backward propagating beams to enjoy the full amplifier aperture.

Double-Pass Amplifier

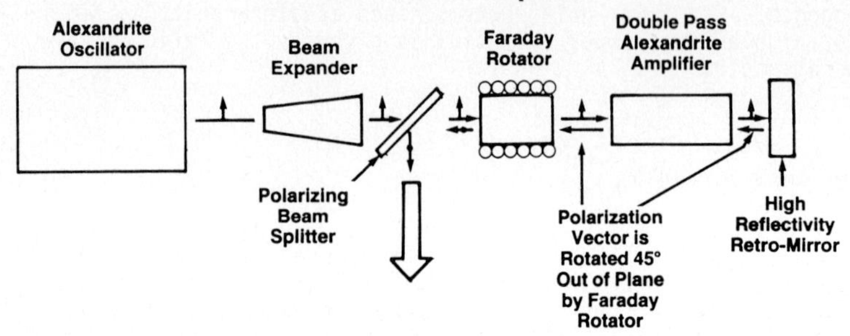

Figure 5: Alexandrite double-pass amplifier: Faraday rotator is of permanent magnet type.

Figure 5 gives a schematic of our alexandrite double-pass amplifier. In this configuration the polarized input is transmitted through a beam splitter, rotated +45° by the Faraday rotator, and amplified by the pumped alexandrite rod. The latter is aligned so that its b axis is parallel to the incident light polarization, i.e., at 45° to the input polarization. The amplified light is retroreflected (by the HR) back through the amplifier rod and into the Faraday rotator. This time the polarization is rotated - 45°. Thus, the backward propagating beam has a 90° polarization shift with respect to the input. This beam is outcoupled by the polarizing beam splitter.

In the past 45° polarization rotation has been achieved by using pulsed Faraday rotators, usually FR-5 Glass surrounded by a Helmholtz coil. The coil was pulsed through a PFN from a charging supply, and synchronized to the laser power supplies. A new approach makes use of a permanent magnet rotator consisting of cored-out Cobalt Samarium magnets (B = 8.5 - 11.0 kG) encapsulated in an aluminum housing. Usually 4 to 6 magnets are necessary (depending on the B) to achieve adequate rotation in FR-5 at 750 nm. A new terbium-based silicate rotator glass, M-16 (Kigre Corp.) gives good rotation at the wavelength of interest, and has a relatively high damage threshold. It does have a somewhat stronger absorption at 750 nm limiting its use at high average power.

In initial experiments, this double-pass amplifier yielded an unsaturated gain per Joule of flashlamp input energy of 0.0036. This compares with a gain/Joule of 0.0023 for a single pass configuration of the same pumping conditions. System losses are still being reduced by improvement of surface coatings. A storage energy of 6.3 J for 450 Joules into the lamps yields a storage efficiency of 1.4%.

Figure 6 details preliminary results for single-pass and double-pass amplification as functions of amplifier input energy. The improved extraction (and consequential gain saturation effects) for the double-pass configuration is evident.

7. Summary

Alexandrite is a robust solid-state laser matieral well suited for use in medium average power applications (1 kW). Energy storage is high but the emission cross-section is relatively low. Fortunately, alexandrite's high damage threshold permits efficient extraction of stored energy.

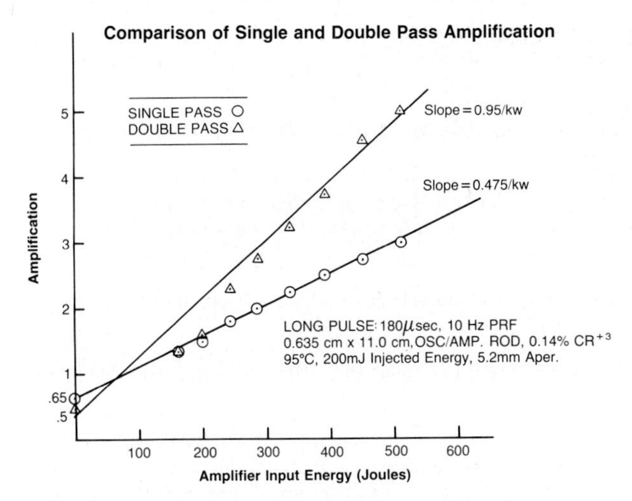

Figure 6: Comparison of alexandrite single- and double-pass amplification as a function of amplifier input energy.

Because alexandrite is a tunable, low cross-section, variable cross-section material, the design of alexandrite lasers does not follow the traditional approaches developed for materials with higher fixed cross sections. Energy storage and energy extraction must be balanced. Under most conditions alexandrite laser designs are driven by energy extraction considerations. Temperature and rod concentration and diameter can be used to balance gain and energy storage. When this is done, high average power operation can be achieved at excellent operating efficiencies. Some performance highlights for alexandrite amplifiers are summarized below [7,8].

Alexandrite Amplifier Performance*

• Amplification	5X
• Energy Stored	6 Joules
• Energy Extracted	2.25 Joules
• Output Energy	3.0 Joules
• Output Power	100 Megawatts

* 0.635 cm x 10 cm rods @95°C, 755 nm.

Acknowledgements

Many helpful and insightful discussions with Drs. Y. Band and J. Walling aided us greatly during the course of this work.

References

1. J.C. Walling, D.F. Heller, H. Samelson, D.J. Harter, J.A. Pete, R.C. Morris, "Tunable Alexandrite Lasers: Development and Performance," IEEE Journal of Quantum Electron., vol. QE-21, No. 10, pp. 1568-1581, Oct. 1985.

2. M.L. Shand, J.C. Walling, and R.C. Morris, "Excited-state absorption in the pump region of alexandrite," J. Appl. Phys. vol. 52, pp. 953-955. 1981.

3. W. Koechner, Solid-State Laser Engineering. New York: Springer-Verlag, 1976.

4. S.C. Seitel, "Alexandrite laser damage testing," Rep. to Allied Corpor. (unpublished) Michelson Lab., Naval Weapons Cen., China Lake, CA 93555, July 2, 1984.

5. D.J. Harter, A. Heiney, and H. Samelson, Allied Corporation, Mt. Bethel, NJ, unpublished results.

6. D.J. Harter and K. Whittle, Allied Corporation, Mt. Bethel, NJ, unpublished results.

7. D.F. Heller, J.J. Barrett, T. Chin, D.J. Harter, J.S. Krasinski, J. Kuper, J.A. Pete, H. Samelson, M.L. Shand, D.R. Siebert, and J.C. Walling, "Advances in alexandrite laser technology," presented at the Topical Meet. on Tunable Solid-State Lasers, Arlington, VA, May 16-17, 1985.

8. O. Kafri, J. Krasinski, and D.F. Heller, "New diagnostics for studying the quality and optical properties of solid-state laser media," presented at the Laser and Electro-Opt., Baltimore, MD, Session ThR6, 1985.

High-Power Injection-Locked Alexandrite Ring Laser

F. de Rougemont, V. Michau, and R. Frey

Laboratoire d'Optique Quantique du C.N.R.S., Ecole Polytechnique, F-91128 Palaiseau Cedex, France

1. INTRODUCTION

During the last decade a lot of research work has been done with alexandrite acting as a laser material [1]. Attractive laser properties, particularly the long lifetime of the excited level of the laser transition (262 μs) and the good thermal conductivity permitted attainment of a high repetition rate tunable solid-state Q-switched laser working at room temperature.

This paper describes the application of injection-locking to an alexandrite laser and also describes the possibility of an extended tunability of this laser source to the infrared region through Raman-induced cavity dumping [2]. This frequency down conversion technique essentially permits achievement of very efficient temporal pulse compression which may be very important for a lot of experiments with alexandrite lasers; by Q-switching an alexandrite laser the pulse durations in the 100 ns-range are too long for many applications.

This relatively long duration is due to the low gain of this laser. For this reason it is difficult to reach high amplification factors : recently a double pass amplification factor of 5 was obtained at a 95°C temperature [1]. On the other hand, because of the good energy storage properties of alexandrite, amplification is more efficient at relatively high input energies, thus achieving a better energy extraction.

In order to obtain high peak powers with a low spectral bandwidth operation an injection-locking scheme seems to be most appropriate; low gain and efficient energy storage are favorable for intracavity amplification. The frequency is selected by means of a low-energy master oscillator containing highly dispersive elements. Energy is extracted from a high-energy slave laser.

This paper is divided in three parts: the master oscillator is described in section 2; the prism tuned ring slave oscillator and its mirror number dependence is described in section 3; and finally, the pulse compression technique by Raman-induced cavity dumping is described in section 4.

2. MASTER OSCILLATOR

The experimental setup is presented in figure 1. The master oscillator is delimited by a dashed line in figure 1. The Apollo Laser double elliptical laser head incorporated two linear xenon flashlamps and the alexandrite rod (5 mm in diameter, 10 cm in length). Both flashlamps and rod were maintained at 60°C by means of a deionized water cooling unit. This temperature was chosen in order to obtain the widest tunability of the master oscillator. The lab-made resonator consisted of an output

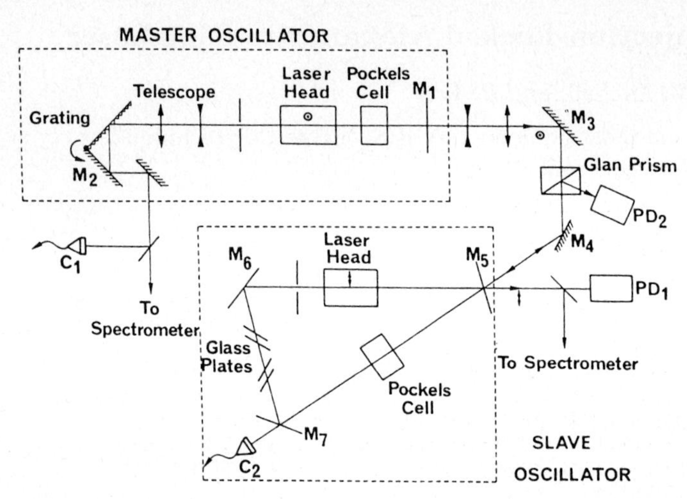

Fig.1 : Experimental setup : the polarizations of the beams are indicated by ⊙ and ↕ for vertically and horizontally polarized beams respectively.

dielectric mirror M_1 (75% reflectivity) and a Littrow-mounted diffraction grating (1224 grooves per mm). The zero-order reflection on the grating and mirror M_2 were used as a second output,thus allowing a permanent control of the wavelength through a spectrometer and the achievement of a trigger pulse by means of the fast photocell C_1. A telescope (f_c=250 mm, f_d= -50 mm) magnified the beam on the grating. Finally the cavity was Q-switched by a Pockels cell by switching off a quarter-wave voltage. A 3.5 mm hole limited the beam to the cleanest spatial region of the rod.

Table 1 : Main characteristics of the master oscillator.(λ : wavelength ; E : energy per pulse ; Δt : FWHM duration,; $\Delta\sigma$: FWHM spectral bandwidth ; θ : total divergence angle at 80 % energy ; ϕ : beam diameter)

λ(μm)	E(mJ)	Δt(ns)	$\Delta\sigma$(cm^{-1})	θ(mrad)	ϕ(mm)
0.725 - 0.775	$\geqslant$ 5	250 - 300	0.03	1	3.5

The main output characteristics are listed in table 1. The tuning range is indicated for a minimum energy of 5 mJ which represented the minimum value for an efficient injection-locking of the slave oscillator. In fact the maximum output energy from the master oscillator was limited to about 20 mJ in order to avoid optical damage inside the cavity. The present limitation of the tuning range on the red side of the spectrum was due to an increase of 15% in losses between 0.76 and 0.78 μm because of inadequate coatings on the optical elements.

3. SLAVE RING OSCILLATOR

3.1. Injection-locking operation

The slave ring oscillator which is delimited by dashed lines in figure 1 consisted of three mirrors : M_6 and M_7 were totally reflecting and M_5 was the output coupler with a 75 % reflectivity. This ring cavity mainly contained a Pockels cell and another laser head from Apollo Laser, Inc. with the same geometrical configuration ; the rod was 9.52 mm in diameter and 10 cm in length. Nevertheless, the beam diameter was restricted to its useful diameter (8mm). Obviously, this cavity provided a laser which could work without injection from the master oscillator : in this case this laser operated on a broadband spectrum around 7515 Å and had two output beams. One of them could travel back to the master oscillator and induce strong optical damage inside it. In order to avoid such damage the cavities were passively decoupled by using a Glan prism. Because of the alexandrite gain dichroism the master oscillator was set to generate a vertically polarized beam which after magnification in a telescope was transmitted through the Glan prism and injected in the ring cavity. Before the Q-switching of the slave oscillator a 2330 V voltage was applied on the Pockels cell, thus acting as a quarter wave plate, which together with the four glass plates at the Brewster angle provided losses greater than the horizontally polarized gain. However, energy was stored at the injected wavelength. Furthermore the rotation sense of the light inside the ring cavity was defined by the injection direction. At a suitable time Q-switching occurred by switching off the 2330 V voltage through the trigger pulse generated by the photodiode C_1. Almost all the horizontally polarized laser energy was actually detected by pyrodetector PD_1. Only less than 1% of the total energy was observed after reflection on the Glan prism on pyrodetector PD_2.

The tunability curve is represented in figure 2. 500 mJ pulses were obtained over a broad tuning range between 7400 Å and 7650 Å [3]. However the total tuning range was not as broad as that of the master oscillator tuning range.

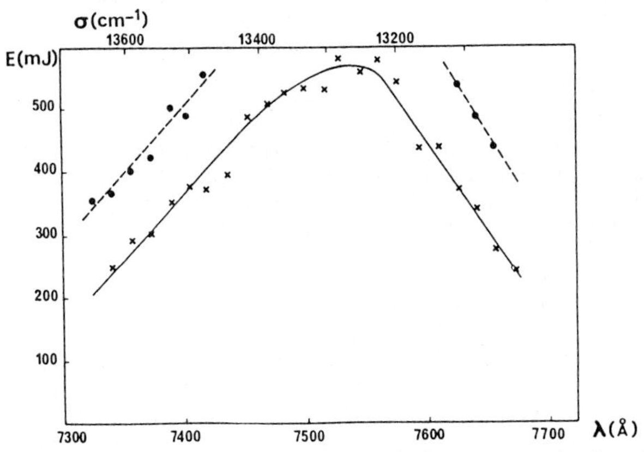

Fig.2 : Energy output for the whole system during injection-locked operation versus wavelength. Solid and dashed lines are for 1.28 and 1.43 times the threshold pumping energy, respectively.

This limitation was due to the dynamical behavior of the slave oscillator. It is known [4] that the buildup time mainly depends on the laser gain and the number of photons inside the cavity at the Q-switching time. As an example, two situations occuring at the same wavelength (7515 Å) are represented in figure 3. Both plots are experimental. In the case (a) the ring cavity was injected and in the case (b) it was not. The buildup time in the first case was shorter because the number of photons at the Q-switching time was larger than in the case (b).

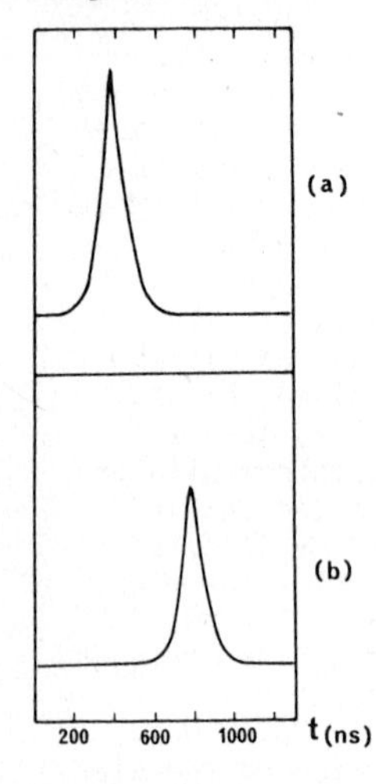

Fig.3 : Buildup time of the pulse in the slave cavity. t=0 is the Q-switching time. (a) Injected cavity at λ=7515 Å (b) Free-running operation at λ=7515 Å

In the situation in which the injected wavelength was not equal to the central wavelength of the alexandrite gain (7515 Å) a competition occured between the injected operation and the non-injected one. Because of the lower gain at a sideband wavelength the buildup time became longer than the central wavelength non-injected operation, thus favouring the non-injected operation on a broadband spectrum. In any case it was necessary to add a dispersive element inside the ring cavity in order to suppress the non-injected behavior.

3.2 Mirror Number Dependence

This section deals with the tunability of ring lasers by prism rotation. As will be demonstrated when the prism is rotated either the beam direction inside the cavity can be modified and the wavelength kept constant, or the wavelength is modified without any change in the output direction, depending on the number of mirrors in the cavity.

A first experiment was done with a three-mirror cavity (fig.4.a). Mirrors M_2 and M_3 were totally reflecting and M_1 was the output coupler ensuring a 75% reflectivity. The cavity contained an alexandrite laser head and two apertures with a variable diameter. A 1 mrad rotation of M_2 around the z rotation axis kept laser energy unchanged but the output beam direction was turned by the same 1 mrad angle. In order to test this three-mirror cavity with a dispersive element, M_2 was replaced by a Pellin-Broca prism. For a 7 mm aperture no significant change of the wavelength was observed with a prism rotation of 6 mrad. However the output direction changed by the same 6 mrad angle. Nevertheless, when the

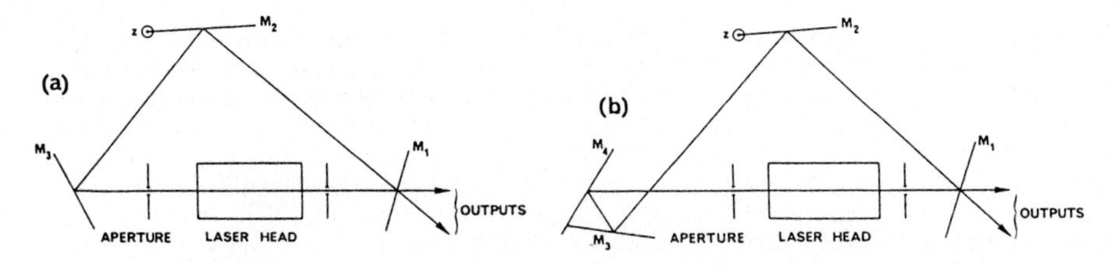

Fig.4 : Experimental setups : (a) Three-mirror cavity (b) Four-mirror cavity

aperture angle of the laser gain medium was reduced a competition between wavelength and direction in the cavity occurred. Indeed a tuning of 25 Å and 88 Å was observed for the same prism rotation in the case of a 5 mm and a 3 mm diameter, respectively.

In a second experiment, the laser head was placed in a four-mirror cavity (fig.4 b). In such a cavity laser operation vanished through a 0.05 mrad rotation of M_2. By replacing M_2 with a Pellin-Broca prism the overall tuning range of the alexandrite laser (0.72 - 0.78 μm) was easily achieved by a 3 mrad rotation of the prism for any aperture in the cavity. No change in the output direction was detected within the experimental accuracy.

These results may be explained and generalized to a n-mirror cavity by using geometrical considerations [5]. Here the demonstration is restricted to the case of three and four-mirror cavities (fig.5).

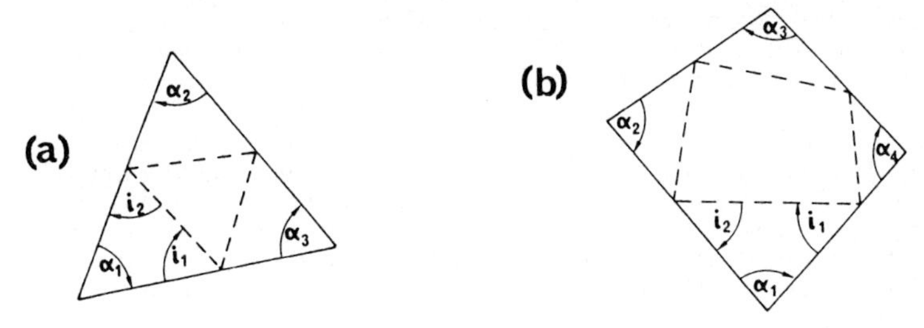

Fig.5 : Geometrical scheme of a ring cavity. (a) Three-mirror case (b) Four-mirror case

A three-mirror cavity may be represented by a triangle as shown by solid lines in figure 5a. The light roundtrip is represented in dashed lines. A similar representation may be done with a quadrilateral for a four-mirror case. The angles i are related to the incidence angles and the α's are the angles between two consecutive mirrors in the cavity. Two sets of equations are easily obtained:

$$
\begin{aligned}
i_1 + i_2 &= \pi - \alpha_1 \\
i_2 + i_3 &= \pi - \alpha_2 \\
i_3 + i_1 &= \pi - \alpha_3
\end{aligned}
\qquad (1a)
$$

$$
\begin{aligned}
i_1 + i_2 &= \pi - \alpha_1 \\
i_2 + i_3 &= \pi - \alpha_2 \\
i_3 + i_4 &= \pi - \alpha_3 \\
i_4 + i_1 &= \pi - \alpha_4
\end{aligned}
\qquad (1b)
$$

Equations (1a) can always be solved. On the other hand, in order to find a set of incidence angles which solve the (1b) equation system with the α parameters, a necessary condition has to be respected in the four-mirror case which is :

$$\alpha_1 + \alpha_3 = \alpha_2 + \alpha_4 . \qquad (2)$$

Consequently, if one mirror is rotated in the four-mirror ring cavity equation (2) is no longer valid, and the quadrilateral does not provide a laser cavity and laser action is suppressed. On the other hand, for the three-mirror cavity, if one mirror is rotated another roundtrip exists according to equation (1a) provided the α's are modified. This other roundtrip is only possible if the gain medium is actually crossed by the beam. In particular these considerations are not true for a point gain medium [6].

In conclusion, tuning a high-energy ring laser through rotation of a dispersive prism is most effective when using a resonator consisting of an even number of mirrors.

3.3 Extended Tunability Alexandrite Setup

According to the previous considerations a new experimental setup for the slave ring oscillator was tested (fig. 6). The operation was the same as described in section 3.1. The four glass plates at the Brewster angle were provided by the input and output sides of the Pockels cell and the Pellin-Broca prism.

With this setup the tunability was the same as in the case of the master oscillator. Main characteristics are given in table 2. Measurements were performed only for energies smaller than 600 mJ in order to avoid optical damage.

4. FREQUENCY DOWN-CONVERSION AND PULSE COMPRESSION BY RAMAN-INDUCED CAVITY DUMPING

Various laboratory investigations require short duration intense infrared pulses. In order to obtain infrared wavelengths from a visible laser, sti-

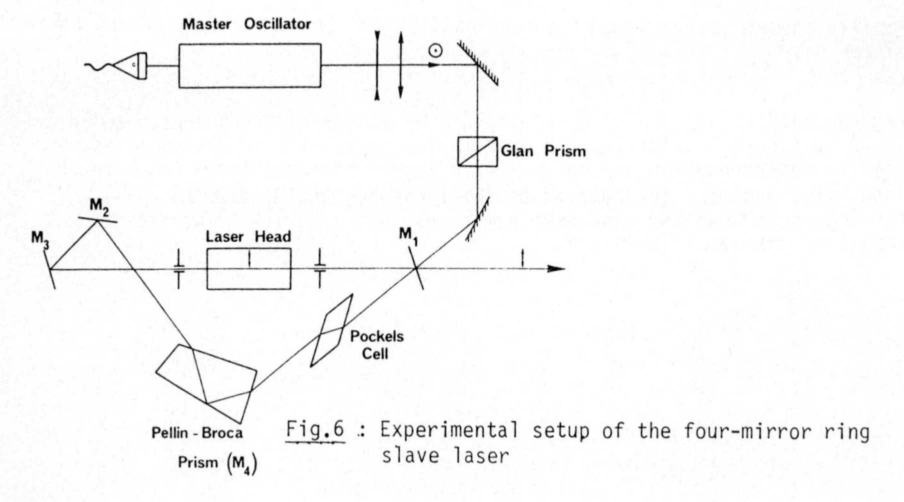

Fig.6 : Experimental setup of the four-mirror ring slave laser

Table 2 : Main characteristics of the slave ring oscillator (notations are given in the table 1 caption)

$\lambda(\mu m)$	E(mJ)	Δt(ns)	$\Delta\sigma(cm^{-1})$	θ(mrad)	ϕ(mm)
0.725 - 0.775	up to 600	$\geqslant 120$	0.03	0.5	7

mulated Raman Scattering [7] has been proven to be a very efficient technique. Since the Raman gain is proportional to the incident beam intensity the efficiency is improved when using high intensities. Because of the long duration of alexandrite pulses (120 ns) such intensities may be increased through a pulse compression technique. Several methods such as Q-switching, pulse slicing, cavity dumping or backward stimulated scattering have been proposed to achieve pulses in the nanosecond range. However more recently an efficient technique called Raman-induced cavity dumping has been demonstrated in our laboratory [2].

The basic idea is to extract the electromagnetic energy stored in a nearly lossless cavity at a storage frequency by using passive cavity dumping induced by stimulated Raman scattering. Indeed the storage cavity is composed of four mirrors (M_1 to M_4) which are totally reflecting at the storage frequency ω_{1S} (fig.7). This cavity contains a Raman medium.

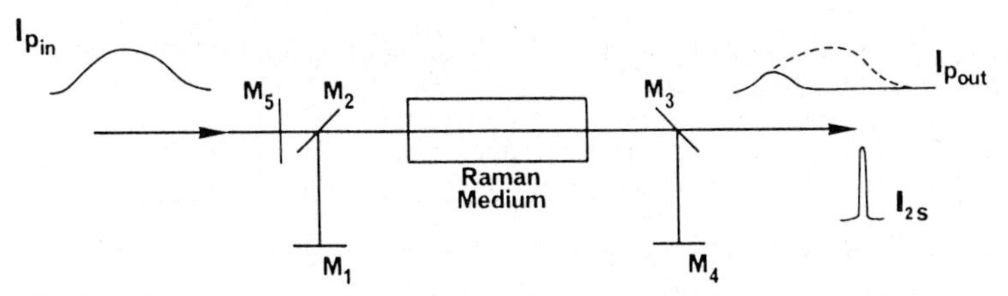

Fig.7 : Schematic arrangement of Raman-induced cavity dumping

The energy storage is achieved through stimulated Raman scattering of an incoming pump pulse at the laser frequency ω_L ; mirrors M_5, M_2 and M_3 are non-reflecting at frequency ω_L. The Raman gain is adjusted in such a way that a second stimulated Raman scattering may occur when the intensity stored at the first Stokes frequency ω_{1S} is maximum. In this case the cavity is dumped by the generation of a second Stokes pulse,the duration of which is about a roundtrip time of the light inside the cavity.

A theoretical study has been done [8] and some experiments with a ruby laser demonstrated pulse compression rates up to 10 [2].

Work is in progress in our laboratory in order to apply Raman-induced cavity dumping to the alexandrite system. A complementary stage is being added to the experimental setup for the pulse compression. It is possible to predict some characteristics for this infrared tunable source by adapting the experimental results obtained with a ruby laser to the alexandrite case. For example, extrapolations predict 90 mJ energy,2ns duration pulses in the 2μm range when using high-pressure hydrogen gas as the Raman medium. It is clear that the tunability of such a system is very broad because of the potential use of several Raman media and the multiplication of the Raman shifts so that most of the infrared region could be reached.

5. CONCLUSION

This paper has demonstrated the possible application of the injection-locking technique to an alexandrite laser thus obtaining a 500 mJ tunable Q-switched laser. A low-energy narrow bandwidth grating tuned alexandrite oscillator was used for injecting a high-energy alexandrite slave ring laser. Use of crossed polarizations for master and slave lasers has also been demonstrated to fully decouple the ring slave oscillator from the master laser. The mirror number dependence of a ring cavity for tuning a high-energy laser has also been emphasized : only a ring cavity consisting of an even number of mirrors is appropriate for the laser tunability. In our experimental setup the ring slave oscillator consisted of three mirrors and one Pellin Broca prism.

Nanosecond duration infrared pulses could be produced by pulse compression through Raman induced cavity dumping. The basic principle of this technique has been explained and predicted results have been given considering experimental results of a previous work with a ruby laser. This technique is now being applied to the alexandrite laser in order to obtain a high peak power laser source tunable in the infrared.

REFERENCES

1. J.C. Walling, D.F. Heller, H. Samelson, D.J. Harter, J.A. Pete, R.C. Morris : IEEE J.Quantum Electron QE 21, 1568 (1985)
2. F. de Rougemont, R. Frey, F. Pradère : Optica Acta 32, 1281 (1985)
3. V. Michau, F. de Rougemont, R. Frey : Appl.Phys. B 39, 219 (1986)
4. O. Svelto : Principles of Lasers, 2nd ed., Plenum Press, New York (1982)
5. V. Michau, F. de Rougemont, R. Frey : to be published in the 1 July issue of Appl.Opt.(1986)
6. A. Fuchs, D. Debelaar, M.M. Salour : Appl.Phys.Lett. 43, 32 (1983)
7. See for example Y.R. Shen : The Principles of Nonlinear Optics, J. Wiley and Sons, New York (1984)
8. R. Frey, F. Pradère : IEEE J.Quantum Electron. QE 20, 786 (1984)

Low Magnification Unstable Resonators Using Radially Varying Birefringent Elements

*D.J. Harter and J.C. Walling**

Allied Corportion, 7 Powder Horn Drive, Mt. Bethel, NJ 07060, USA

1. Introduction

It has been shown theoretically that a gaussian gain profile can confine an unstable resonator mode so that such a mode has a finite but large beam radius [1-4]. The stabilization of a large beam radius by a radially dependent gain profile has also been experimentally verified [5]. A gaussian loss element (i.e., a gaussian reflector) can also affect the beam profile in a similar manner [6-8]. More recently, Eggleston et al [9] have pointed out that a radially birefringent reflector can be approximated as a gaussian reflector, and have shown that such an element can support a large unstable resonator mode which does not suffer Fresnel fringing due to sharp changes in the reflectivity typical for an unstable resonator. This fringing can be a source of regions of high intensities which can cause optical damage. Giuliani et al [10] have shown that such a radial birefringent element (RBE) unstable resonator can be applied to a pulsed Nd:YAG laser. Ruby and alexandrite are usually oscillated at lower gain configuration than Nd:YAG so that low magnification unstable resonators are necessary to support an unstable resonator mode. Low magnification unstable resonators have higher internal standing wave power densities and have a greater problem with Fresnel fringing and, hence, are more susceptible to optical damage. Low magnification unstable resonators are less stable against perturbations in the unstable resonator mode which can also lead to damage [4,10]. All experiments with "Polka Dot" unstable resonators with alexandrite lasers have led to optical damage at relatively low output powers due to a stable mode oscillating from the central high reflectivity "Polka Dot" [11]. In this paper a low magnification unstable resonator using radially birefringent elements which is less susceptible to damage is described. This resonator is less susceptible to damage for two reasons. One reason is the radially birefringent element is a soft aperture without Fresnel fringing effect. Also, in the configuration used in our experiments, the Q for the unstable resonator mode was comparable to the Q for stable mode, so accidental lasing in lower order modes did not cause optical damage. In this paper we will also describe our results with this resonator for alexandrite and ruby.

2. Description of RBE Reflector

Figure 1 illustrates the radially birefringent element reflector which was used in the experiments. The reflector consists of two lenses (elements 3 and 4) which have equal and opposite curvature, and which are mounted in close proximity to yield a very long focal length. The reflector also has two compensator plates (elements 2 and 5), a polarizer (element 1) and a flat mirror (element 6). The radially birefringent elements

* Present address is: 5833 Lomond Drive, San Diego, CA 92120

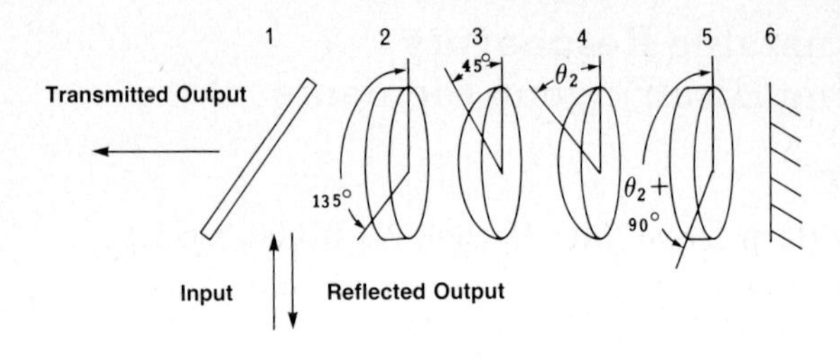

Fig. 1 The elements which constitute the RBE output coupler are shown as well as the angle between the fast axis of each element and vertical polarization. Elements 3 and 4 are birefringent lenses, 2 and 5 are compensator plates, 1 is a polarizer, and 6 is a mirror.

in this reflector are fabricated from two birefringent crystal quartz lenses where the fast and slow axes of the crystal are perpendicular to the optical axis of the lens. One of the lenses (element 2) is a multiple zero ($0 + n\pi$) waveplate for 750 nm at its center and the curvature of the concave surface is chosen so that the lens is a multiple quarter ($1/4 + n\pi$) waveplate at a selected distance, r_o, from the center of the lens. This lens is rotated so that the angle between its fast axis and the vertical polarization of the incoming laser light is 45°. After double passing through this element, the polarization of light transmitted through the center of the lens is not changed and all of this light is reflected back into the laser cavity at the polarizer. The polarization of light transmitted through the lens at radius r_o is rotated 90° and all of this light is reflected out of the resonator at the polarizer.

The second lens (element 3) is a multiple 1/4 waveplate ($1/4 + m\pi$) at its center and the curvature of the lens is equal and opposite to the curvature of the first lens (element 2) so this lens is a multiple zero waveplate ($0 + m\pi$) at radius, r_o. The angle between the fast axis of this lens and the verticle polarization of the incoming laser light is a variable (θ_2) so that the fraction of light which is transmitted out of the laser cavity can be continuously varied. The reflectance, R, of this radially birefringent element reflector is shown in Fig. 2 as a function of a radius for various values of θ_2 and is given by [12]

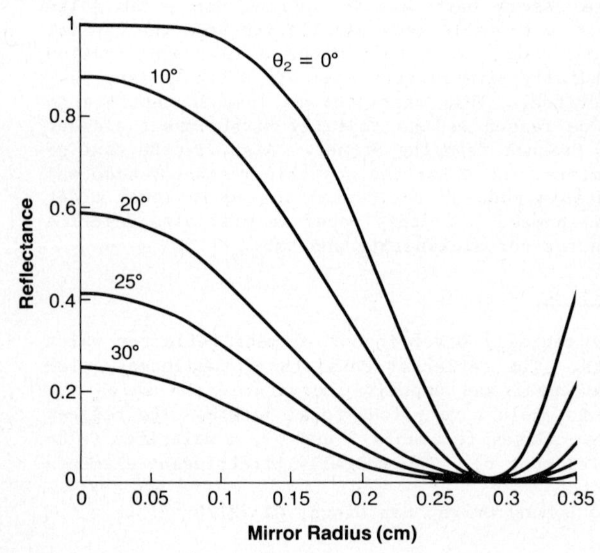

Fig. 2 The reflectance of the radially birefringent filter used in the experiments is plotted as a function of radius. The zero of the reflectance is approximately at the aperture of the 1/4" rod. The filter consists of five elements: polarizer, -21 and +21 cm radius quartz lens with compensator plates. θ_2 is the angle between the direction of polarization and the fast axis of the quartz lens which is a quarter wave plate in its center.

$$R = \cos^2(2\alpha_r)[\cos^2(2\theta_2) + \sin^2(2\alpha_r)\ (\sin 2\theta_2 - 1)^2], \tag{1}$$

where

$$\alpha_r = \frac{-\pi \Delta n r^2}{2\lambda\rho} = \frac{\pi}{4}\frac{r^2}{r_o^2} \tag{2}$$

and Δn is the difference in refractive index between the fast and slow axes (0.00891), λ the lasing wavelength (750 nm), ρ is the curvature of element 3 (21 cm), and r_o is the first zero of the reflectance (3.0 mm).

Each of the lenses (3 and 4) is rigidly mounted with an additional plate (elements 2 and 5). These plates are also made of crystal quartz with thicknesses chosen to give the same birefringence as the multiple zero order (n or $m\pi$) of the lens it is mounted with. Therefore, by mounting the compensator plate so that its fast axis is parallel to the slow axis of the lens, the plate compensates for the birefringence due to the finite thickness of the lens and the pair of elements is a radially birefringent element with single order 0 to 1/4 wave birefringence. The primary effect of these compensator plates is to make the reflectance of this radially birefringent reflector relatively insensitive to wavelength differences between the ruby lasing wavelength at 694.3 nm and the lasing wavelengths of alexandrite (700-800 nm).

3. RBE Resonator Design

For the design of an unstable resonator, it is important to know the output coupling or loss, L, per pass. We will derive the loss of a gaussian beam from passing through a gaussian and a RBE reflector with and without apertures. The resonator mode for the RBE resonator is approximated as having a gaussian profile.

The output coupling, L, for a gaussian beam of radius r through a reflectance, R(r), with an aperture next to the reflector with radius r_c is

$$L = 1 - \frac{\int_0^{2\pi}\int_0^{r_c} \exp(-2r'^2/r^2)\ R(r')r'dr'd\theta}{\int_0^{2\pi}\int_0^{\infty} \exp(-2r'^2/r^2)r'dr'd\theta}. \tag{3}$$

For a gaussian reflector with reflectance

$$R(r') = \exp(-2r'^2/r_m^2)$$

the loss is

$$L = 1 - r_e^2/r^2\ (1 - \exp\ (-2r_c^2/r_e^2)), \tag{4}$$

where

$$1/r_e^2 = 1/r_m^2 + 1/r^2 \quad . \tag{5}$$

Here r_e is the beam radius of the exiting gaussian beam profile. When $r_c \to \infty$, (4) reduces to the more common equation for the loss, L, [9,13] for an unstable resonator which is given by $1-1/M^2$ where M, the magnification of the resonator, is equal to the ratio of the incoming beam radius to that of the exiting beam radius (r/r_e).

For the case of a radially birefringent output coupler, with the reflectance given in (1) and $\theta_2 = 0$, with an aperture that has a radius r_c equal to the first zero of the RBE element, r_o, the loss can be found again by integrating (3) to give

$$L = 1 - \frac{(1-\exp(-P\pi))(2P^2 + 5)}{2(P^2 + 4)(P^2 + 1)} - \frac{P^2}{P^2 + 1} \tag{6}$$

with $P = 2r_c^2/\pi r^2$.

The loss for the RBE with no aperture is

$$L = 3/[2(Q^2 + 4)(Q^2 + 1)] \quad , \tag{7}$$

with $Q = 2r_o^2/\pi r^2$.

Figure 3 shows the output coupling for the RBE output coupler with an aperture of radius $r_o=r_c$ and without an aperture ((6) and (7) and Figs. 3(a) and (d), respectively) and the output coupling for the gaussian output coupler with an aperture ((5) and Fig. 3(b)). In Fig. 3 it can be seen that the output coupling of the RBE does not exceed 0.5 as the radius of the incident beam increases. The reason for this is because the reflectance of the RBE reflector begins to oscillate rapidly between 0 and 1 beyond the first zero of the reflectance, r_o. It should be noted that when r/r_m or r/r_o is increased, which is necessary for high magnification operation (large incident beam radius over existing beam radius), then the aperture loss becomes significant. The output coupling, L, for an RBE output coupler with an aperture is the sum of the output coupling from the aperture and the RBE output coupler. From Fig. 3(b) it is clear that as the ratio r/r_o becomes greater than 1, then the output coupling from the aperture (Fig. 3(c)) becomes larger than the output coupling from the RBE output coupler (Fig. 3(d)). Therefore, the most efficient way to operate the RBE unstable resonator is at low magnification. The output coupling can be increased by varying θ_2 in (1) or by another means.

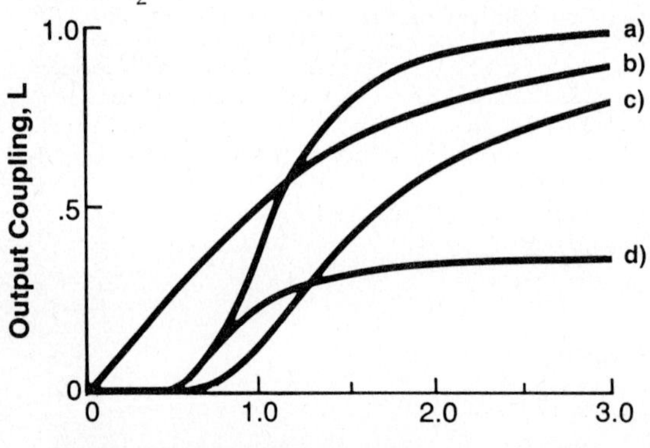

Fig. 3 The output coupling of a gaussian reflector is plotted as a function of the ratio of the incident gaussian beam radius to the radius of the gaussian reflector, r_m, with an aperture $r_c=r_m$ (b). The output coupling is also plotted for a radial birefringent filter without an aperture (d) and with an aperature (a) where $r_c=r_o$. The loss from the aperture, R(r)=1, is also plotted (c).

When the RBE output coupler is approximated as a gaussian output coupler with $r_o=r_m$, then the intracavity beam radius and the magnification of the resonator can be determined by the use of ray matrices where the matrix element for the gaussian loss element is [4,13]

$$\begin{vmatrix} 1 & 0 \\ \frac{-i\lambda}{\pi r_m^2} & 1 \end{vmatrix} . \qquad (8)$$

Using this matrix element with the other ABCD matrices for the resonator and applying the self-consistency requirement for the complex beam parameter after a round trip, two solutions for the quadratic equation for the complex beam parameter, q, can be found. For conditions where there is no radially dependent gain or loss all the matrix elements are real and the negative root solution for q gives positive, real roots for the beam radius when $-1 < (A+D)/2 < 1$ [14]. However, with complex matrix elements, the positive root of the quadratic equation can give positive real roots for the beam radius without satisfying the above inequality. This is the unstable resonator mode that we are interested in.

The beam radius has been calculated as a function of position in a resonator with a RBE output coupler and is shown in Fig. 4. The RBE output coupler causes a step function in the beam radius since it transmits more of the beam at the aperture than on axis. After the beam radius is reduced from passing through the RBE output coupler, the beam expands until it reaches the output coupler again. The resonator parameters used in Fig. 4 are the parameters for the RBE unstable resonator used with alexandrite. The set-up for ruby was similar except the curvature of M_2 was 10 meters since there is less thermal lensing from ruby because it is operated at lower repetition rates (~10 ppm).

4. Experimental Results

During the experiments with the alexandrite unstable resonator, the thermal lensing of the alexandrite rod could be continuously varied by changing the pulse repetition rate (1-20 pps). In this manner the resonator could be changed from the stable to the unstable resonator mode. The stability of the resonator is checked by the method described by Eggleston et al.[9] where a card is placed in the cavity which clips the beam. If the

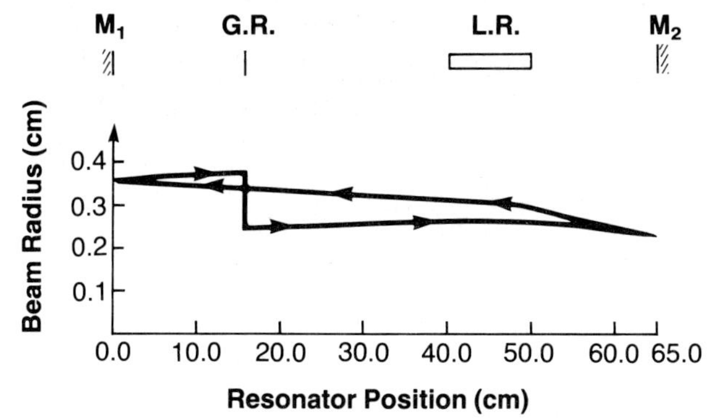

Fig. 4 The beam radius is plotted as a function of position for the alexandrite RBE unstable resonator laser using ABCD matrices. M_1 is a flat reflector, M_2 is a -80 cm reflector, the thermal lensing of the rod is 1.1 diopters and the radius of the gaussian reflector is 0.3 cm.

laser is stable, the card clips both sides of the beam. However, if it is unstable, the card only clips the same side of the beam as the card. The laser is usually operated well beyond the changeover point since there is a large region where the laser operated in a combination of stable and unstable modes.

As the thermal focusing of the alexandrite laser rod is decreased, the magnification of the unstable resonator increases, but the efficiency of the laser decreases. This is most likely due to the loss due to the aperture, as is illustrated in Fig. 3. The ratio of the output coupling from the RBE and M_1 has been measured. M_1 has a known reflectivity so the output coupling of the RBE is thus known. By changing θ_2, large output coupling could not be obtained as is predicted from (1) no matter what the repetition rate, gain, or laser wavelength used. The greatest output coupling that has been obtained is 25%. One possible explanation may be that the thicknesses of the birefringent elements are not exactly correct. However, when these elements were tested extracavity with a cw alexandrite laser, the reflectance qualitatively followed the predicted curves shown in Fig. 1. The elements were thought to be not correct since tuning of the wavelength of the pulsed alexandrite laser with these elements intracavity varied the output coupling. Newer elements of the same design were used which did not show a change of output coupling with wavelength. To obtain sufficient output coupling from the first elements, the reflectivity of M_1 was lowered to 60%. The RBE filter was not used for output coupling but only for stabilizing the unstable resonator mode. Approximately 10-15% light was output coupled through the RBE filter, which is a loss. However, this configuration has the additional benefit in that the Q for the stable resonator modes is not much higher than the unstable resonator modes, so that accidental lasing in a lower order mode due to misalignment is not catastrophic.

The divergence of the laser was measured by magnifying and imaging the magnified focal spot of the beam after being focused by a 10 cm lens. The size of the focal spot is compared to the diffraction-limited spot for an helium-neon beam of the same initial diameter. Figure 5(a) shows the imaged spot from a multi-transverse mode alexandrite resonator operating with approximately 6-8x the diffraction limit. The magnified image of the focal spot is shown in a photograph 5(b) and in a 3-dimensional plot for the alexandrite RBE laser (5c). The output of the alexandrite laser is ~2.5 times the diffraction limit. The laser output at the time of this measurement was 400 mJ per pulse in a 40 nanosecond pulse at a 8 Hz repetition rate. Figures 5(e&f) shows the focusing ruby RBE output at 250 mJ and 5(d) shows the diffraction-limited helium-neon laser focal spot for the imaging system used with the ruby laser. As can be seen, the majority of the output is in a spot which is close to the diffraction limit.

There are two reasons why the divergence of the alexandrite is greater than that for ruby. One reason is that the alexandrite was operating at higher powers, and it was observed that at approximately 250 mJ output the alexandrite laser is also close to diffraction limited. Also, the alexandrite laser rod chosen was one of poorer optical quality (0.9 waves distortion at 632 nm across the rod face). This rod was chosen to show the ability of the unstable resonator to give high-quality output even with a relatively poor optical quality rod.

5. Conclusion

In conclusion, both ruby and alexandrite have oscillated in low magnification unstable resonators using radially birefringent elements. Usually

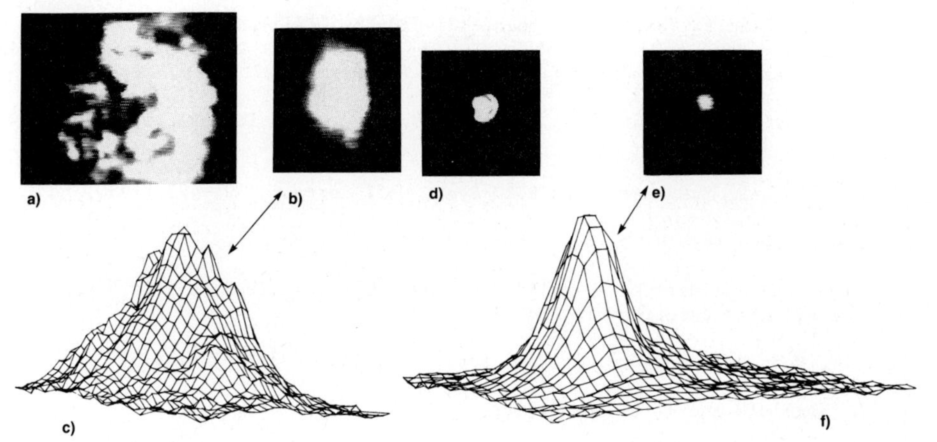

Fig. 5 The magnified image of the focal spot from a multi-transverse mode alexandrite laser which is 6-8x diffraction limited is shown in 5(a). The focal spot from an alexandrite RBE laser at 400 mJ, Q-switched is shown in a photograph 5(b) as well as a 3-dimensional plot 5(c). The alexandrite beam in 5(b&c) is ~2.5 diffraction limited, 5(e&f) show the image of the focal spot from a 250 mJ, Q-switched RBE ruby laser. Fig. 5(d) is a diffraction-limited spot at 632 nm for the imaging system used with the ruby laser.

low magnification unstable resonators have considerable Fresnel fringing and little stability from high Q, stable mode operation, and thus a considerable problem with optical damage. The smooth quasi-gaussian filtering solves the problem of the Fresnel fringing. High-power Q-switched output, which fills the active volume of the rod with near-diffraction limited output, has been obtained using the low magnification unstable resonator with radially birefringent elements.

Acknowledgements

The authors would like to thank Drs. R.L. Byer and E.P. Ippen for the helpful and stimulating discussions concerning this work and K. Whittle and G. Shipp for their technical assistance in the experiments. This research has been partially supported by Apollo Lasers Inc.

References

1. H. Kogelnik, Appl. Opt. 4, 1562-1569 (1965).

2. L.W. Casperson and A. Yariv, Appl. Opt. 11, 462-466 (1972).

3. G.J. Ernst and W.J. Witteman, IEEE J. Quantum Electron., QE-9, 911-918, (1973).

4. L.W. Casperson, IEEE J. Quantum Electron., QE-10, 629-634 (1974).

5. L.W. Casperson and A. Yariv, Appl. Phys. Lett., 12, 355-357 (1968).

6. H. Zucker, Bell Syst. Tech. J. 49, 2343-2376 (1970).

7. A. Yariv and P. Yeh, Opt. Commun. 13, 370-374 (1975).

8. A.N. Chester, Appl. Opt. 11, 2584-2590 (1972).

9. J.M. Eggleston, G. Giuliani, and R.L. Byer, J. Opt. Soc. Am. 71, 1264-1272 (1981).

10. G. Giuliani, Y.K. Park, and R.L. Byer, Opt. Lett. 5, 491-493 (1980).

11. Unpublished results by D. Harter and also by L. Horowitz.

12. Eqn. 1 can be derived by using Jones matrices described in Ref. 9 for the elements described in Fig. 1.

13. L.W. Casperson and S.D. Lunnam, Appl. Opt. 14, 1193-1199. (1975).

14. H. Kogelnik and T. Li, Appl. Opt. 5, 1550-1567 (1966).

Injection Locking Broadly Tunable, Q-Switched Alexandrite Lasers Using Semiconductor Laser Diodes

J. Krasinski, P. Papanestor, J.A. Pete, and D.F. Heller

Allied Corporation, 7 Powder Horn Drive, Mt. Bethel, NJ 07060, USA

We have shown that the frequency of a broadly tunable, high-power, alexandrite laser can be critically controlled by injection from a 1 μW single frequency cw laser diode. The physics, tuning, and performance of this laser system are described.

1. Introduction

Traditionally, the frequency control of broadly tunable lasers has been accomplished using loss control techniques: most commonly prisms, gratings, etalons, Fabry Perot interferometers or other forms of intracavity filters. Unfortunately, all loss control techniques introduce some losses at the selected as well as at the deselected frequencies. This makes for some obvious disadvantages, particularly when loss control elements are used in conjunction with low-gain lasers. Intracavity insertion losses increase with the number of intracavity tuning elements. Consequently, the output power and efficiency of a single frequency (single longitudinal mode) laser are usually only fractions of their free-running values. Also, optical damage to intracavity frequency control elements can be limiting to laser performance.

An alternative and more efficient method for tuning and line narrowing pulsed lasers utilizes injection locking. The laser cavity is injected (seeded) with photons from another source, e.g., a low-power narrow linewidth laser. Injection occurs prior to pulse build-up in the injection locked oscillator (regenerative amplifier). Thus, at the onset of amplification, the intracavity power density at the injected frequency is greater than that at other (non-injected) frequencies. Longitudinal modes that overlap these injected frequencies build up sooner and saturate the gain, while other modes are just beginning to build up from the noise (see Fig. 1). As a result, most of the energy stored in the active (gain) medium is extracted at the injected frequency. If the laser's gain bandwidth permits amplification across many modes, the energy generated in the unwanted (background) modes increases proportionally with the number of these modes. This can reduce the spectral purity (signal-to-noise ratio) for a given injected power unless additional measures are used to bandwidth limit the gain. In the absence of additional measures, the signal-to-noise ratio can be increased by raising the power of the injecting beam. Figure 1 shows (schematically) the time dependence of the intracavity power density when pulse generation starts from spontaneous emission (noise) (upper curve) and when generation is seeded by injection from an external signal with intensity only 100 times greater than the noise (lower curve). In our experiments a birefringent tuner was used to bandwidth limit the gain to a few Ångstrom region centered near the injected frequency. Many requirements for frequency selectivity in an injection seeded laser cavity are less critical than are those for a loss-tuned

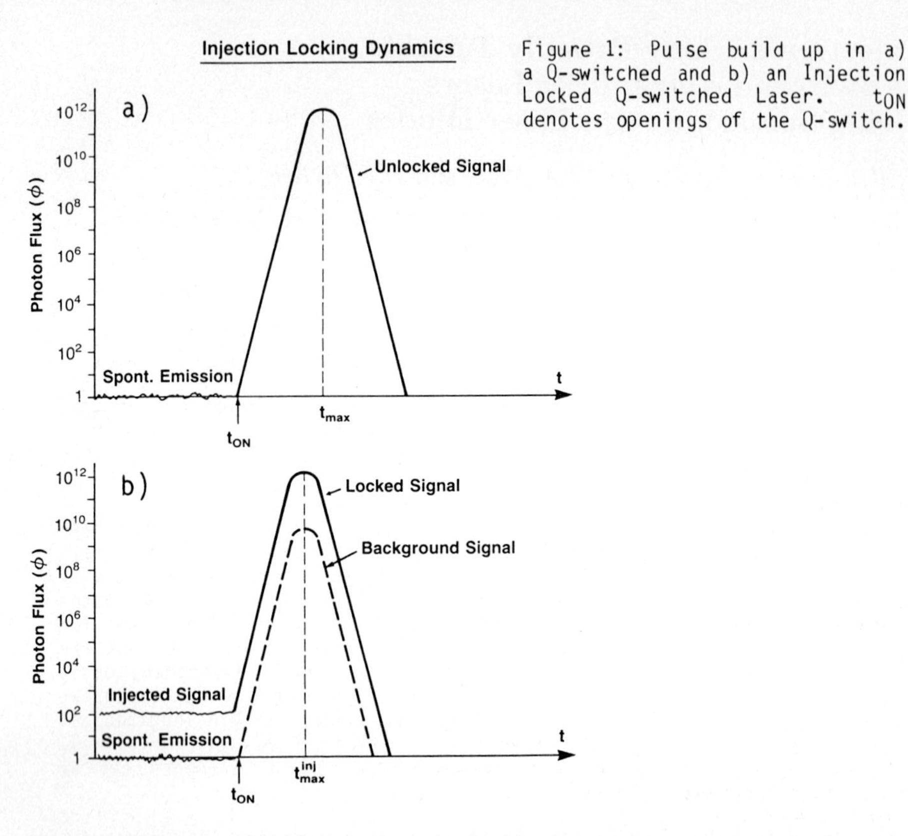

Figure 1: Pulse build up in a) a Q-switched and b) an Injection Locked Q-switched Laser. t_{ON} denotes openings of the Q-switch.

laser. It is sufficient to have the gain maximized near the desired output frequency. Since cavity losses are reduced, laser output is increased. Cavity length stability is important for stable single mode operation.

2. Experimental Configuration

In typical configurations for injection-locked oscillators, the injecting beam is introduced into the cavity through the oscillator's high reflector[1]. Only a small part of the incident power is transmitted into the laser cavity. Here, we describe a system that makes use of a nonreciprocal optical element, a (3-port) Faraday circulator (Fig. 2), in which passive injection can occur through the output coupler of the injection locked oscillator. The optical circulator consists of a 45° Faraday rotator and a thin film polarizer. The Faraday rotator used a Hoya FR-5 glass rod (11 mm diameter x 50 mm long). This was placed in a cylindrical channel drilled along the symmetry axis of a stack of four rare earth (SmCo) magnets, each 50 mm diameter and 12.5 mm length. Two extra magnets with reversed polarization were placed at the ends of the stack increasing the field inside the channel. The thin film polarizer had a selectivity of better than 100:1 at the injected frequency. Due to the nonreciprocal operation of the circulator, a beam delivered into port P1 of the circulator is transferred exclusively to port P2. Port P2 is coupled to port P3, and P3 to P1.

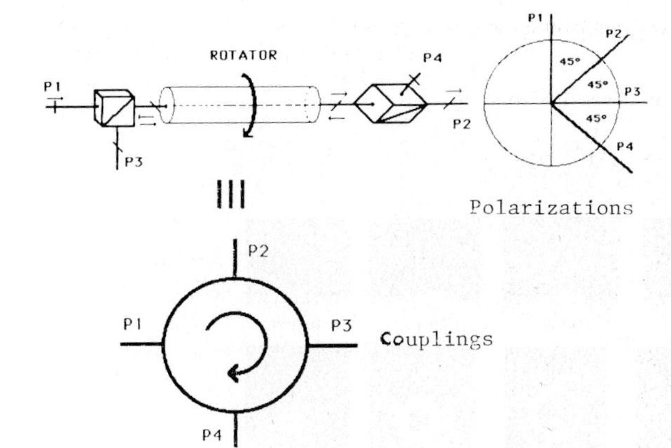

Figure 2: A 4-port optical circulator consisting of two polarizing beam splitters and a Faraday rotator. Beam polarizations and couplings are indicated. In our experiments one of the beam splitters was replaced by the injection-locked oscillator effecting 3-port circulation.

Injection Locking by a Semiconductor Diode Laser

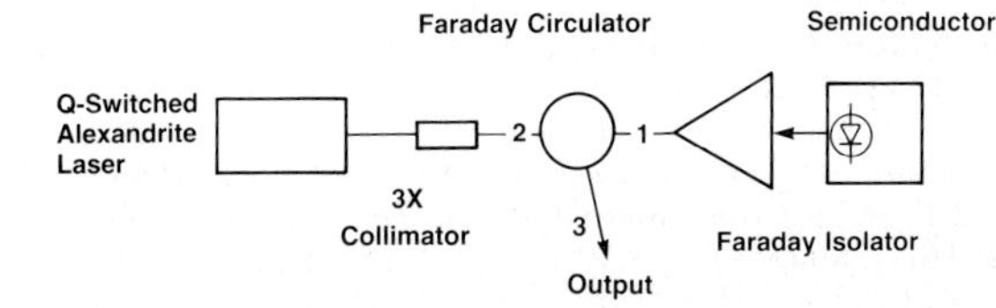

Figure 3: Schematic of Experimental Apparatus

Figure 3 shows the experimental configuration. The narrowband injecting beam was generated by a single frequency temperature tunable diode laser (Mitsubishi 6411, $\Delta\nu_{diode} \approx 50$ MHz) that was directed into the Faraday circulator. The light was transmitted through the output coupler (T=.5) into the cavity of the alexandrite injection-locked oscillator. The alexandrite laser output beam was directed back into the circulator and, due to the circulator's nonreciprocal properties, was coupled out of the system. Thus, the diode laser was protected from damage by the high-power output beam. Because the polarizer used in the Faraday circulator had some leakage at the undesired polarization, an optical (Faraday) isolator was placed in front of the laser diode for additional protection.

3. Performance

Figure 4 shows the output spectrum of the alexandrite injection-locked oscillator (ILO) as a function of the output power of the injecting diode laser. The output power of the ILO was independent of the diode laser power. As can be seen, reliable frequency locking was obtained even when the injected power was as low as 1 μW. At this power level ~50% of the laser power was emitted in the seeded modes and the rest was in a non-locked background. For cw injecting powers above 10 μW, there was no detectable spectral background and practically all energy was emitted in the seeded modes. At low injecting power a rather precise matching of the injecting beam to the ILO cavity modes was required, but at higher injecting power this sensitivity was greatly reduced. No attempt was made to stabilize the length of the ILO cavity or to critically match the diode frequency to a single cavity mode. Consequently, the injecting frequency

Injection Locking by a Semiconductor Laser Diode

Alexandrite Laser Spectrum

$\lambda = 755$ nm

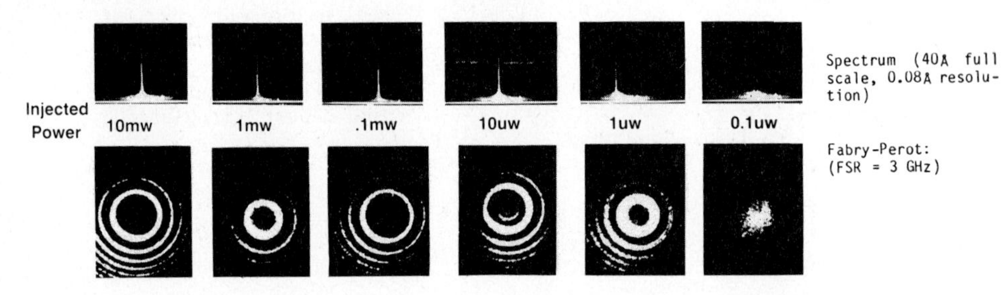

Figure 4: Spectral Analysis of Injection-Locked Laser Output

typically overlapped two cavity modes of the ILO. As a result, the time dependent (Q-switched) output pulse from the ILO (Fig. 5) showed some modulation corresponding to the relative amplitudes of the two seeded modes. The Q-switched output power of the alexandrite laser was about 10.0 MW (300 mJ, 30 ns). For an injecting power of 10 μW the apparent power gain of the system was 10^{12}. However, only energy injected during the pulse build-up time was effective in locking the output (the cavity lifetime was ~10 ns). Because the emission cross-section and thus the spontaneous emission background of alexandrite is polarized and low, alexandrite oscillators are particularly suitable for high-fidelity frequency control by injection locking to low-power (diode laser) sources. Locking can be achieved by seeding the cavity with only $\sim 10^5$ photons.

4. Conclusions

We report reliable frequency control of diode laser injection-locked, Q-switched, alexandrite lasers using incident cw injecting powers as low as 1 μW. This performance was made possible by alexandrite's low emission

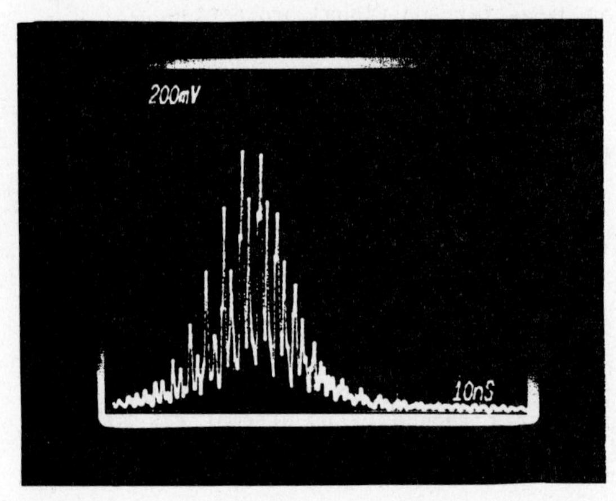

Figure 5: Time-Dependent Output Pulse From Injection-Locked Alexandrite Laser

cross-section, that allows excellent mode discrimination, and by use of a Faraday circulator that permitted injection through the locked oscillator's output coupler. We anticipate that this technique will permit the development of narrowband pulsed laser sources with transform limited linewidths that are tunable over much of the near IR. Output from these sources can be converted by nonlinear processes[2] to provide easily controlled high-quality spectral sources throughout the IR, visible, and UV.

References

1. W.R. Rappaport, J.J. Yeh, and R.C. Sam, "High-Efficiency Injection Locking of Q-switched Alexandrite Laser," CLEO '83 Technical Digest, 206 (1983).

2. J. Krasinski, P. Papanestor, and D.F. Heller, "Tunable Alexandrite Laser Pumped Intracavity Raman Laser," in Proceedings of SPIE, "High Power and Solid State Lasers," V. 622, 105 (1986).

Panel Discussion on Chromium Tunable Lasers

J.C. Walling

Light Age, Inc., San Diego, CA 92120, USA

Panel Membership

NAME	AFFILIATION
Dr. L. Andrews	GTE-Laboratories, Waltham, MA
Dr. J. Caird	Lawrence Livermore National Lab, CA
Dr. D. F. Heller	Allied Corporation, Mt. Bethel, NJ
Prof. G. Huber	University of Hamburg, Hamburg, FRG
Dr. H. P. Jenssen	Massachusetts Institute of Technology
Dr. B. H. T. Chai	Allied Corporation, Morristown, NJ
Prof. R. C. Powell	Oklahoma State University, OK
Dr. J. C. Walling (Panel Leader)	Light Age, Inc., San Diego, CA

1. Status of Cr^{3+} and V^{3+} doped tunable lasers.

At present at least 14 distinct laser materials have been demonstrated in this series. These exhibit wide variations in their suitability as laser media. The search for new materials continues and it is anticipated that many new materials will be found. The search continues for improvements in the important parameters: wavelength coverage, energy storage, fluorescence lifetime, radiative quantum yield, emission cross-section, and excited state absorption. One product of this effort, $Cr:ScBO_3$, was reported for the first time at this conference.

The wavelength coverage of these lasers, as now known, is presented in Table 1. At present wavelength coverage extends from about 710 to 1340 nm

Table 1: MATERIALS

	COMMON NAME OR ACRONYM	CHEMICAL FORMULA	WAVELENGTH (PEAK/RANGE)* nm
1.	ALEXANDRITE	$BeAl_2O_4:Cr^{3+}$	710-820
2.	EMERALD	$Be_3Al_2(SiO_3)_6:Cr^{3+}$	720-842
3.	Cr:GSGG	$Gd_3(Sc,Ga)_2Ga_3O_{12}:Cr^{3+}$	742-842
4.	Cr:GSAG	$Gd_3Sc_2Al_3O_{12}:Cr^{3+}$	735-820
5.	Cr:GGG	$Gd_3Ga_5O_{12}:Cr^{3+}$	745

6.	Cr:LLGG	$(La,Lu)_3(Lu,Ga)_2Ga_3O_{12}:Cr^{3+}$	850
7.	Cr:YSGG	$Y_3Sc_2Ga_3O_{12}:Cr^{3+}$	750
8.	Cr:YGG	$Y_3Ga_3O_{12}:Cr^{3+}$	730
9.	$Cr:KZnF_3$	$KZnF_3:Cr^{3+}$	825
10.	$Cr:ScBO_3$	$ScBO_3:Cr^{3+}$	790-900
11.	$Cr:ZnWO_4$	$ZnWO_4:Cr^{3+}$	980-1090
12.	$Cr:SrAlF_5$	$SrAlF_5:Cr^{3+}$	825-1010
13.	$V:MgF_2$	$MgF_2:V^{2+}$	1070-1150
14.	$V:CsCaF_3$	$CsCaF_3:V^{2+}$	1240-1340

*Values are compiled from various literature and direct sources. These involve noncommensurate conditions of observation, and should be considered approximate.

with one gap near 1200 nm. The performance of these materials varies substantially, and the materials covering certain regions are not particularly gifted, particularly for the longer wavelength. $Cr:SrAlF_5$ has not performed with expected efficiency; it is suspected of harboring so called "dark sites," parasitic Cr sites that absorb pump light but relax nonradiatively. On the long wavelength side, beyond 900 nm, none of the materials are strong performers at least in their present state of crystal development. Beyond 1 μm, V^{2+} doped in fluoride hosts has been the most widely researched lasing ion. In principle, there should be no particular problem in perfecting the crystal growth of $V:MgF_2$ and $V:CsCaF_3$. Even so, intrinsic problems may well limit the utility of these sources.

In contrast, the performance of the short wavelength materials continues to be strong, led by alexandrite, the archetype and first member of this series to be lased in a tunable mode. Alexandrite is the most advanced material of this series as a result of extensive material growth development and improved resonators to deal with the low (compared to ruby or Nd:YAG) emission cross - section and high - energy storage. Alexandrite has demonstrated 100 W in Q-switched operation, 1 J at 100 Hz, from two 6.34 mm diamter 11 cm long rods in an oscillator-amplifier configuration. The performance of single longitudinal-single transverse mode advanced alexandrite oscillators has reached 200 mJ per pulse.

While not as advanced in material development as alexandrite, emerald has some attractive intrinsic properties. The relative energy of the ^2E and $^4T_2$ multiplets, which establishes the balance between emission cross-section and fluorescence lifetime in this series of materials, is almost ideal in emerald for laser operation. At the same time emerald is believed to have superior thermal lensing properties. On the other hand, because of a crystallographic phase transition during cool down, emerald cannot be produced by Czochralski methods; it is grown hydrothermally instead. Even though hydrothermal growth is a difficult technique to develop, progress suggests that suitable laser material will be produced as efforts continue.

The performance of the Cr doped garnets, particularly GSAG, has improved also. While Cr:GSGG grows well by Czochralski methods, color center formation, most likely associated with a gallium deficiency, has been a continuing problem. Cr:GSAG, on the other hand, not only grows well by Czochralski methods, but is essentially free of obvious color center formation. Still, the laser efficiency of Cr:GSAG has been significantly less than for Cr:GSGG for reasons not yet fully understood. A major advantage for Cr:GSAG is that the Cr distribution coefficient, a measure of the relative concentration of Cr in the boule and in the melt, is near unity. It is generally expected that continued research will yield satisfactory growth methods for both materials.

Finally, Cr:$KZnF_3$ has become the second tunable solid state laser material (after alexandrite) to be offered commercially. This material operates from 785 to 865 nm, with good slope efficiency when laser pumped (<14% has been reported). While this efficiency is substantially lower than the 30 to 50% range found for emerald and alexandrite, its wavelength coverage does extend further into the IR.

2. Evaluation

For general research applications, a broadly tunable laser has high utility and is important. For many "fielded" applications, however, the ability to target specific wavelengths or to operate anywhere within a specified frequency range is more important than broad tunability. To address these requirements, it is desirable to find a set of good materials that together cover the broad wavelength range of interest.

In these fielded applications, small size, efficiency, and reliability are critical requirements. The $3d^3$ series of materials offer the advantage of simplicity of design and size reduction. For the near term, this is based on direct lamp-pumped operation, obviating more complex laser-pumped laser techniques. Progress in developing shorter wavelength semiconductor diode laser pumps for these solid-state lasers are expected to lead to eventual system simplification and size reduction. For flashlamp excitation the energetics of many of these materials are close to the ideal: They absorb in the visible where the lamps emit; and the storage time is long enough and the energy storage great enough to permit efficient pumping within the constraint of manageable lamp life. In this property, these lasers are distinguished from other important and/or promising tunable lasers, including color center lasers and Ti:Sapphire, which have high cross sections but short fluorescence lifetime.

Frequency conversion, particularly harmonic up-conversion of these lasers, can be expected to enhance their range of application considerably and make them competitive as tunable ultraviolet solid-state laser sources. Using nonlinear fixed frequency conversion techniques, such as Raman shifting, these lasers permit complete tunable coverage of the visible and near IR.

3. Future research

From the standpoint of new materials, Cr^{3+} doped metaphosphates, $(APO_3)_3$ where A could be Sc, In, Lu, or Y, offer interesting possibilities for achieving desired laser characteristics with peak emission in the 880 to 890 nm range. From a more basic research perspective, other candidate materials might be found among the well-known phosphors when produced as single crystals. Promising candidates include the tungstates, vanadates and molybdates that form interesting charge transfer complexes. For the longer wavelength regime one looks to V^{2+} in fluoride hosts, since it appears doubtful that Cr^{3+} will be able to extend beyond about 1 μm.

Excited state absorption is an important limitation in these materials. Like the laser emission itself, excited state absorption is vibrationally assisted by the laser emission itself and may produce strong broad bands. Different vibrational modes may couple with the ground to excited state transition and the ground to doubly excited state transition. It is, therefore, not possible to predict the linewidth, strength and exact position of the excited state absorption transition (between excited and doubly excited states) level energies and widths derived from ground state absorption data alone. It is unfortunately the case that inadequate attention has been paid to the theoretical foundations of this problem. It would be helpful if theory were developed to predict optimal crystal structures, and developed to predict figure of merit parameters to guide selection of promising candidates.

Due to the comparative weakness of the materials in the long wavelength range, it seems appropriate that further effort be applied to materials with long wavelength potential. Finally, the importance of operation at wavelengths outside the near infrared, where these lasers are "in band," necessitates the development of suitable frequency conversion methods.

Part VI

Titanium Sapphire Lasers

Room-Temperature cw Operation of the Ti:Al_2O_3 Laser

A. Sanchez, R.E. Fahey, A.J. Strauss, and R.L. Aggarwal

Lincoln Laboratory, Massachusetts Institute of Technology,
Lexington, MA 02173, USA

This paper reports on the room-temperature cw operation of several Ti:Al_2O_3 laser rods pumped by an Ar-ion laser. For one rod, a threshold of 2.3 W, maximum output of 1.6 W, and slope power efficiency of 19% were obtained at 770 nm, and continuous tuning was observed from 693 to 948 nm.

1. Introduction

In this paper we report on the room-temperature cw operation of Ti:Al_2O_3 lasers with output power at the one-watt level. MOULTON [1] reported cw operation at liquid nitrogen temperature, but at room temperature he obtained only quasi-cw operation by chopping the pump beam at a 16% duty cycle. ALBERS, STARK and HUBER [2] obtained cw operation at room temperature, but the maximum output power was less than 10 mW. In the present investigation, by using an Ar-ion laser operating in all the blue-green lines as the pump source, maximum power of 1.6 W has been obtained at 770 nm, near the peak of the gain profile, and continuously tunable cw operation has been achieved from 693 to 948 nm. In addition, cw lasing has been observed at temperatures up to 120 C. We have also obtained conclusive evidence for the homogeneous nature of the lasing transition.

2. Experimental Procedure

Six Ti:Al_2O_3 laser rods, containing between 0.024 and 0.099 wt.% Ti_2O_3, were fabricated from crystals that were grown at Lincoln Laboratory by a thermal-gradient-freeze technique [3]. The end faces of the rods, which were 6 mm^2 in cross-section, were cut either normal or at the Brewster angle (60°) and polished optically flat. The c axis was perpendicular to the rod axis, and in the Brewster case it was also in the plane of incidence. Antireflection coatings were applied to the normally cut faces.

For laser operation the Ti:Al_2O_3 rod, supported by a water-cooled heat sink, was placed at the waist of an astigmatically compensated three-mirror cavity as shown in Fig. 1. The pump radiation from the Ar-ion laser was injected into the cavity through the folding mirror and focused into the laser by a lens placed outside the cavity.

3. Experimental Results

Initial experiments were performed on a Brewster-cut rod with length (ℓ) of 1.8 cm containing 0.024 wt.% Ti_2O_3. Figure 2 shows the laser out-

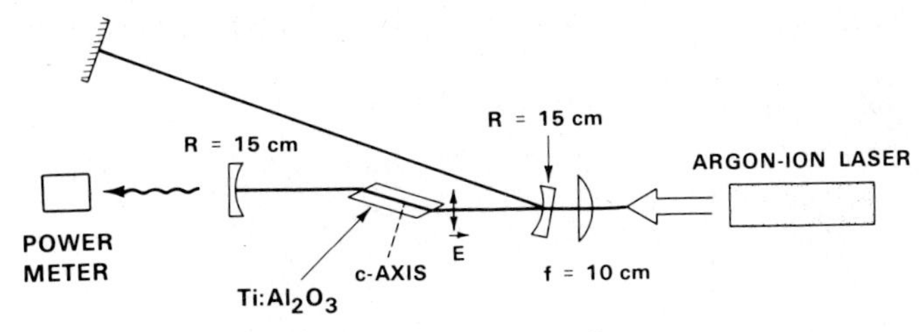

Fig. 1 Schematic of the experimental setup

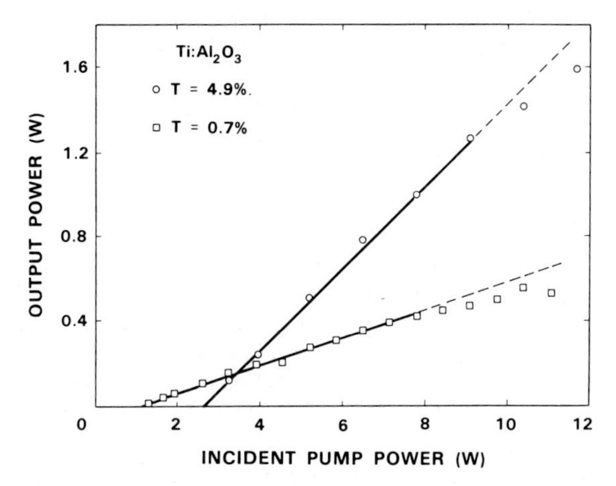

Fig. 2 Output power at 770 nm from a cw room-temperature $Ti:Al_2O_3$ laser vs incident pump power from an all-line Ar-ion laser

put vs incident pump power for two coupling mirrors with 0.7% and 4.9% transmittance. The maximum power output was 1.6 W. The slope efficiencies (η) obtained with these couplers were 6.5% and 19%, respectively. Using these values of η and the fraction of incident power absorbed (0.7), we obtain values of (64 ± 10)% for the internal quantum efficiency and (2.4 ± 0.5)% for the round-trip cavity loss (L). This value of L, which is consistent with the dependence of the threshold on coupling mirror transmittance, implies an upper bound of $\alpha = L/(2\ell) = 0.007\ cm^{-1}$ on the absorption coefficient of the $Ti:Al_2O_3$ rod at the laser wavelength in the π polarization ($E \parallel c$). This value of α is 100 times smaller than the absorption coefficient (α_m) for the pump radiation at 490 nm.

In subsequent experiments we investigated two additional Brewster rods cut from the same crystal and three normally cut rods from a different crystal with a higher doping level. Because of the concentration gradient along the growth axis in crystals grown by the thermal-gradient-freeze technique [3] the six rods had different doping levels. Room-temperature cw operation was obtained for each of the rods. The $Ti:Al_2O_3$ laser radiation was coupled from the cavity using a flat mirror with a transmittance of either 0.7, 2.5 or 4.9%. From the slope efficiency values obtained for the different transmittances, the round-trip cavity loss L was obtained for each rod. The normalized loss, $L/2\ell$, is plotted in Fig. 3 vs α_m. If the residual absorption of the $Ti:Al_2O_3$ [5] is entirely re-

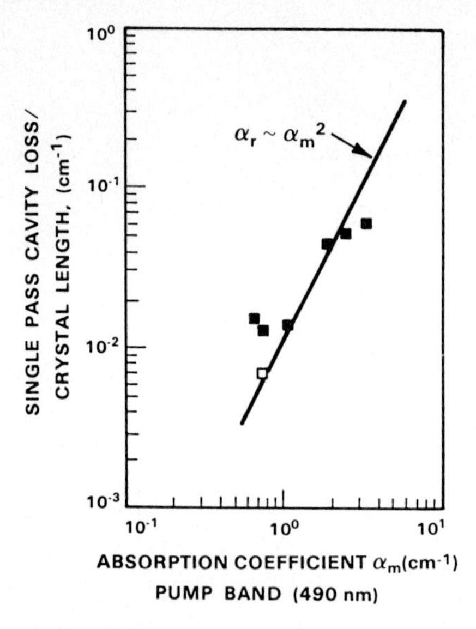

Fig. 3 Normalized cavity loss vs pump band absorption coefficient (α_m) for six laser rods. The two data points with the same α_m were obtained for the same rod in two different cavities, and their separation indicates the uncertainty of the data.

sponsible for the cavity loss, the normalized loss will be equal to the residual absorption coefficient. The solid line with a quadratic slope is representative of the residual absorption of our as-grown crystals as measured with a spectrophotometer [3]. The two types of data are generally consistent, although those obtained from the laser measurements exhibit more scatter. The residual absorption as measured by a spectrophotometer can therefore be used to predict laser performance.

Tuning studies were made on the laser rod of Fig. 2. In this case all three cavity mirrors had broadband, highly reflecting dielectric coatings. The laser radiation was coupled out by a near-Brewster intracavity plate and observed with a 0.1 nm resolution grating spectrophotometer. Tuning was accomplished with a three-plate intracavity Lyot filter. In the initial experiments, tuning was limited to the range from 705 to 915 nm due to branch jumping of the filter. When an intracavity prism was added to prevent branch jumping, laser emission was obtained from 693 to 948 nm. In this case the tuning range was limited by the bandwidth of the cavity mirrors. Figure 4 shows the laser threshold as a function of wavelength. The spikes at long wavelengths can be attributed to the characteristics of the cavity mirrors.

Because of the high efficiency of the Ti:Al_2O_3 laser [1], it has generally been assumed that the laser transition is homogeneously broadened. We have verified this assumption by measuring the spectral distribution of the spontaneous fluorescence emitted by the laser rod of Fig. 2 in a direction perpendicular to the axis of the rod. The results are shown in Fig. 5, where the spectra obtained for three different operating conditions are plotted on the same intensity scale. Trace (a) is the spectrum obtained when the pump power was about 1.7 times the threshold

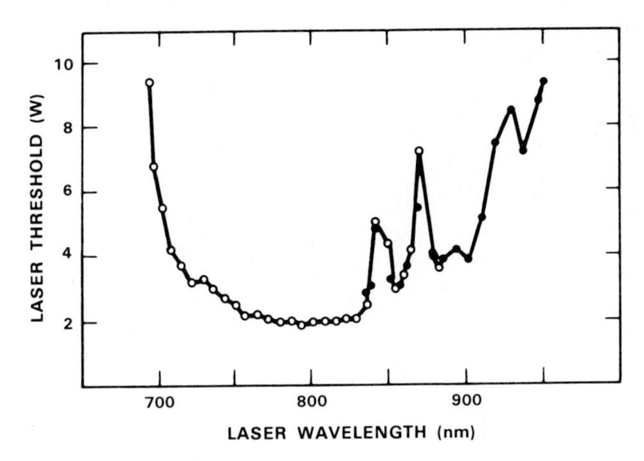

Fig. 4 Tuning characteristics of the $Ti:Al_2O_3$ laser. The threshold power (incident on the rod) is plotted vs laser wavelength.

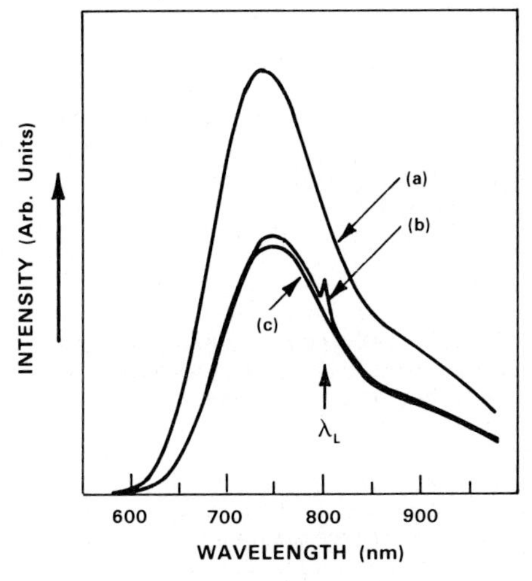

Fig. 5 Spontaneous fluorescence spectra for the $Ti:Al_2O_3$ laser recorded when the laser is a) off, b) on, and c) at threshold.

value but laser action was prevented by inserting an absorber in the laser cavity. When the absorber was removed, laser action took place at λ_L = 800 nm, and the fluorescence intensity was reduced by the same factor at all wavelengths, as shown by trace (b). (The small spike at λ_L results from scattering of the laser radiation.) The uniform decrease in intensity establishes that the $Ti:Al_2O_3$ transition is homogeneously broadened. Trace (c) is the fluorescence spectrum when the pump power was decreased to the threshold value. As expected, this trace is almost coincident with trace (b).

Laser operation of the rod of Fig. 2 has also been studied as a function of temperature. Figure 6(a) shows the threshold pump power P_{th} for

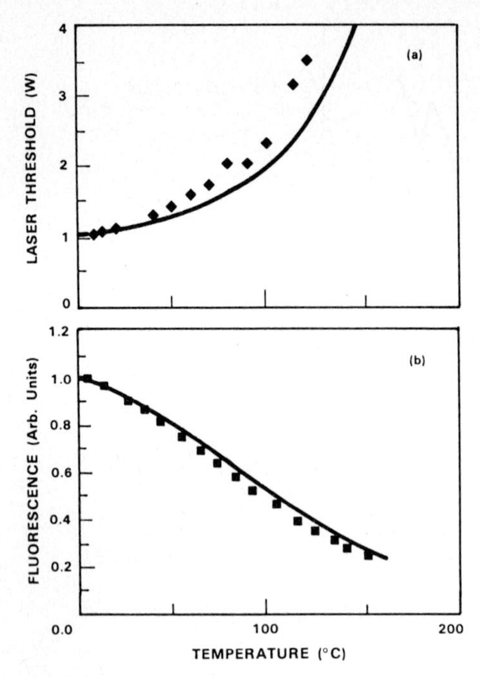

Fig. 6 Temperature dependence of a) threshold pump power, and b) fluorescence intensity. The circles represent the data points, and the curves are the predictions based on known lifetime data.

laser emission at 780 nm as a function of heat-sink temperature. Continuous operation was obtained at temperatures up to 120 C. The experimental points are compared with a normalized curve that was calculated from the relationship $P_{th} \simeq \gamma_t$, where γ_t is the total relaxation rate for spontaneous fluorescence, by using the published data [1,2] for the temperature dependence of γ_t. It is believed that the deviations observed at higher temperatures occur because the temperature of the lasing filament becomes greater than the measured heat-sink temperature.

In addition, it has been verified that the increase in total relaxation rate with increasing temperature is entirely due to the contribution of nonradiative relaxation processes. The intensity I_f of the spontaneous fluorescence emitted by the laser rod at a fixed power below threshold has been measured as a function of temperature. The experimental points, shown in Fig. 6(b), are fit well by the normalized curve shown, which was calculated from the relationship $I_f \simeq \gamma_r/\gamma_t$, where γ_r is the radiative relaxation rate, by again using the published data [1,2] for the temperature dependence of γ_t and also making the assumption that γ_r is independent of temperature. The good fit verifies this assumption. Since $\gamma_t = \gamma_r + \gamma_{nr}$, where γ_{nr} is the nonradiative relaxation rate, it can be concluded that the increase in γ_t and P_{th} with temperature is due entirely to the increase in γ_{nr}. It should be noted that although the increase in laser threshold is a direct consequence of the nonradiative processes, the laser slope efficiency remains constant over the investigated temperature range.

4. Conclusion

Room-temperature cw operation of a $Ti:Al_2O_3$ laser has been demonstrated at an output power level of 1.6 W at 770 nm. The laser has been tuned from 693 to 948 nm. Lasing was observed up to 120°C. Evidence was obtained for the homogeneous nature of the laser transition. The laser characteristics of rods with doping levels ranging from 0.024 to 0.099 wt% Ti_2O_3 are consistent with the values of the residual absorption obtained from spectrophotometer measurements.

Acknowledgements

The authors would like to thank G. D. Silva for polishing the laser rods and D. J. Sullivan for technical assistance. This research was supported by the Defense Advanced Research Projects Agency.

References

1. P. F. Moulton, J. Opt. Soc. Am. B3, 125 (1986).
2. P. Albers, E. Stark and G. Huber, J. Opt. Soc. Am. B3, 134 (1986).
3. R. E. Fahey, A. J. Strauss, A. Sanchez and R. L. Aggarwal, this conference, Paper ThA3.
4. A. Sanchez, R. E. Fahey, A. Strauss and R. L. Aggarwal, Opt. Lett. 11, 363 (1986).
5. P. Lacovara, L. Esterowitz and M. Kokta, IEEE J. Quantum Electron. QE-21, 1614 (1985).

Continuous Wave Tunable Laser Operation of Ti^{3+}-Doped Sapphire at 300 K

P. Albers[1], *H.P. Jenssen*[1], *G. Huber*[2], *and M. Kokta*[3]

[1]Massachusetts Institute of Technology, Crystal Physics and Optical Electronics Laboratory, Cambridge, MA 02139, USA
[2]Institut für Angewandte Physik, Universität Hamburg, Jungiusstr. 11, D-2000 Hamburg 36, Fed. Rep. of Germany
[3]Union Carbide, Inc., Wahougal, WA 98671, USA

The $Ti^{3+}:Al_2O_3$ laser operated cw at room temperature. The slope efficiency is 15.2% with a maximum output of 64 mW at 720 nm. Continuous tuning was possible from 706 nm to 1024 nm. Extreme long wavelength operation was obtained in a 1.06 μm Nd-laser cavity.

1. Introduction

Sapphire is a very promising Ti^{3+} host material for broadly tunable laser applications [1]. It is isostructual to Ti_2O_3 with a trigonal distorted octahedral lattice site for the dopant. The strong crystal field of Al_2O_3 results in a wide splitting of the ^2D free ion level. This simple level splitting of Ti^{3+} and an Al_2O_3 bandgap of about 10 eV prevents excited state absorption both for pump wavelengths and emission wavelengths as well as any resonant absorption of the Stokes shifted emission. Ti^{3+} has, besides its crystal field split (d^1) ^2D free ion level, no levels of any other configurations for energies below 10 eV.

2. Absorption and Gain Curve

The absorption spectrum of a typical $Ti^{3+}:Al_2O_3$ crystal is shown in Fig. 1. The broad double band in the range 400 nm to 600 nm is due to

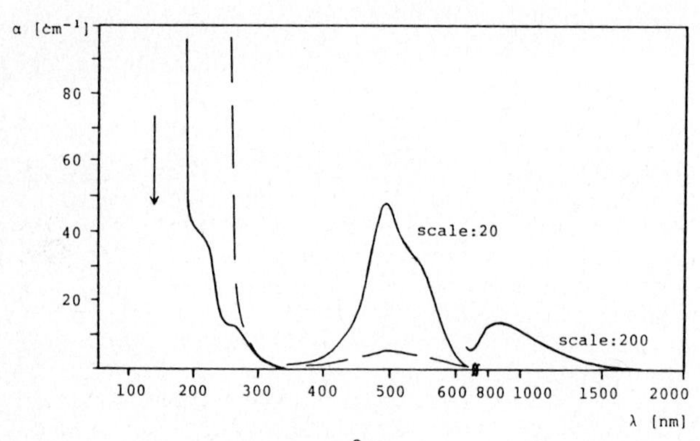

Fig. 1. Absorption of $Ti^{3+}:Al_2O_3$ at 300K with E||c

the crystal field split ^2D levels including the influence of the dynamic Jahn-Teller effect [2,3]. The two peaks at 485 nm and 550 nm have absorption cross-sections for $E \parallel c$ of $\sigma = 6.1 \cdot 10^{-20}$ cm^2 and $\sigma = 2.1 \cdot 10^{-20}$ cm^2 based on the nominal 0.1 wt % Ti^{3+} concentration in the crystal. In addition to the Ti^{3+} absorption band, the spectrum shows a weak absorption in the near infrared and a very strong absorption in the UV below 300 nm.

The infrared absorption is centered at 800 nm with a long tail to beyond 2000 nm. This absorption depends on the crystal growth conditions and on the additional annealing processes. The absorption strength is in the order of 0.15 cm^{-1} for as grown samples and close to zero for annealed crystals. The long tail of the absorption band is mostly due to scattering effects. A careful polishing and cleaning of the surfaces of the crystals reduces the absorption between 1100 nm and 2000 nm almost completely.

The UV absorption bands are still under investigation. As grown Ti^{3+} doped sapphire crystals have an apparent band edge at about 280 nm compared to that of undoped sapphire at 140 nm, indicated by an arrow. Preliminary results show that this "shift" of the band edge depends on the growth conditions and can be influenced by different annealing processes. With oxidizing atmosphere one gets a small Ti^{3+} absorption and an increase of the charge transfer bands (dashed curve). In reducing atmosphere the concentration of trivalent titanium and the Ti^{3+}-bands increases. At the same time the UV absorption between 200 nm and 300 nm decreases. Therefore the results of the annealing processes indicate that the charge transfer bands of Ti^{4+} are responsible for this absorption. The remaining bands in the UV must be partly due to Ti^{3+} charge transfer bands. The effect of the UV bands on the laser properties of Ti^{3+} doped sapphire is under investigation.

The gain curve of $Ti^{3+}:Al_2O_3$ extends from 600 nm to beyond 1100 nm with a half-width of 2910 cm^{-1} [Fig. 2] and with a maximum at 790 nm. Both in absorption and fluorescence the transitions for $E \perp c$ are about 60% weaker than for $E \parallel c$. Using the theory of McCumber [4], one can calculate the effective emission peak cross-section $\sigma_e = 4.5 \cdot 10^{-19}$ cm^2, which should yield a maximum gain of $\sigma_e \cdot N = 15$ cm^{-1} at full inversion for $\lambda = 790$ nm.

3. Free-running cw laser

The laser experiments were realized with 0.1 wt % Ti^{3+} doped sapphire grown by the Czochralski method. The pump source was an argon laser at 488 nm/514 nm. The 4.2 mm long laser crystal absorbed up to 465 mW. The resonator was nearly concentric with mirror radii of 5 cm. With 0.8% output coupling we obtained the input/output curve shown in Fig. 3. The maximum output is 60 mW near 720 nm; this wavelength was fixed by the mirrors used. The laser threshold is less than 100 mW and the slope efficiency is 15.2%. From this value the internal losses in the resonator were calculated to be L = 2.9%. A weak infrared absorption band and scattering losses (observable with the pump beam) can account for most of these losses.

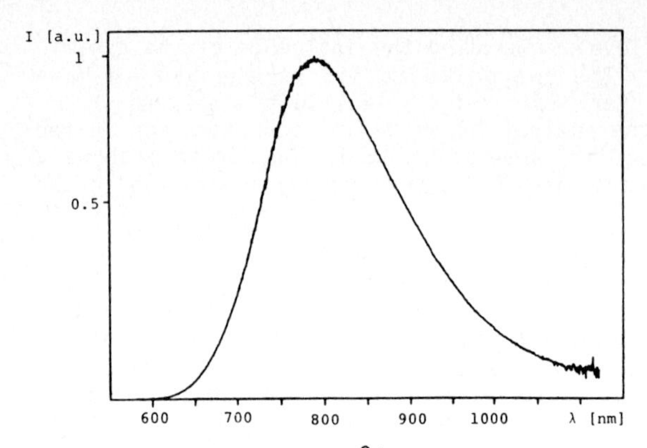

Fig. 2. Gain curve of $Ti^{3+}:Al_2O_3$ at 300K with E||c. The spectrum is corrected for the spectral sensitivity of the detection system

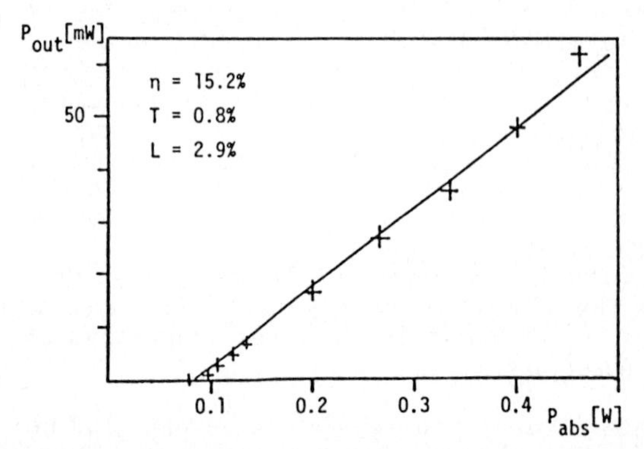

Fig. 3. Output power versus input power of $Ti^{3+}:Al_2O_3$

4. Tuning of the $Ti^{3+}:Al_2O_3$ laser

Continuous tuning of the laser could be achieved in a nearly concentric resonator with a 5 cm-radius mirror (input coupler), an antireflection coated internal lens, a Brewster angle prism and a plain output coupler with $T \leq 1\%$ within the tuning range of Fig. 4. The pump beam was focused with a 5 cm-lens onto the crystal. The tuning curve up to 1024 nm was obtained with the argon pump laser at 13W, a repetition rate of 60 Hz, and a duty cycle of 10%. The wavelength of the $Ti^{3+}:Al_2O_3$ laser was measured with a 0.25 m-Spex. It was even possible to operate the laser crystal inside a concentric 1.06 μm-Nd^{3+}-resonator with 99.9% reflectivity. This demonstrates that even a wider tuning range can be expected.

In conclusion, we have achieved cw operation of the $Ti^{3+}:Al_2O_3$ laser at room temperature with a high slope efficiency of 15.2%. We could demonstrate an extremely broad tuning range, which extends beyond 1 μm wavelength.

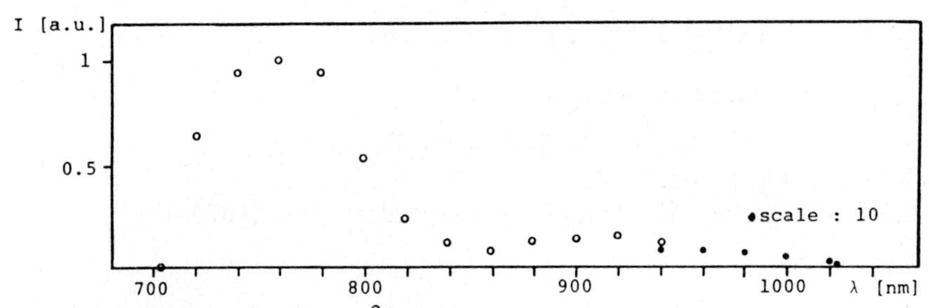

Fig. 4. Tuning curve of $Ti^{3+}:Al_2O_3$

Acknowledgements

We thank R. Horn at the University of Hamburg for measuring the long wavelength tuning in Fig. 4. The work at M. I. T. was funded by the Lawrence Livermore National Laboratories, Subcontract No. 5960505.

References

1. P. Moulton, Opt. News, 8, 9, 1982

2. P. Albers, E. Stark, and G. Huber, J. Opt. Soc. Am B., **3**, 134, 1986

3. P. Albers, Dissertation, Universität Hamburg, 1985

4. D. E. McCumber, Phys. Rev., **134**, A299, 1964

Efficient, Tunable Ti:Sapphire Laser

W.R. Rapoport [1] *and C.P. Khattak* [2]

[1] General Electric Company, P.O. Box 50 00, Room 289, Binghamton, NY 13902, USA

[2] Crystal Systems, Inc., 27 Congress Street, Salem, MA 01970, USA

Titanium-doped Sapphire ($Ti{:}Al_2O_3$) is a promising candidate for an efficient, broadly tunable solid-state laser source. The utility of $Ti{:}Al_2O_3$ was first pointed out by Moulton [1] and has since been studied at several other facilities [2]. Efficient operation is critically dependent upon the quality of the laser material. Until recently, few samples of high-grade material with low loss coefficients were available.

Several samples were grown by the Czochralski process and by the heat exchanger method (HEM). These samples were scanned in transmission for both π and σ polarizations using a Perkin-Elmer Lambda 9 Spectrophotometer. Due to the long length of the crystals, a cross-check was performed by scanning the samples with a tunable $Ti{:}Al_2O_3$ laser with <3Å bandwidth. Absorption coefficients ranged from <0.003/cm to >0.11/cm at the peak gain wavelength (800 nm).

Absorption spectra for two samples are shown in Fig. 1A (HEM) and 1B (Czochralski). Figure 2A shows a laser scan for both π and σ polarizations for an HEM crystal, while 2B shows a typical Czochralski-grown crystal. The most interesting feature of these traces is the small absorption in the σ polarization for Fig. 1A versus the large absorptive peak for the same polarization in Fig. 1B near 800 nm.

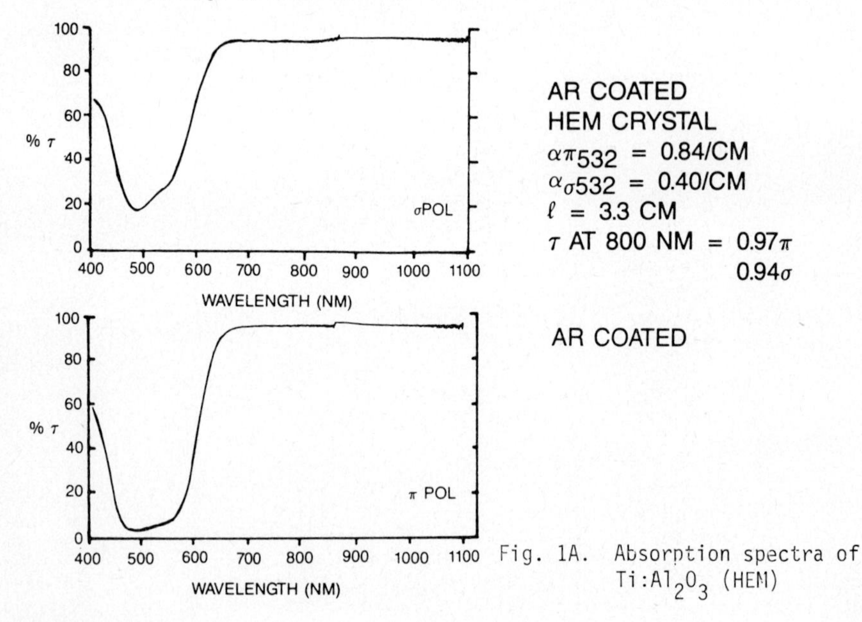

Fig. 1A. Absorption spectra of $Ti{:}Al_2O_3$ (HEM)

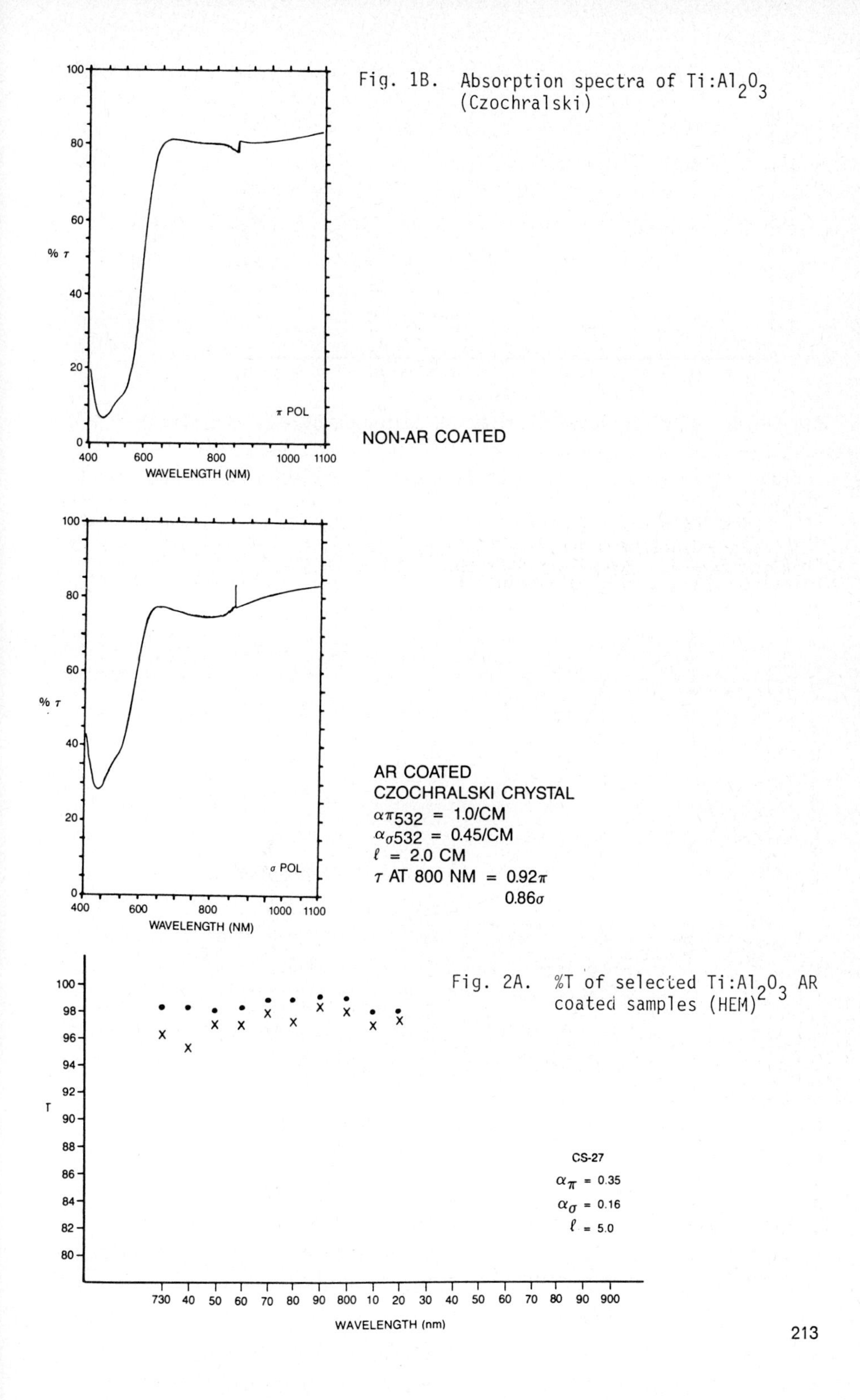

Fig. 1B. Absorption spectra of $Ti:Al_2O_3$ (Czochralski)

Fig. 2A. %T of selected $Ti:Al_2O_3$ AR coated samples (HEM)

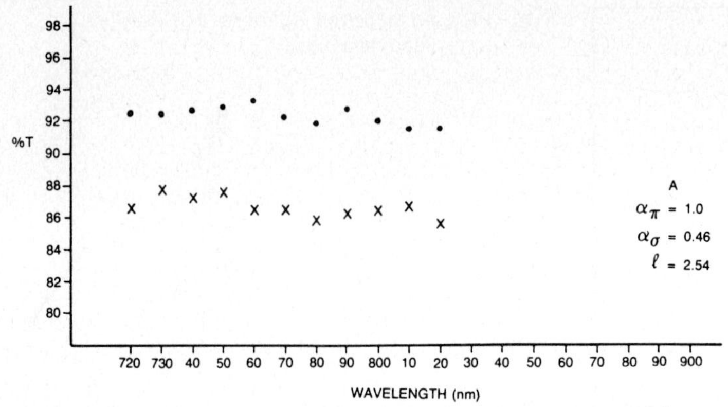

Figure 2B. %T of Selected Ti:Al_2O_3 AR Coated Samples (Czochralski)

Fluorescence spectra were taken for several samples using an EG&G OMA III. HEM and Czochralski spectra were quite similar with the exception of a large Cr^{3+} contamination in several of the Czochralski samples in the σ polarization (Fig. 3). Figure 4 plots the relative cross-section data obtained by multiplying the fluorescence curve by λ^5 [3]. Included in the figure are relative cross-section data points taken from laser energy threshold values vs. a peak cross-section of ~3.8 x 10^{-19} ± 20%.

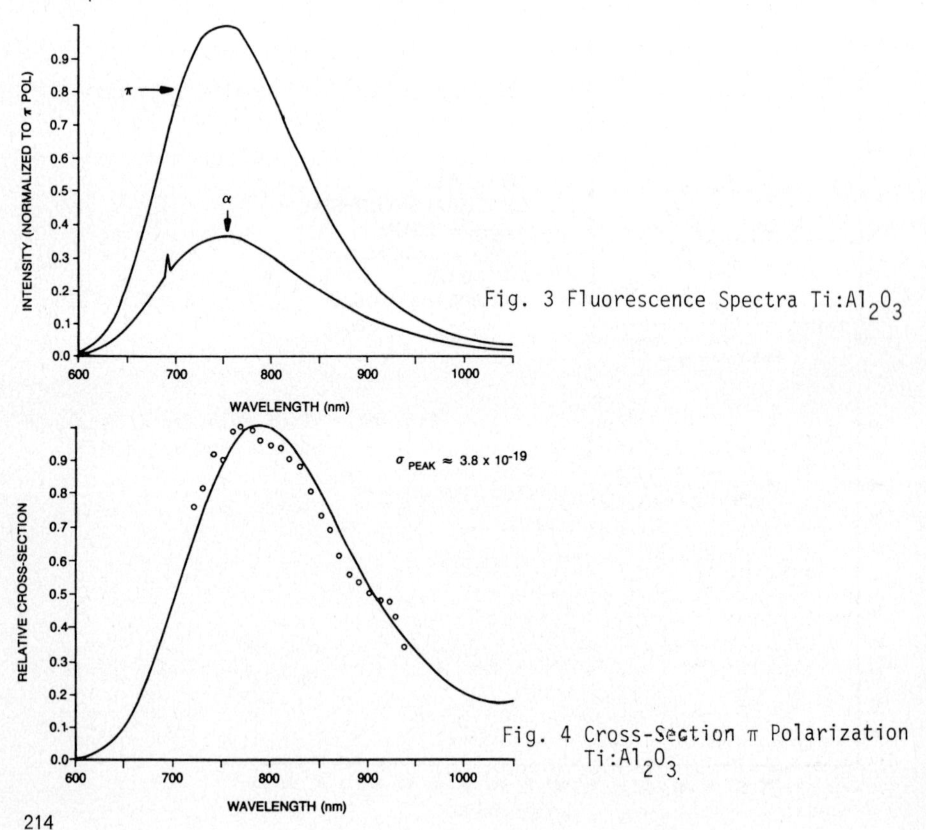

Fig. 3 Fluorescence Spectra Ti:Al_2O_3

Fig. 4 Cross-Section π Polarization Ti:Al_2O_3

The peak value for the cross-section was substantiated by performing a small signal gain measurement. A single mode CW laser diode lasing at 787 nm was directed through the central area of a longitudinally pumped $Ti:Al_2O_3$ crystal. The 532 nm pump was generated by using a telescopic resonator producing a near diffraction-limited TEMoo mode beam. This was then amplified in a single-pass saturated amplifier producing 350 mj/pulse at 5 Hz. The output was doubled to 532 nm using KD*P generating ~175 mj of high-quality green energy. The spatial profile of the pulse was nearly a flat top. The diode laser was directed through the rod, passing through several apertures in order to minimize the spontaneous emission, then through a high index prism for greater selectivity. The detector was located an appreciable distance from the pumped rod. A plot of the small-signal gain can be seen in Fig. 5. The stored energy, which is related to the ln G, decayed with a time constant of 3.2 µsec. The fit yielded a peak cross-section of 3.8×10^{-19} at 787 nm.

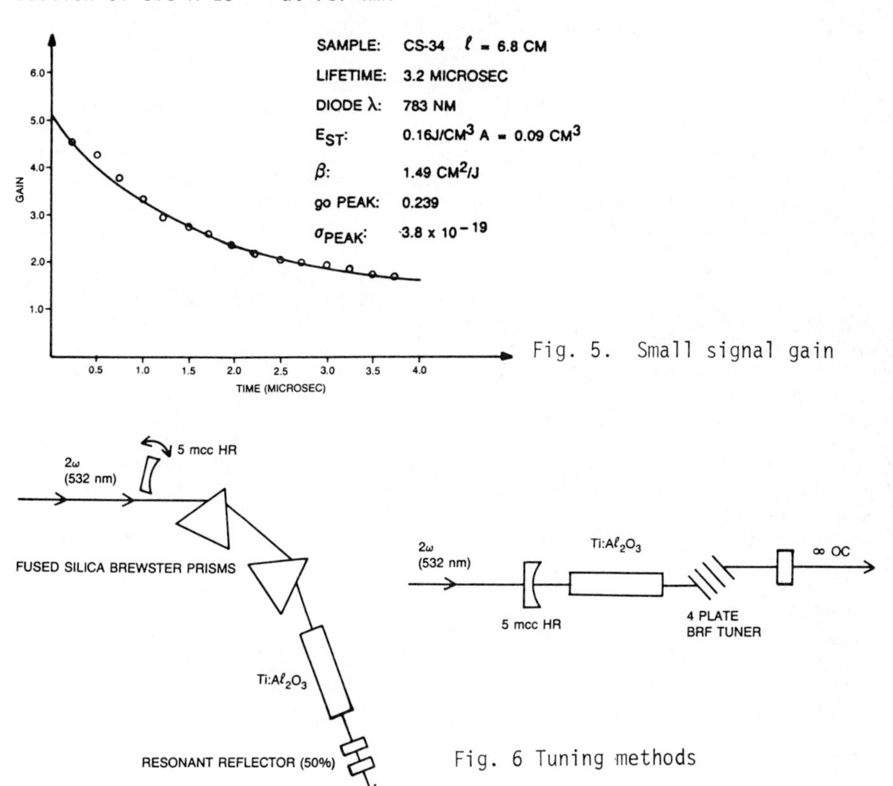

Fig. 5. Small signal gain

Fig. 6 Tuning methods

The $Ti:Al_2O_3$ experimental lasers were tuned two ways as seen in Fig. 6. Prism tuning offered the benefit of being able to pump the crystal without the pump beam having to go through the rear mirror. The output coupler was typically a resonant reflector with a FSR of ~ 1Å and a peak reflectivity of ~ 50%. This enabled the laser to be tuned over very large ranges without having to change mirrors. The lasing bandwidth was approximately 3 nm. The second method used a four-plate BRF tuner. A lasing bandwidth of 3 - 7Å was easily achieved with almost no insertion loss. The major disadvantages were having to pump through the rear mirror and the finite tuning range of the four-plate tuner.

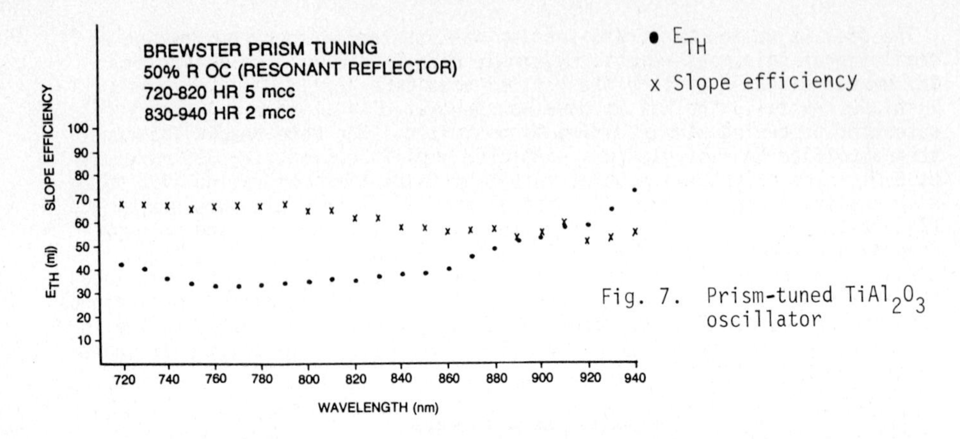

Fig. 7. Prism-tuned $TiAl_2O_3$ oscillator

Figure 7 shows a tuning run for a dual prism-tuned $Ti:Al_2O_3$ laser. Threshold and slope efficiency are plotted vs. wavelength. This data was taken for a high-quality HEM grown sample. The high slope efficiency is consistent with a quantum efficiency near unity. This sample exhibited < 0.5%/cm loss and had a figure of merit (FOM) defined as $\alpha 532/\alpha 800$ of $\sim$ 120. Similar runs made for other samples yielded results consistent with their measured loss values. The material was pumped longitudinally at 532 nm (15 ns FWHM) at 0.5 to 2.2 J/cm^2.

Figure 8 is an example of high slope efficiency, low threshold and high conversion efficiency (> 50%) that was measured for several HEM grown crystals. FOM's of up to 175 have been measured with the laser performance commensurate with that low loss figure. The HEM grown crystals grown to date are typically lower doped than the Czochralski samples and therefore must be longer in length in order to achieve 95% absorption of the pump energy. The long lengths of these crystals have a positive effect on thermal dissipation and have a more favorable energy deposition profile. Since these crystals are being grown with absorption in the lasing area of < 0.5/cm at 800 nm, the lengths don't affect the lasing performance to a great degree. Samples from four different boules have all shown $\sim$ 50% conversion (red out at 800 nm to green absorbed).

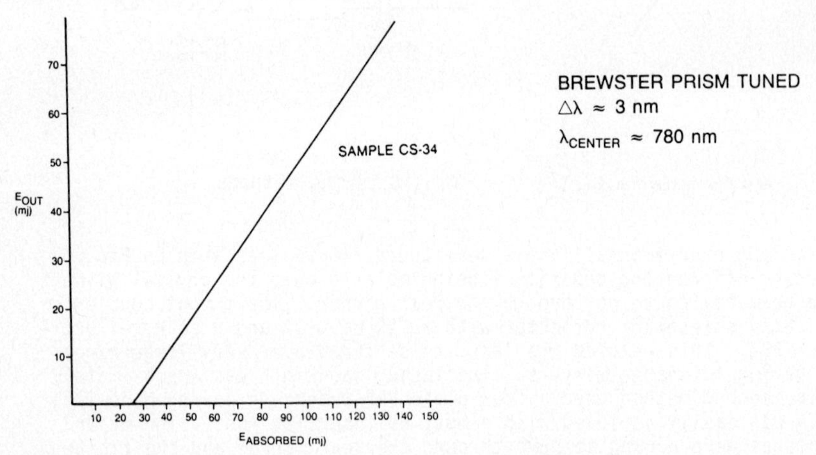

Fig. 8. $Ti:Al_2O_3$ oscillator output vs. absorbed input

Strides still have to be made in the areas of low damage threshold (4 - 5 J/cm^2) for large spot laser operation; also a background absorption that appears to increase with increasing Ti^{3+} concentration that extends out past 1000 nm and exists in both polarizations. Boules are now being produced that are 4" in diameter and 4" in length. The material currently being grown by the HEM is a large step forward in reaching the quality needed for reproducible and high-quality $Ti:Al_2O_3$ lasers.

[1] Crystal growth supported by NSF grant #DMR-8313257

References

1 P. Moulton, Optics News (Nov./Dec. 1982) P. 9

2 P. Moulton, Laser Focus (May 1983) P. 83
L. DeShazer, in: Laser Technology Group Meeting, P. Moulton, moderator, OSA Annual Meeting, New Orleans, 1983. Leon Esterowitz, LaJolla Conference on Tunable Lasers, LaJolla, 1984, Proc. published by Springer-Verlag, Vol. 47.

3 P. Moulton, Spectroscopic and Laser Characteristics of $Ti:Al_2O_3$, JOSA B, Vol. 3, No. 1 (Jan. 1986).

Amplifier and Line-Narrowed Oscillator Performance of $Ti:Al_2O_3$

N.P. Barnes and D.K. Remelius*

Los Alamos National Laboratory, Los Alamos, NM 87545, USA

Laser-pumped $Ti:Al_2O_3$ has been operated as an amplifier and as a line narrowed oscillator. Amplifier data is presented as a function of pump geometry, delay between pump and probe, probe bandwidth, and probe energy.

1. Introduction

Amplifier and line-narrowed oscillator performance of $Ti:Al_2O_3$ has been investigated to determine the applicability of this material to refined applications such as isotope separation and communications. Oscillators employing $Ti:Al_2O_3$ are well known for their wide tuning ranges [1-5]. However, for many applications, a narrow linewidth at a precise wavelength is also essential. To date, the energy output of laser-pumped $Ti:Al_2O_3$ has been somewhat limited. An obvious method of increasing the energy output is through the use of amplifiers. It is toward these ends that the investigation was directed.

2. Experimental Arrangement

Two frequency-doubled Nd:YAG lasers, utilizing a common amplifier, were used to pump two pieces of $Ti:Al_2O_3$, one operating as an oscillator and the other acting as an amplifier. An optical schematic of the experimental arrangement appears in Figure 1. Each of the Nd:YAG oscillators utilizes a 6.3 x 76 mm Nd:YAG laser rod in a single elliptical cavity and a cylindrical ring electrode KD*P Q-switch to produce about 50 mJ in a Gaussian beam. Beam quality is somewhat less that twice diffraction limited. By employing two independent oscillators, the timing between the pump beams for the two pieces of $Ti:Al_2O_3$ could be readily varied. Outputs from these two oscillators were combined using a Glan prism and amplified in a common amplifier. As the amplifier is operated somewhat below saturation, substantial amplification of both beams could be obtained. Outputs from the amplifier were then separated using a second Glan prism and frequency doubled in a Type II KD*P crystal. One of the frequency-doubled beams was used to pump a $Ti:Al_2O_3$ oscillator. A SF6 prism was used

*Present address: Langley Research Center, Hampton, VA 23665-5225

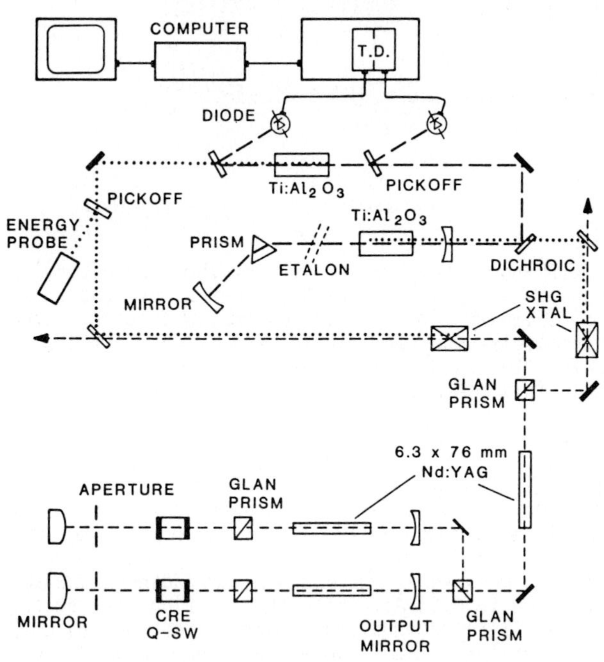

Fig.1. Experimental arrangement for measuring gain

to tune the oscillator and various etalons could be inserted into the resonator for further line narrowing. Output from the oscillator could be used as a probe beam for the amplifier. The pump energy from the second frequency-doubled beam was used to pump the amplifier and its energy was continuously monitored. The probe beam was also monitored before and after transversing the pumped region of the $Ti:Al_2O_3$. Outputs from the detectors were recorded on a transient digitizer which was subsequently read by an on-line computer. Gain was measured by suppressing the Q-switch trigger to the Nd:YAG laser delivering the pump energy to the $Ti:Al_2O_3$ amplifier on alternate pulses. Gain was then interpreted as the ratio of the transmission through the amplifier with and without the pump beam, using suitable averaging.

3. Oscillator Performance

Narrow-line performance of the $Ti:Al_2O_3$ laser oscillator was determined by measuring the threshold, E_{TH}, and the slope efficiency, η_S, of the oscillator as a function of output bandwidth. Threshold and slope efficiency are calculated using incident pump energy to the oscillator rather than absorbed energy. Results of these measurements appear in Table 1 which displays the comparative performance of the various options. For these experiments, a 0.60 reflecting output mirror was utilized. The operating wavelength of the $Ti:Al_2O_3$ laser was maintained at about 0.78 μm

Table 1. Narrow-line $Ti:Al_2O_3$ performance

CONFIGURATION	η_S	E_{TH}(mJ)	$\Delta\lambda$(Å)
SF6 PRISM	0.20	14.7	3.2
0.25 mm UNCOATED	0.21	15.9	0.66
0.86 mm UNCOATED	0.15	14.2	0.08

throughout the tests. With only the SF6 prism in the resonator, a threshold of 14.7 mJ and a slope efficiency of 0.20 were observed. As successive line-narrowing elements are added to the resonator, the threshold remains approximately constant while the slope efficiency decreased somewhat. However, the variation of these parameters is not far beyond experimental error. This behavior is characteristic of a high gain system.

The linewidth of the $Ti:Al_2O_3$ oscillator was determined by using a monochromator and a diode array and eventually a Fabry-Perot etalon. The full width at half-maximum linewidth of the $Ti:Al_2O_3$ laser with only the SF6 prism in the resonator was measured to be 3.2 Å. By adding a 0.25-mm etalon, the linewidth decreased to 0.66 Å. With the addition of a second etalon, 0.86 mm thick, the linewidth was too narrow to be measured accurately with the monochromator. Linewidth was then measured by determining the width of the Fabry-Perot rings and was determined to be 0.08 Å. Further narrowing could be achieved with higher finesse etalons if desired. Since the $Ti:Al_2O_3$ laser can be a high-gain system, further

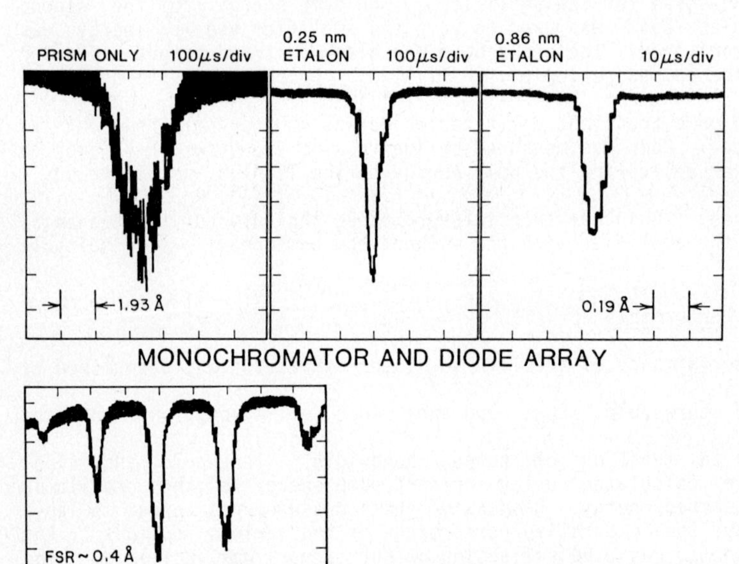

Fig.2. Linewidth measurement of $Ti:Al_2O_3$

narrowing of the laser output is not expected to significantly decrease the laser output.

4. Amplifier Performance

Either a circular or a Gaussian beam profile could be used to pump the laser amplifier. Beam profiles were obtained using a reticon diode array. Typical profiles for the two beams are shown in Figure 3. The first beam profile can be well approximated by a Gaussian while the second beam profile can be approximated by a circular beam profile. Numerical values for the beam profiles, displayed by the reticon array, were verified using a translating straight edge or a variable aperture and determining the transmission as a function of straight-edge position or variable aperture diameter. Appropriate curve-fitting techniques were then used to determine a numerical value for the beam radius to be used in analyzing the data.

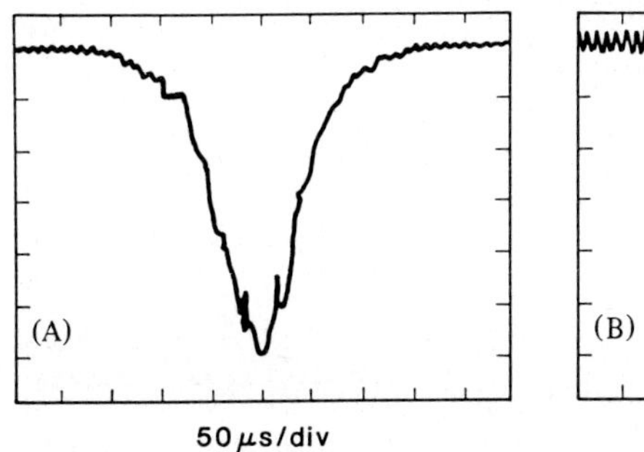

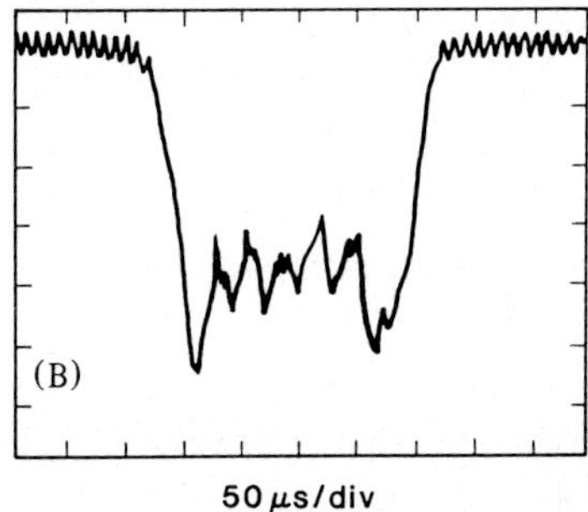

Fig.3. Beam profile for nearly Gaussian (A) and nearly circular beam profiles (B)

The logarithm of the gain of the $Ti:Al_2O_3$ amplifier is plotted versus the pump energy with the Gaussian pump beam profile in Figure 4. Gain of the amplifier is measured 200 ns after the peak of the pump pulse. Gain is representative of the small signal regime with probe energies on the order of 0.1 mJ in a 1.0 mm beam diameter. Gains on the order of exp(1) are observed with a 0.018 J pump input. A straight line represents the theoretically derived gain for a Gaussian beam enjoying a Gaussian gain profile [3]. An average gain can be calculated by evaluating the integral

$$G_A = \int_0^\infty \frac{2}{\pi w_o^2} \exp\left(-\frac{2\rho^2}{w_o^2}\right) \exp\left(\sigma_e \ell N_R \exp\left(-\frac{2\rho^2}{w_g}\right)\right) 2\pi\rho d\rho \,, \quad (1)$$

where

$$N_R = \frac{E_p \lambda_p}{hc} \quad \frac{2}{\pi w_g^2 \ell} \quad (1 - T_c) . \quad (2)$$

In this expression w_o is the beam radius of the $Ti:Al_2O_3$ laser, σ_e is the effective stimulated emission cross-section, ℓ is the length of the $Ti:Al_2O_3$ crystal, E_p is the pump energy, λ_p is the pump wavelength, h is

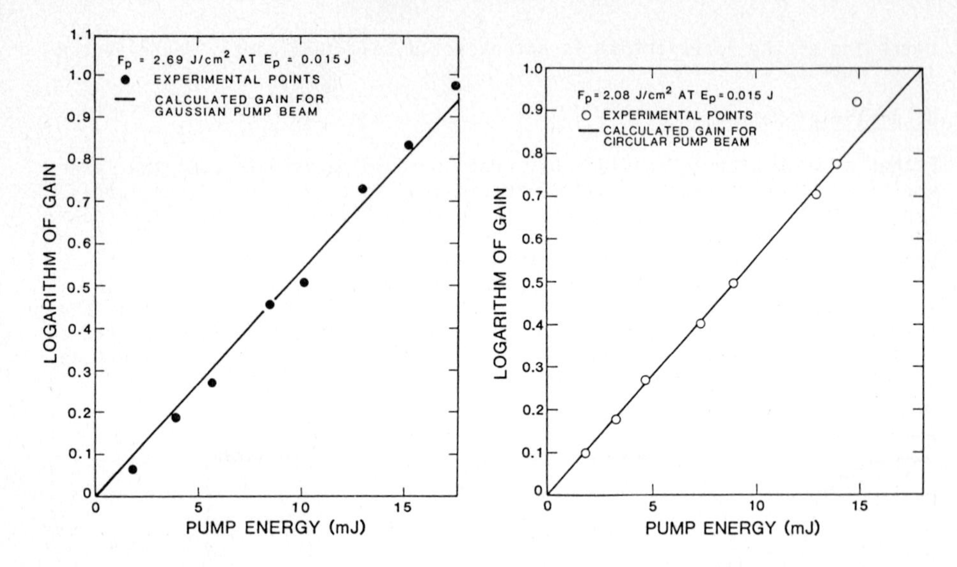

Fig.4. Logarithm of gain of $Ti:Al_2O_3$ vs. pump energy (Gaussian pump beam)

Fig.5. Logarithm of gain of $Ti:Al_2O_3$ vs. pump energy (circular pump beam)

Planck's constant, c is the speed of light, w_g is the beam radius of the Gaussian pump beam, and T_c is the transmissivity of the pump beam through the $Ti:Al_2O_3$ crystal. Good agreement between the experimental data and the theoretical model can be observed. It may be noted that with a 0.015 J input, the peak energy density, F_p, is about 2.7 J/cm^2.

The logarithm of the gain of the $Ti:Al_2O_3$ amplifier is plotted versus the pump energy with the circular pump beam profile in Figure 5. Again a gain of exp(1) can be observed with approximately 0.018 J pump. A straight line represents the theoretically derived gain but for a circular beam profile. Small signal gain in this case is given by:

$$G_A = \exp\left(\sigma_e N_R \ell\right) \left(1 - \exp\left(-\frac{2w_o^2}{w_c^2}\right)\right) + \exp\left(-\frac{2w_o^2}{w_c^2}\right). \qquad (3)$$

Parameters in this expression are the same as those above except w_c is the beam radius of the circular pump beam. Good agreement can be observed between the theoretically derived results and the experimental data. With this gain profile, however, a 0.015 J input produces a peak energy density of about 2.1 J/cm^2. Thus, while gains are comparable, the peak pump energy density is considerably reduced.

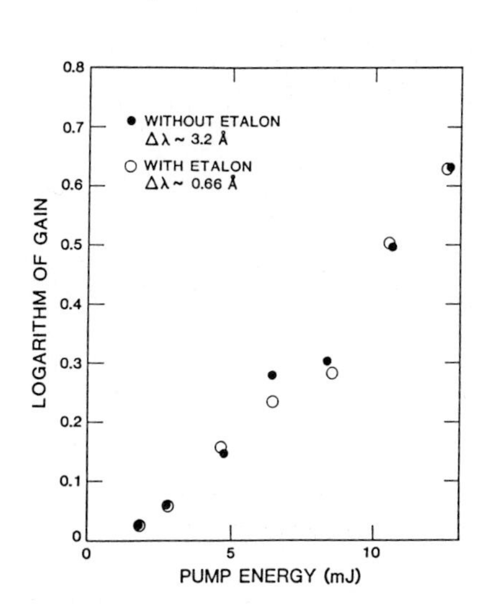

Fig.6. Gain of $Ti{:}Al_2O_3$ vs. pump energy

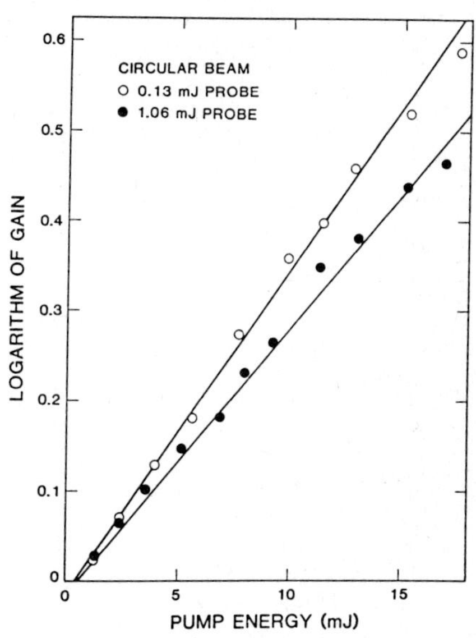

Fig.7. Logarithm of gain of $Ti{:}Al_2O_3$ vs. pump energy (for two-probe energies)

Gain of a $Ti{:}Al_2O_3$ amplifier appears to be relatively independent of the spectral bandwidth of the probe beam. Experimental gain curves for two probe beams with significantly different spectral bandwidths are shown in Figure 6. The probe beams were generated in the $Ti{:}Al_2O_3$ oscillator by using the prism only and the prism in conjunction with an etalon. As can be seen from the figure, gain as a function of the pump energy appears to be independent of the spectral bandwidth of the probe beam. A gain independent of spectral bandwidth of the probe beam would be expected of a homogeneously broadened laser.

Gain has been measured as a function of the probe energy as well as the pump energy. In Figure 7 are plotted the logarithms of the gain as a function of the pump energy, parametrically for two different probe energies. Additional data was taken for a third probe beam energy intermediate to the two energies shown in Figure 7. However, since these points lie between the two curves already displayed, it was omitted for sake of clarity. Data such as this allows a determination of the effective stimulated emission cross-section by two different methods.

Since presumably the energy stored in the upper laser level and the gain profile of the amplifier are known, a measurement of the gain can yield a value for the effective stimulated cross-section. Toward this end, the gain curve with the lowest probe energy was used. A value for the effective stimulated emission cross-section was determined by minimizing the root mean square deviation between the experimental points and the gain predicted by the above equation. When corrected for the delay between the pump beam and the probe beam by using the measured upper laser level lifetime, this procedure yields a value for the effective stimulated

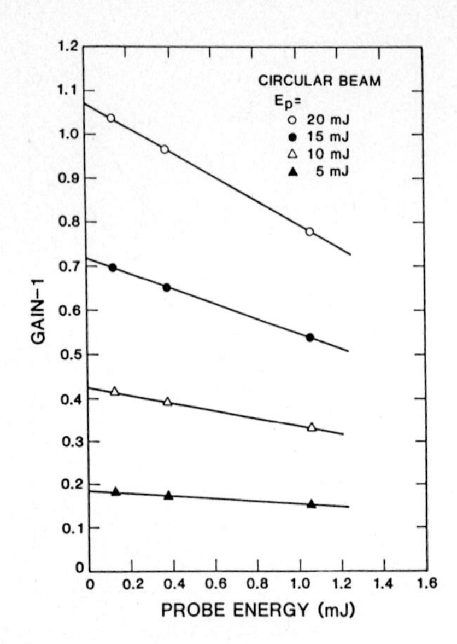

Fig.8. Gain of $Ti:Al_2O_3$ vs probe energy

emission cross-section of 2.5 x 10^{-23} m^2. This value is only slightly above the value for the parameter reported in the literature [1].

A value for the effective stimulated emission cross-section can also be obtained by plotting the gain as a function of the probe energy as shown in Figure 8. Gain was calculated for selected pump energies by interpolating between the experimental data points at a given probe energy. An amplifier model [6] predicts that the initial slope of the gain versus probe energy should be proportional to the product $G_o(G_o-1)/2\ E_{SAT}$ where G_o is the small signal gain and E_{SAT} is the saturation energy for the particular beam profile. Using the saturation energy obtained from the slope of the lines and a knowledge of the beam profile, a value for the effective stimulated emission cross-section can be obtained. Experimental values of the probe energy do fall on a straight line. Furthermore, the slope of these lines do increase with increasing gain. Using the lowest gain versus probe energy curve, a value for the effective stimulated emission cross-section was obtained. However, it was approximately a factor of 2 larger than the value obtained by considering the gain as a function of the pump energy. But, the larger effective stimulated emission cross-section is more in keeping with observed pulse evolution time intervals for the $Ti:Al_2O_3$ oscillator. Pulse evolution time intervals reported previously [2], also indicate that gain is higher than would be expected using the lower value for the effective stimulated emission cross-section and the inversion profile deduced from the measured absorbed pump energy, and beam radius.

5. Discussion

These differences may be resolved if the upper laser level population decayed more rapidly than expected from the measurement of the upper laser level lifetime. Gain data has been corrected for the loss in gain suffered over a 200-ns delay. However, an effective upper laser level lifetime of

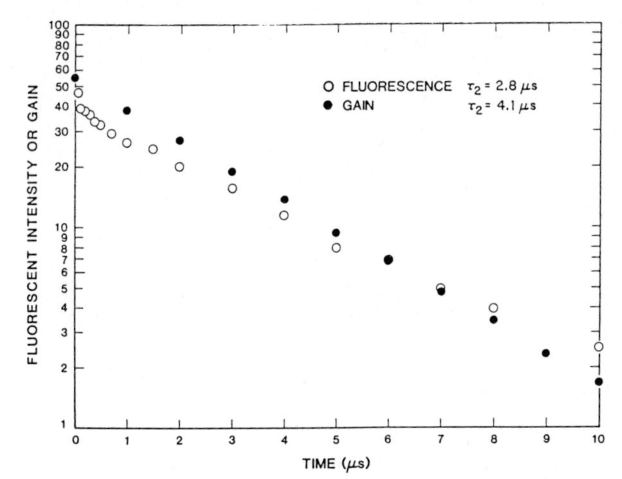

Fig.9. Decay of upper laser level of $Ti:Al_2O_3$ vs. time

about 300 ns instead of about 3.0 μsec, would reduce the upper laser level population by a factor of approximately 2, necessary to reconcile the two derived values for the effective stimulated emission cross-section. To determine if this were the case, the gain was measured as a function of delay between the pump and the probe pulses. Figure 9 shows gain data taken using a Gaussian pump profile as a function of delay. Also plotted is the decay of the fluorescent intensity at low pump levels. On the semilogarithmic plot, the fluorescent intensity at low pump level decays exponentially. Deviations from the exponential decay are slight. However, the gain does not appear to display the same behavior. While not obvious on the logarithmic plot, the initial decay does appear to be more rapid than the final decay. Between the first two points, the gain appears to decay by approximately a factor of 0.7. Unfortunately, there was enough jitter in the system that the gain could not be determined at short time intervals with more resolution. Several possibilities exist to explain the relatively-fast initial decay of the gain, the following among them.

Amplified spontaneous emission could occur which would tend to deplete the population inversion. An estimate of the single pass gain can be obtained by evaluation of:

$$\sigma_e N_o \ell \cong \sigma_e \frac{\lambda_p E_p}{hc} \frac{2}{\pi W_g^2} (1 - T_c), \qquad (4)$$

where N_0 is the upper laser level population density. If the derived value of the effective stimulated emission cross-section is accurate, single pass gains a little in excess of 2.0 could occur at amplifier pump energies used in these experiments. A large effective stimulated emission cross-section would, of course, lead to greater amplified spontaneous emission loss. However, the aspect ratio of the excited volume is significantly less than encountered with laser rods such as Nd:YAG. Consequently, it is not certain that longitudinal amplified spontaneous emission could account for all of the observed loss in gain.

Transverse amplified spontaneous emission may also tend to deplete the upper laser level population. Although the gain in the transverse

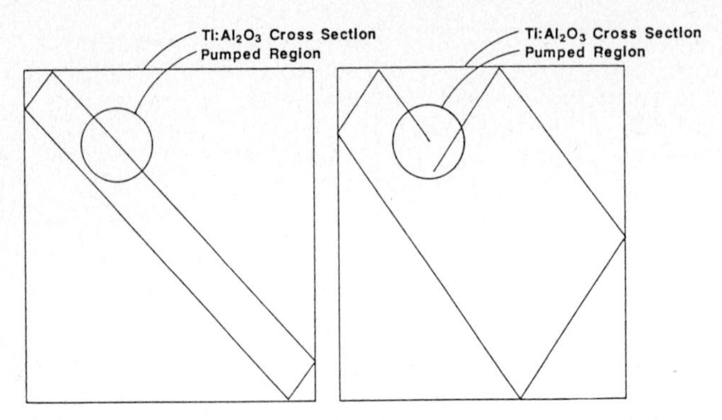

Fig.10. Amplified spontaneous emission normal to pump beam

direction is an order of magnitude smaller than in the longitudinal direction, feedback of the amplified spontaneous emission may occur if the emission occurs in certain directions. An example of feedback appears in Figure 10. Opposite corners of the Ti:Al_2O_3 rectangular solid could act as Porro prisms to return the spontaneously emitted radiation for amplification. While only a relatively small fraction of the radiation could return to the gain cross-section after only 4 internal reflections, other directions of emission could enjoy gain after more internal reflections. One such direction is also shown in Figure 10. In this case, the radiation is returned to the gain region after 12 internal reflections. While the gain per pass of transverse amplified spontaneous emission is smaller than longitudinal amplified spontaneous emission, at least some of the radiation would be returned for amplification after only a modest loss. The loss in the Ti:Al_2O_3 at the lasing wavelength as well as the crystal dimensions would be significant parameters if this were a significant loss mechanism.

Multiple sites for the Ti atom could explain the nonexponential decay. If each site had a different decay time, it would yield a nonexponential decay. An explanation such as this tends to be supported by the fact that apparently Ti atoms can form pairs [7]. However, the fact that the fluorescent decay at low excitation levels appears to be exponential does not tend to support this explanation.

Higher phonon densities resulting from high excitation densities could result in an increased nonradiative decay rate. On the average, absorption of a single photon would yield about 8 phonons as the excited state relaxed. While this would yield a negligible increase in the temperature, it would significantly increase the phonon occupation factor at the phonon frequencies needed to cause an increased nonradiative decay rate. Normally the phonon distribution function would thermalize rapidly. However, because of the high Debye frequency of Al_2O_3, the density of phonon states to which phonons could decay is low. Consequently, thermalization time constants above tens of picoseconds may be possible. While not appearing likely, if the phonons could exist for time intervals on the order of 100 ns, an increased nonradiative decay rate could result. Time intervals on this order are comparable to the time interval required for a phonon to propagate out of the excited volume of the laser material.

Further experiments need to be performed to determine which of these loss mechanisms is dominant. Until the loss is identified and quantified, determination of the effective stimulated emission by measuring the gain is somewhat uncertain. However, it does appear that the value derived and presented here is a lower limit to the effective stimulated emission cross section.

References

1. P. F. Moulton, J. Opt. Soc. Am. B. (in press) and "Solid-State Research Report," Lincoln Laboratory, MIT, 15, (1982:83).

2. N. P. Barnes and D. K. Remelius: "Ti:Al_2O_3 Laser Pumped By a Frequency Doubled Nd:YAG Laser," Presented at CLEO 1985, Baltimore, Maryland, May, (1985).

3. N. P. Barnes and D. K. Remelius: "Gain, Lasing and Tuning Characteristics of Ti:Al_2O_3," Presented at CLEO 1985, Baltimore, Maryland, May, (1985).

4. C. E. Byvik and A. M. Buoncristiani: "Analysis of Vibronic Transitions in Titanium-Doped Sapphire Using the Temperature of the Fluorescence Spectra," IEEE J. Quant. Elect. QE-21, 1619-1623, (1985).

5. P. Lacovara, L. Esterowitz, and M. Kokta: "Growth, Spectroscopy, and Lasing of Titanium-Doped Sapphire," IEEE J. Quant. Elect. QE-21, 1614-1618, (1985).

6. N. P. Barnes, V. J. Corcoran, I. A. Crabbe, L. L. Harper, R. W. Williams, and J. W. Wragg: "Solid-State Laser Technology," IEEE J. Quant. Elect. QE-10, 195-201, (1974).

7. R. L. Aggarwal, A. Sanchez, R. E. Fahey, and A. J. Strauss: "Magnetic and Optical Measurements on Ti:Al_2O_3 Crystals for Laser Applications: Concentration and Absorption Cross Section of Ti^{3+} Ions," Appl. Phys. Lett. 48, 1345-1347, (1986).

Oscillator and Amplifier Performance of Ti:Sapphire

L.G. DeShazer, J.M. Eggleston, and K.W. Kangas

Spectra Technology Inc., 2755 Northup Way, Bellevue, WA 98004, USA

Doubled Nd:YAG lasers pumping Ti:Sapphire show promise as a tunable, all-solid-state laser system. Experimental efforts are directed at developing efficient systems for converting 532 nm pump light into the near-IR laser emission by Ti:Sapphire. The oscillator performance of Ti:Sapphire is well predicted by numerical models. Preliminary amplification experiments are examined and modeled. Reduction in efficiency due to parasitic oscillations is observed, modeled, and suppressed.

1. Introduction

Ti:Sapphire is a broadly tunable laser material [1] whose basic laser properties are currently under experimental examination. The research reviewed in this paper covers experiments and modeling intended to better determine the performance characteristics of Ti:Sapphire lasers.

2. Emission Cross-Section

The cross-section curve for a transition is related to the fluorescence intensity per unit wavelength curve for the transition, by a factor of the wavelength (λ) to the fifth power [2]. In particular,

$$\sigma_{\pi}(\lambda) = \frac{3}{2} \frac{\eta \lambda^5}{\tau_f \left[\int \lambda I(\lambda) d\lambda \right] 8\pi n^2 c} I(\lambda), \qquad [1]$$

where η is the fluorescence efficiency, τ_f is the observed fluorescence lifetime, and all other parameters have their customary meaning. The factor of 3/2 arises because the π emission ($E \parallel C$) is twice as intense as the σ emission ($E \perp C$). The predicted line shape for a ligand field transition is a Poisson distribution [3] given by

$$\sigma(\nu) = \sigma_s S^P / P! \quad , \qquad [2]$$

where $P = (\nu_o - \nu)/\nu_p$

ν_o = energy of zero–phonon line at 626 nm,

ν_p = phonon energy, a fitting parameter,

S = expectation value of the number of photons, a fitting parameter,

σ_s = the scale factor, a fitting parameter.

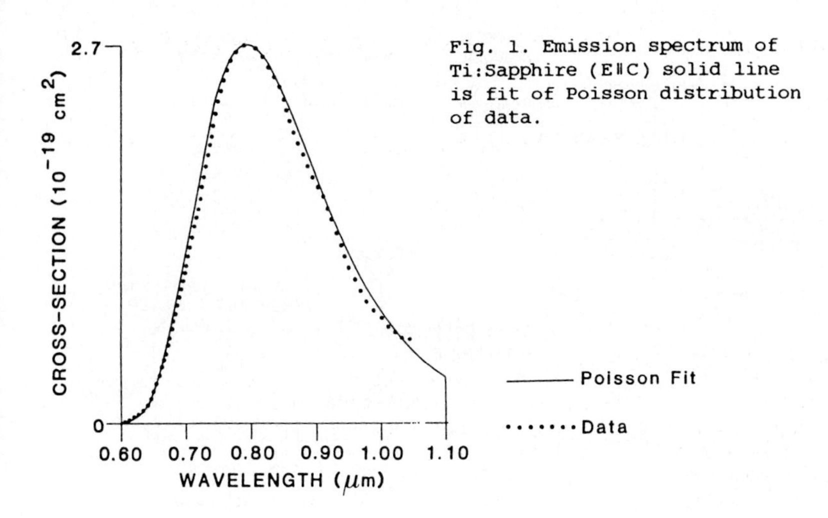

Fig. 1. Emission spectrum of Ti:Sapphire (E∥C) solid line is fit of Poisson distribution of data.

The fit shown in Figure 1 has parameters ν_p = 611.19 cm^{-1}, ν_o = 15968 cm^{-1}, S = 6.02, and σ_s = 3.98 x 10^{-21} cm^2, yielding a peak cross-section of 2.7 x 10^{-19} cm^2. In deriving this fit, the ratio of τ_f/η was assumed to be 4.2 μsec, which is slightly higher than the currently accepted zero-temperature lifetime limit of 3.9 μsec. This choice is based on our laboratory measurements of lifetime and results in a slightly low estimation of cross-section. A scale parameter (σ_s) of 4.28 x 10^{-21} cm^2 and peak cross-section of 2.9 x 10^{-19} cm^2 results if the 3.9 μsec lifetime is used. The zero-phonon line was allowed to vary in this fit, and the resulting value is reasonable. The cross-section predicted by the fit with the 4.2 μsec radiative lifetime was used in all the modeling discussed in this paper.

3. Oscillator Experiment

Figure 2 illustrates the typical experimental configuration for pumping Ti:Sapphire with a green laser, and it is the same setup used for previous experiments [4–7]. End pumping has been used for Ti:Sapphire due to the combination of relatively low green pump absorption coefficient (≈1 cm^{-1}), high infrared passive loss (≈0.1 cm^{-1}), and high laser saturation fluence (≈1 J/cm^2). Both oscillator and amplifier modules were end-pumped by doubled Nd:YAG lasers (Quanta-Ray Model DCR-2) at 532 nm in 10 ns at 1 Hz repetition rate. The green pump beam was focused through the high reflectance rear mirror with a pump beam diameter near 3 mm, with up to 160 mJ absorbed per pulse. The oscillator was constructed with a flat-flat resonator. The Brewster-angle prism was used as a tuning element in some experiments. Tuning with the prism was achieved from 720 to 1030 nm with a 4 nm bandwidth. Narrower bandwidths of 0.5 A were produced by using a birefringent filter and etalons in conjunction with the prism. For the amplifier, the green pump beam and the input beam from the Ti:Sapphire oscillator were directed into the amplifier crystal using a dichroic beamsplitter.

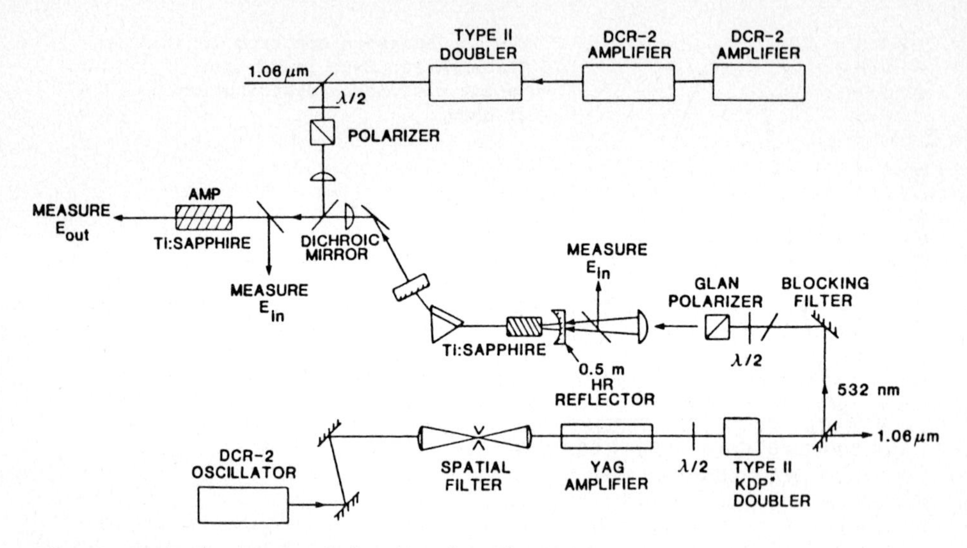

Fig. 2. Typical experimental setup for Ti:Sapphire oscillator/amplifier.

Figure 3 shows the performance of a Ti:Sapphire oscillator and the results of numerical modeling. The peak conversion efficiency in this experiment is 42 percent. The slope efficiency varies during the pulse due to atomic and optical kinetics. The single-pass crystal transmission in the oscillator was measured externally to be 81 percent per pass. The pump beam spot size was measured to be about 2.7 mm.

4. Oscillator Model for Ti:Sapphire

The performance of Ti:Sapphire oscillators is completely understood from the simple material and optics parameters. While Ti:Sapphire has only two electronic states, it has a very large number of vibronic states (forming a vibronic continuum), which allows the laser to be treated as a four-level laser. The numerical model used in fitting the data modeled the Ti:Sapphire as a four-level system, with a 3.2 μsec lifetime for the upper laser level, and a very fast relaxation between both the pump level and upper laser level and between the lower laser level and the ground state.

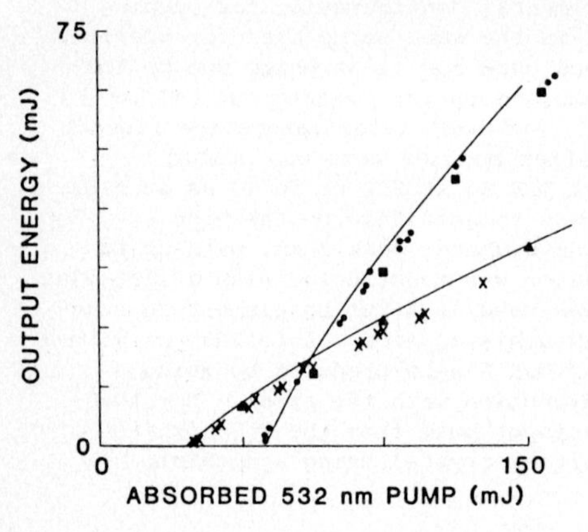

Fig. 3. Ti:Sapphire oscillator performance compared to theoretical model. X are data and ▲ are model results for 20% output coupler, and solid dots are data and ■ are model results for 60% output coupler.

The cross-section was determined from the Poisson fit to the fluorescence spectrum, and the number of excited states from the number of pump photons absorbed. The model did not include any saturation by the pump, two-photon absorption of either the pump or laser photons, or excited state absorption. The optical kinetics included three-dimensional physical optics propagation, which interacted with a 'thin sheet' gain medium. The initial conditions in the gain media were determined by the pumping geometry. The initial optical field was modeled as a spatially uncorrelated random noise source. The noise intensity was adjusted to be reasonably close to theoretical quantum noise levels.

Adjusting only the spot size within the ±10% experimental uncertainty, this model has accurately predicted the absolute performance of a Ti:Sapphire oscillator. An example of this fit is shown in Figure 3 for laser operation at 800 nm with 40% and 80% output reflectors. Calculations using both two-dimensional and three-dimensional codes are displayed. The three-dimensional code is more accurate but requires more computation time. The oscillator pulse build-up time and pulse width are also well predicted by this model, illustrated in Figure 4. The build-up time was defined as the delay between the peak of the pump pulse and the peak of the output pulse. The pulse width is the full width at half maximum of the output. Since the YAG pump pulse was much shorter than the build-up time of the output pulse, the Ti:Sapphire laser pulse was a single gain-switched (Q-switched) spike of order 10 ns FWHM.

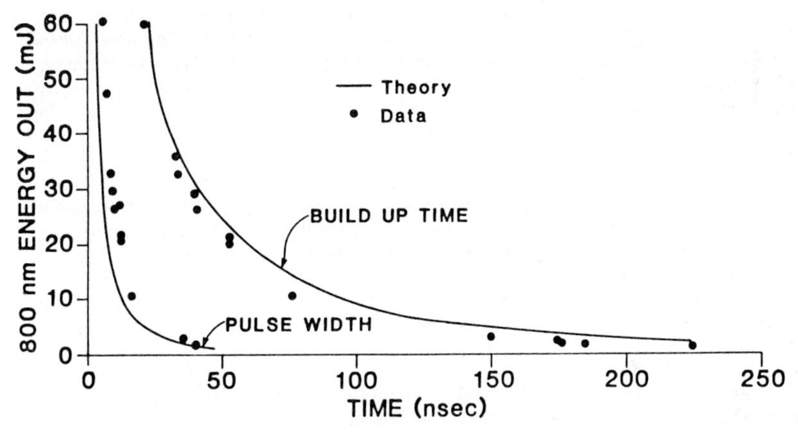

Fig. 4. Buildup and pulse width of gain-switched oscillator pulse from Ti:Sapphire: comparison of theory to experiment.

5. Parasitic Oscillations

Because of the relatively high gain in Ti:Sapphire when pumped by doubled Nd:YAG pulses, parasitic oscillations within the laser crystal can be a problem. Parasitics due to total-internal reflections from polished crystal faces were observed in both oscillator and amplifier operation. Figure 5 illustrates the effect of parasitic suppression in an 800-nm oscillator where the only change between data runs was the suppression of parasitic modes. This suppression entailed adding frosted glass, with an index matching fluid, to the unused polished sides of the gain medium. These sides were originally polished for spectroscopic purposes.

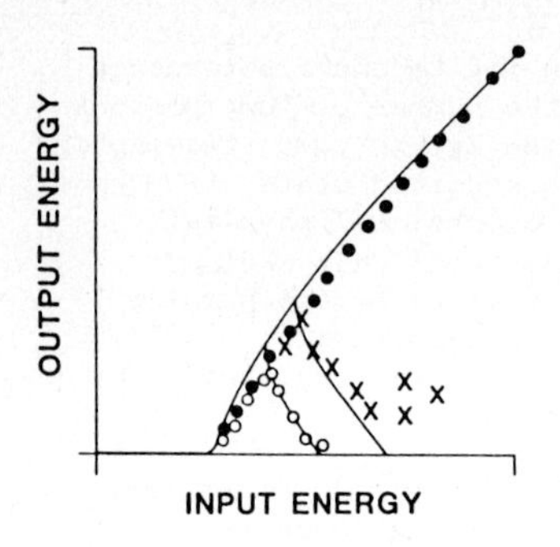

Fig. 5. Experimental and numerical results on gain-switched parasitics in Ti:Sapphire oscillators.

Identification of the 'roll-over' in the oscillator output as due to the onset of parasitics took some study, since parasitics normally clamp the output energy, not decrease it [8]. However, roll-over in the input-output curve occurred because the pumping time was short compared to the parasitic build-up time, which resulted in gain-switched parasitic modes. These gain-switched parasitics also have been modeled and are well described by theory.

Figure 5 also shows the output from a one-dimensional model of an oscillator with an internal parasitic. The inputs to the model were the optical cavity round-trip time and losses, the parasitic path round-trip time, the pump time, and the ratio of thresholds between the parasitic and desired optical mode. The parasitic path round-trip time was assumed to be the time required to travel once around the Ti:Sapphire crystal, which was substantially shorter than the cavity round-trip time. The initial rise in the output curve occurs because the optical mode is above threshold and the parasitic is not. This rise continues even beyond the parasitic 'threshold' because the build-up time of the cavity mode, when it is well above threshold, is shorter than the build-up time of the parasitic, when it is just barely above threshold. Increased pumping causes the parasitic build-up time to drop faster than the desired mode build-up time. The sharp drop in the output occurs as the build-up time of the parasitic changes from being slightly longer than the cavity mode build-up time (no parasitic loss) to slightly shorter (significant parasitic loss). The slow decay of the cavity output with pumping arises from the kinetics of a gain-switched (Q-switched) pulse.

Unlike CW lasers, gain-switched and Q-switched lasers leave a cavity below threshold, not at threshold [9]. In fact, the peak in the intracavity power of a gain-switched laser occurs when the gain equals the loss. Beyond this point, the intracavity power continues to circulate in the cavity and the gain medium continues to amplify it, albeit not as fast as the loss depletes it. This continued amplification depletes the gain medium below the gain-equals-loss (threshold) condition. This effect allows a gain-switched parasitic to completely extinguish the output from a cavity mode which has a lower threshold but longer round-trip time. This effect also contributes to the roll over in the output curves in Figure 3.

It should be noted that in Figure 5 the threshold ratio was adjusted to achieve roll over at the appropriate point. Discrepancies between modeling and experiment occur because the model assumed single optical and parasitic modes that were completely overlapped in the gain media. In reality, there is multi-mode operation with significant, but not complete, overlap in the gain media.

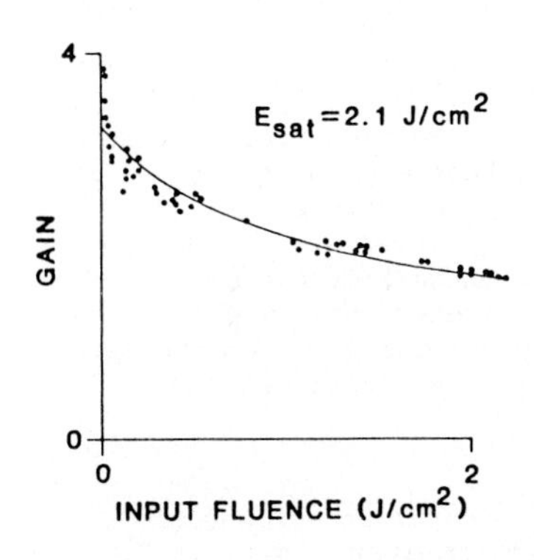

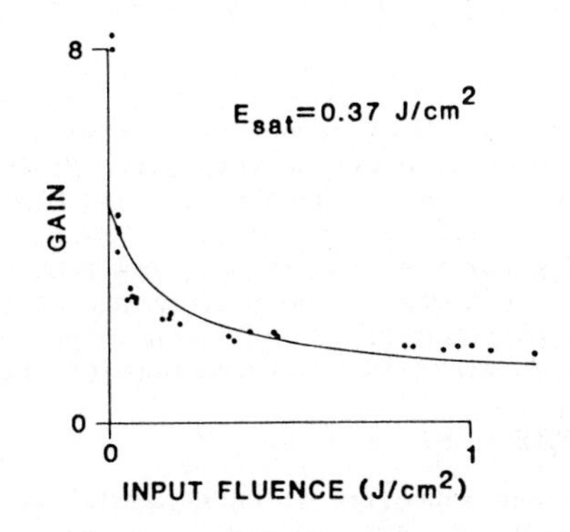

Fig. 6. Amplifier Saturation Data

6. Amplifier Experiment

Figure 6 shows two examples of amplifier performance at 800 nm, illustrating the saturation of gain for Ti:Sapphire crystal 2 cm long. The two curves are at the same wavelength but for different oscillator setups. The variation in saturation fluences arises from a factor of three uncertainty in the effective spot area. This uncertainty arises from the small beam area used and from the fact that the beam shape was not smooth. The values do, however, straddle the theoretical value of 0.9 J/cm^2. A more detailed fit would require larger input energies and carefully measured (smooth) pump and extracting beam profiles.

The solid curves in this figure are a least squares fit of a numerical amplifier model to the measured data. The model was a one-dimensional, four-level amplifier model based on a rate equation analysis. Background loss and the gain distribution due to absorption of the pump light were included. Upper state fluorescence was ignored, as the extraction occurs on time scales that are short compared to the upper state lifetime. The Fresnel losses are already compensated for in the data. The fitted parameters are the lossless gain coefficient (G_o) and the saturation flux ($E_{sat} = h\nu/\sigma$). The role of these parameters can be seen in the standard lossless saturation equation for four-level storage amplifiers

$$G = \ell n \left\{ 1 + G_o \left[\exp\left[\frac{E_{in}}{E_{sat}} \right] - 1 \right] \right\}, \qquad [3]$$

which the model reduces to in the absence of background loss. The fitting routine adjusted the lossless gain (G_o) and the saturation fluence to achieve a least squares deviation between the logarithm of the measured and predicted gain.

The deviation in the small signal gain from the fitted curve is assumed to arise from hot spots in the pump beam. Since the amplification is exponentially dependent on the stored energy, small regions that are intensely pumped can have a significant impact on the small signal amplification, with very little total energy content.

7. Conclusion

In conclusion, the agreement between measured and modeled performance of Ti:Sapphire oscillators is very good, indicating that excited state absorption, two-photon absorption, and other nonlinear effects are unimportant for powers of order 500 MW/cm^2. A four-level laser model is appropriate for the transition. The behavior of parasitics can be understood in terms of the performance of gain-switched optical modes. The amplifier performance is not inconsistent with predictions, given the large experimental error in spot size predictions of initial experiments.

Acknowledgements

This work was supported by NASA-Langley Research Center under contract NAS1-17960 and by STI independent research and development. The Ti:Sapphire crystals were grown with the Czochralski method by Union Carbide Corporation.

1. P.F. Moulton: Opt. News 8(6), 9 (1982). Also P.F. Moulton: "Spectroscopic and laser characteristics of Ti:Al_2O_3," J. Opt. Soc. Am. B 3, 125 (1986).

2. See for instance, M. Planck: Theory of Heat Radiation (Dover, NY, 1959). Also, B.F. Aull and H.P. Jenssen: "Vibronic interactions in Nd:YAG resulting in nonreciprocity of absorption and stimulated emission cross sections," IEEE J. Quan. Elect. QE-18, 925 (1982).

3. D. Curie: "Absorption and Emission Spectra," in Optical Properties of Ions in Solids, ed. by B. Di Bartolo (Plenum Press, NY, 1975) p. 89.

4. L.G. DeShazer, G.F. Albrecht, and J.F. Seamans: "Tunable Titanium Sapphire Lasers," presented at SPIE Conference on High Power and Solid-State Lasers, Los Angeles, CA, January 1986.

5. G.F. Albrecht, J.M. Eggleston, and J.J. Ewing: "Measurements on Ti^{3+}:Al_2O_3 as a Lasing Material," Tunable Solid-State Lasers, Proceedings of the First International Conference, La Jolla, CA, 13-15 June 1984, P. Hammerling, A.B. Budgor, and A. Pinto, Editors (Springer-Verlag, Berlin, 1985).

6. G.F. Albrecht, J.M. Eggleston, and J.J. Ewing: "An Evaluation of the Ti:Sapphire Tunable Laser System," Tunable Solid-State Lasers for Remote Sensing, Proceedings of the N.A.S.A. Conference, Stanford University, Stanford, CA, 1-3 October 1984, R.L. Byer, E.K. Gustafson, and R. Trebino, Editors (Springer-Verlag, Berlin).

7. G.F. Albrecht, J.M. Eggleston, and J.J. Ewing: "Measurements of Ti^{3+}:Al_2O_3 as a Lasing Material," Opt. Comm. 52, 401 (1985).

8. W.R. Sooy et al.: "Dynamic Limitations on the Attainable Inversion in Ruby Lasers," in Quantum Electronics Conference Paris 1963, Vol. 2, ed. by Grivet, p. 1103.

9. W. Wagner and B. Lengyel: "Evaluation of the Giant Pulse in a Laser," J. Appl. Phys. 34, 2040 (1963).

Laser Performance and Temperature-Dependent Spectroscopy of Titanium-Doped Crystals

K.L. Schepler

Air Force Wright Aeronautical Laboratories, AFWAL/AADO-1, WPAFB OH 45433, USA

1. Laser Performance of Ti:sapphire

Ti:sapphire ($Ti^{3+}:Al_2O_3$) is of strong interest because of its ability to lase over a continuously tunable, broadband, spectral region [1]. However, crystal quality has continued to be plagued by problems with the presence of scattering centers and spurious absorption losses in the spectral region where lasing is expected. To assess the current state-of-the-art in Ti:sapphire crystal growth, laser crystals were procured from two crystal growth facilities. A crystal 48 mm long by 5 mm in diameter with Brewster angle ends was cut from a boule grown by the Czochralski (CZ) method. It had a nominal doping of 0.1 wt% titanium. Another crystal, 35 mm long and 7 mm in diameter, was grown using the Heat Exchanger Method (HEM); it also had Brewster angle ends. Based on relative absorption of 532 nm laser radiation in the two crystals, the titanium doping level in the HEM crystal was calculated to be 0.03 wt%.

Laser performance of the two crystals was measured under identical pumping and laser resonator conditions. Figure 1 is a diagram of the experimental setup. The pump laser was a frequency-doubled Nd:YAG laser with 10 ns pulses. A 254 mm lens focused the pump beam onto the Ti:sapphire crystal. The flat input mirror was coated to transmit 532 nm light but was highly reflective in the 730-810 nm region -- greater than 99%. The output mirror was 50% reflective in the 730-810 region and had a 3 m radius of curvature. A prism was used for tuning experiments to provide wavelength discrimination.

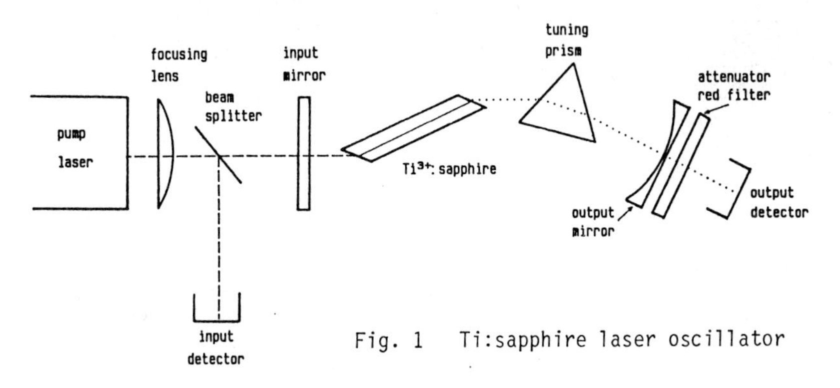

Fig. 1 Ti:sapphire laser oscillator

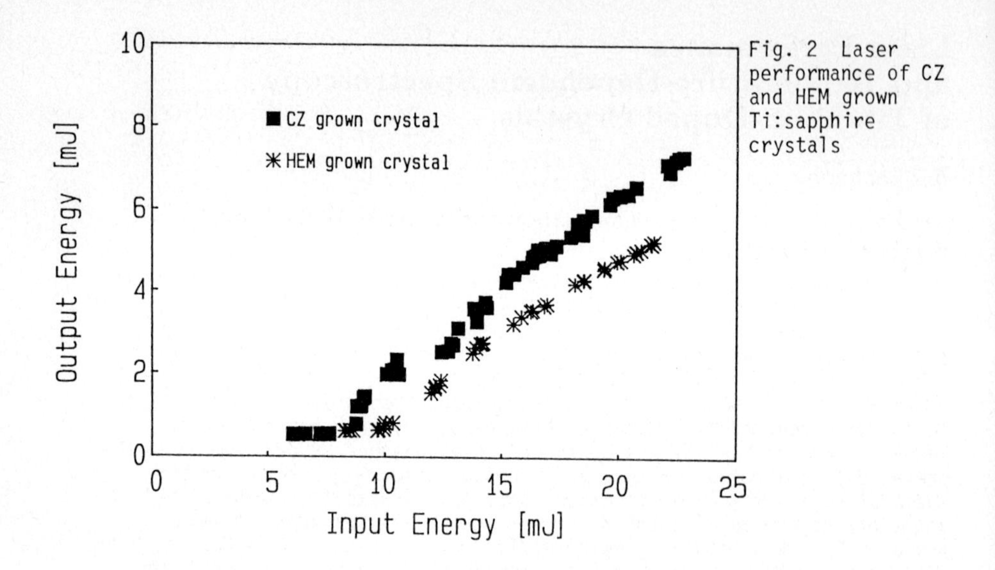

Fig. 2 Laser performance of CZ and HEM grown Ti:sapphire crystals

Output energy per pulse as a function of input energy is shown in Fig. 2. In this case no tuning element was present. Input energy was energy incident on the Ti:sapphire crystal surface. Threshold was at 8 mJ for the CZ crystal and at 10 mJ for the HEM crystal. Energy slope efficiencies were 50% and 42% for the CZ and HEM crystals respectively. But the HEM crystal, because of its lighter doping, absorbed only 60% of the incident energy, whereas the CZ crystal absorbed 99% of the incident light. Quantum slope efficiency based on photons absorbed was almost 100% for the HEM crystal and 72% for the CZ crystal. Both crystals discussed here are improvements over previously available material.

The tuning range for the HEM crystal is shown in Fig. 3. Only one pair of mirrors was used and tuning was limited by bandwidths of the mirror coatings and by surface damage of the laser crystal at input energies greater than 25 mJ.

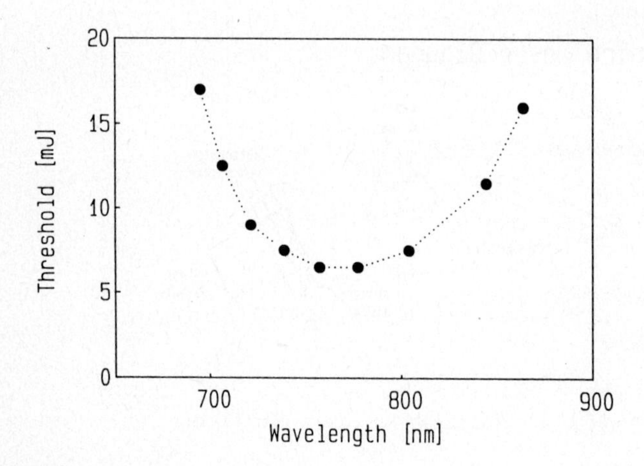

Fig. 3 Ti:sapphire lasing threshold for the HEM crystal and one set of mirrors

Direct measurement of absorption and scattering losses was made by observing the pre-crystal and post-crystal energies of 755 nm, 8 ns laser pulses. The CZ crystal had a loss of 2.5%/cm and the HEM crystal was slightly better with a measured loss of 2.2%/cm. Loss in a more heavily doped HEM-grown crystal (not tested for laser performance) was found to be 1.8%/cm which is encouraging for the prospect of increasing the titanium-ion doping without incurring additional loss.

2. Spectroscopic Comparison of Ti:sapphire and Ti:YALO

Ti^{3+} ions differ in their optical properties when placed in different host crystals because of the sensitivity of d-orbitals to the surrounding crystal lattice. New host materials could be useful for improving titanium laser performance or changing the laser tuning range. Reported here is a comparison of the spectroscopic properties of Ti:sapphire and Ti:YALO ($Ti^{3+}:YAlO_3$). Figure 4 shows the polarized fluorescence spectra of Ti:sapphire at room temperature and at 11K; these are similar to previous measurements [2,3]. The c-axis and a-axis spectra are shown to scale, i.e., c-axis fluorescence is about twice the intensity of the a-axis fluorescence. Also, room-temperature fluorescence intensity (assuming the absorption coefficient is not a strong function of temperature) decreases to approximately half of what it was at 11K; room-temperature spectra in Fig. 4 have been multiplied by a factor of two to show relative lineshape changes. The inset shows detail at the high-energy side of the fluorescence spectrum at 11K. Here the zero-phonon lines and single-phonon lines are clearly visible.

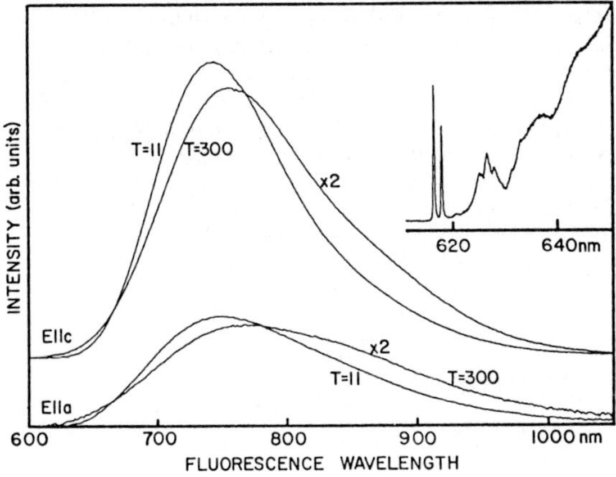

Fig. 4 Ti^{3+} polarized fluorescence in Al_2O_3

Ti:YALO polarized fluorescence spectra at 11K and 300K are shown in Fig. 5. As first shown by MOULTON [3], Ti:YALO fluorescence is shifted to shorter wavelengths when compared to Ti:sapphire. But only small changes were noted in relative intensity and peak position for different polarizations. Also, Ti:YALO shows no significant changes in total fluorescence intensity; this was true even at temperatures as high as 400K. The inset shows the zero-phonon, single-phonon and multiple-phonon lines. There is one zero-phonon line at 540 nm but other peaks have not been positively identified.

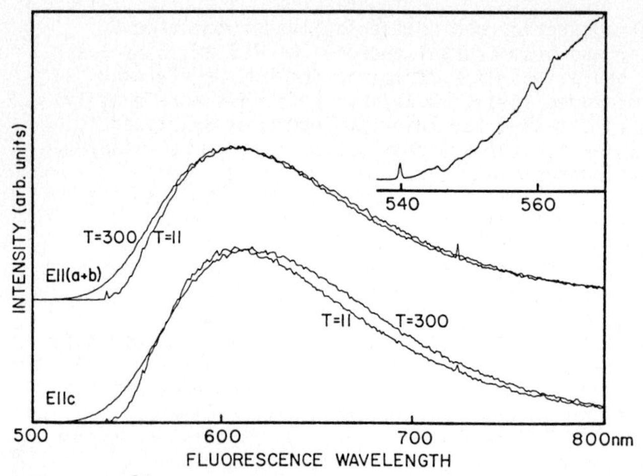

Fig. 5 Ti^{3+} polarized fluorescence in $YAlO_3$

In contrast to Ti:sapphire the linewidth of Ti:YALO changed little with temperature. Ti:sapphire linewidth (FWHM) increased 600 cm^{-1} between 11K and 300K. But Ti:YALO linewidth increased by less that 200 cm^{-1} over the same temperature span. Ti:YALO did show an increased rate of broadening at temperatures above 300K.

Temperature dependence of fluorescence lifetime was also measured in Ti:sapphire and Ti:YALO. Laser pulses 10 ns in duration and at 532 nm were used to pump both crystals. The fluorescence was collected, passed through a monochrometer set at the fluorescence peak and detected with a photomultiplier. Fluorescence pulses were signal averaged and lifetimes were calculated using least square fits. The results are shown in Fig. 6. The Ti:sapphire results are similar to results published previously [4-7]. The points are experimental measurements and the solid curve is a calculation based on thermal population of a two-level model where the higher level has a very short lifetime due to rapid nonradiative relaxation from the upper level to the ground state.

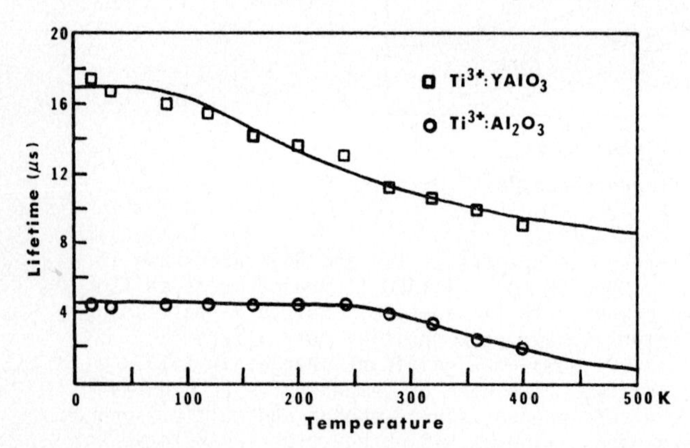

Fig. 6 Temperature dependence of fluorescence lifetime

In Ti:YALO the lifetime is longer than in Ti:sapphire and it has a different dependence on temperature. The two-level model fit to the Ti:YALO data is not as satisfying. It requires that the energy separation between the levels be small (280 cm^{-1}) and that the lifetime of the upper level be similar to that of the lower level. The lower level lifetime was taken to be 17 ms with the upper level fit being 8 ms. The poor fit to the curve may be related to the poor optical quality of the crystal; numerous cracks were present. Lifetime data showed a distribution of the lifetimes, i.e., plots of ln(intensity) versus time were not straight lines. A possible explanation of the small two-level splitting is that in Ti:YALO the two-fold degeneracy of the ^2E level is lifted by a crystal field distortion or some other mechanism.

Fluorescence intensity and fluorescence lifetime measurements indicate that nonradiative transitions become increasingly significant in Ti:sapphire as temperature rises above 250K. However, Ti:YALO does not exhibit nonradiative transitions even at temperatures up to 400K. The absence of nonradiative loss in Ti:YALO should benefit its potential use as a cw laser.

3. Summary

Laser performance measurements of Ti:sapphire crystals grown using two different growth techniques have been made. The measurements show that both growth techniques are capable of producing high-quality crystals with near theoretical performance. Also, measurements of the spectroscopic properties of Ti:sapphire and Ti:YALO have been made. The longer lifetime, the absence of nonradiative loss, and the shift of the fluorescence of Ti:YALO to shorter wavelengths in comparison to Ti:sapphire make Ti:YALO an interesting candidate for laser operation.

References

1. P. F. Moulton, Optics News 8, 9 (1982)
2. B. F. Gachter and J. A. Koningstein, J. Chem. Phys. 60, 2003 (1974)
3. P. F. Moulton: In Springer Series in Optical Sciences 47, 7 (Springer-Verlag, Berlin, Heidelberg, New York, Tokyo 1985)
4. C. E. Byvik and A. M. Buoncristiani, IEEE J. Quantum Electron. QE-21, 1619 (1985)
5. R. C. Powell, G. E. Venikouas, L. Xi, J. K. Tyminski and M. R. Kokta, J. Chem. Phys. 84, 662 (1986)
6. P. F. Moulton, J. Opt. Soc. Am. B 3, 125 (1986)
7. P. Albers, E. Stark and G. Huber, J. Opt. Soc. Am. B 3, 134 (1986)

Flashlamp-Pumped Titanium-Doped Sapphire Laser

P. Lacovara and L. Esterowitz*

Naval Research Laboratory, Washington, DC 20375, USA

At CLEO '85 we reported an efficient flashlamp-pumped titanium-doped sapphire laser which is capable of output energies of several hundred millijoules. Despite advances in efficiency, however, high power operation of the laser was still limited in part by short flashlamp life (approximately 10^3 shots). We now report a system with greatly extended flashlamp life (10^5-10^6 shot) which for the first time makes possible a flashlamp-pumped system with multi-watt average power.

The operation of xenon flashlamps, even those designed for dye laser use, at 2-10 joules per centimeter input into the lamp in a microsecond pulse generally results in explosion of the lamp after a few thousand shots. Using a shock loading, and a high current simmer (3-5 Amp CW) to provide a centered, low inductance arc channel, we have obtained lamp lifetimes of at least 250,000 shots. Most importantly, lamp failure was by gradual decrease in output due to opaque deposits inside the envelope rather than by catastrophic failure.

We have investigated other aspects of flashlamp-pumped titanium-sapphire which merit discussion. Consistently we have found that the laser slope efficiency using a small eccentricity aluminum plated ellipse is about twice as high as that for a diffuse cavity (this has been tested for 4 different diffuse cavities, 3 commercial and one made in-house). This appears to be due to the small pump light absorption of presently available low loss (i.e., low doped) titanium-sapphire laser material. While an elliptical cavity directs light from the flashlamp into the rod, a diffuse cavity randomizes pump light, so the efficiency is strongly dependent on the ratio of laser rod absorption to all other (flashlamp and cavity walls) absorption in the cavity. Efficiency calculations comparing these two configurations agree closely with experiment.

We have estimated the effect on laser performance of the temperature-dependent quantum efficiency in titanium-doped sapphire. This effect could conceivably distort the parabolic temperature profile in the laser rod, making thermal lensing compensation non-trivial, and if it were strong enough that the temperature inside of the rod rose greatly, it could dramatically alter the transverse gain profile. In reality, the effect is not large enough to be a factor for moderate (few hundred watt/cm^3) pump loadings. Inspection of the heat equation with a positive temperature-dependent heating term shows that it can be reduced to Bessel's equation of zeroth order, with solutions that are Bessel functions of zeroth order. When these are expanded for small arguments

* Unified Industries, Springfield, Va.

we find that for expected pump loadings the distortion of the temperature profile is negligible, and that thermal runaway (where the temperature in the hottest part of the rod increases continually to some high value) is only expected for very high loadings, for example, one kilowatt/cm^3 in a 1 cm diameter laser rod.

In our first attempts at high average power operation of titanium-doped sapphire we have obtained 2.5 watts (125 mJ at 20 Hz). We expect that with higher doped, lower loss laser material, and refined pumping techniques, higher average powers will be obtained from flashlamp-pumped titanium-doped sapphire.

Optical and Mass Spectroscopic Analyses of Titanium-Doped Sapphire Crystals

C.E. Byvik[1], *A.M. Buoncristiani*[1], *S.J. McMurray*[1], *and M. Kokta*[2]

[1]NASA Langley Research Center, Hampton, VA 23665, USA
[2]Union Carbide, Washougal, WA 98671, USA

Remote sensing of atmospheric species from space-based platforms using differential absorption LIDAR (DIAL) will require a high-energy, long lifetime, efficient, and tunable laser transmitter. Tunable solid state laser materials such as titanium-doped sapphire (Ti:Sapphire) and alexandrite are currently candidates for DIAL laser transmitters. The interest in the development of laser quality Ti:Sapphire is based on its broad tuning range which extends into the near IR (700 to 1100 nm) MOULTON [1]. However, initial growths of Ti:Sapphire exhibited significant and deleterious absorption in the lasing region as shown in Figure 1. One of the hypotheses suggested for the origin of this parasitic absorption is the presence of impurities incorporated into the laser crystal at growth. Evidence for the presence of such an impurity ion appears in the fluorescence spectrum of Ti:Sapphire excited by 514.5 nm radiation from an argon ion laser (Figure 2). The sharp peak at 690 nm is attributed to the R-lines of chromium.

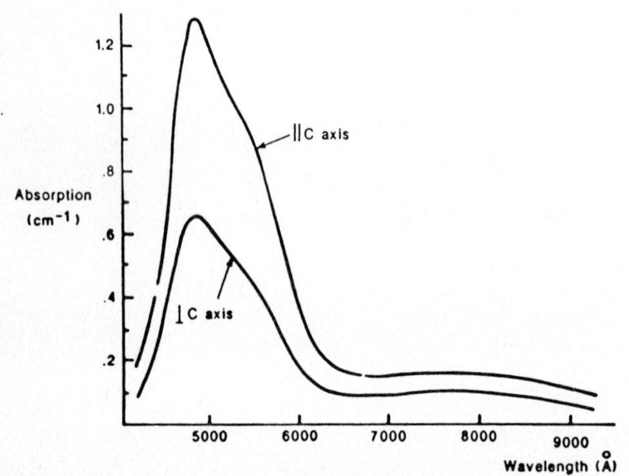

Figure 1. Absorption spectra for Ti:Sapphire.

In order to identify other impurities in Ti:Sapphire laser crystals as they are currently grown, we performed an elemental mass spectroscopic analysis on samples of the materials used in the melt and on samples of the crystal grown subsequently by both the Czochralski and the Heat Exchange methods. Data from this analysis (see Table 1) indicate that, in fact, large concentrations of transition metal ions

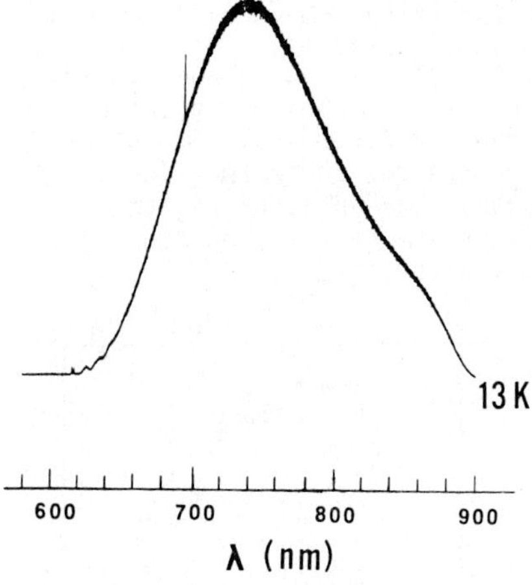

Figure 2. Fluorescence Spectrum of Ti:Sapphire, showing the R-line peak attributed to Chromium.

Table 1

MASS SPECTROGRAPHIC ANALYSIS OF TI:SAPPHIRE SAMPLES. CONCENTRATIONS PPM

Sample/Element	Ti	V	Cr	Mn	Fe	Ni
Crackle	<0.04	0.1	1.0	0.2	3.0	<0.04
Top	10	1.0	25	1.0	15	0.1
Middle	35	1.0	50	2.0	20	0.1
Bottom	20	0.5	75	0.5	15	0.1
TiO(2)		0.3	1.0	0.06	2	1
Fiber growth	100	0.3	65	0.1	7	1
HEM	100	0.3	75	0.2	5	0.4

are incorporated into the crystal during the growth process. The data confirms the presence of significant concentrations of chromium, iron and vanadium. The chromium fluorescence indicated in Figure 2 can occur either through direct excitation of the chromium by the argon ion laser or by indirect excitation through a coupling of the titanium with the chromium. The fluorescence due to the chromium represents a loss mechanism resulting in a potentially lower quantum efficiency for Ti:Sapphire lasing.

The fluorescence spectrum of the as grown Ti:Sapphire crystals shows no evidence of the presence of iron. Weak iron fluoresence has been observed previously in the spectra of iron doped with TiO_2 and $SrTiO_3$

BYVIK et al [2]. Thus the effect of unintentionally doped chromium and iron ions in Ti:Sapphire is understood qualitatively. In order to examine the influence of vanadium a crystal containing 50 ppm vanadium and 0.05% titanium was grown by the Czochralski method. The absorption spectrum of this co-doped sample is shown in Figure 3; it shows a significant overlap in the absorption spectra for these two ions. To assess the effect of the presence of vanadium as an impurity ion on the lasing efficiency of Ti:Sapphire, the fluorescence lifetime and quantum efficiency of this co-doped sample was measured.

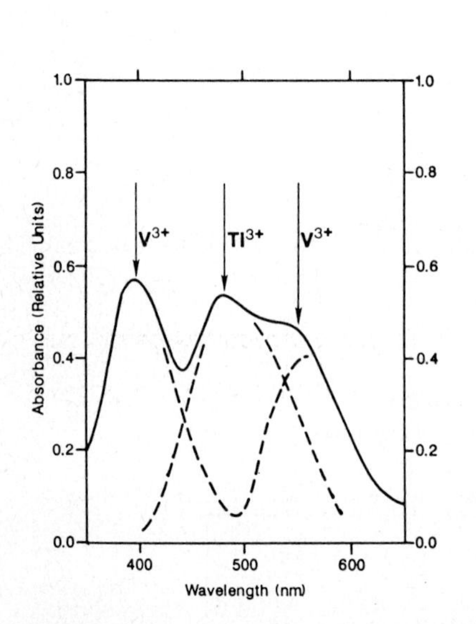

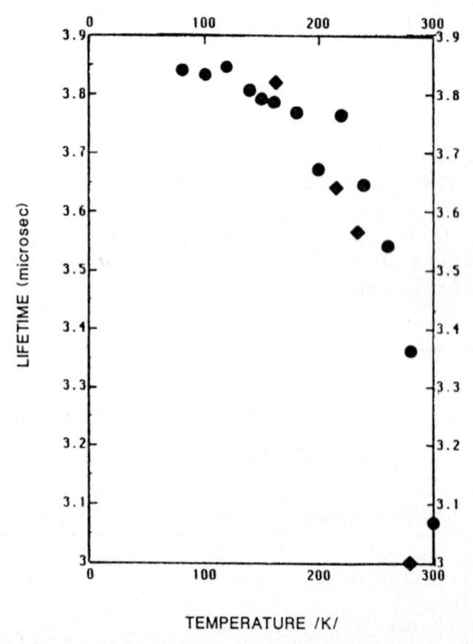

Figure 3. Absorption spectrum of Sapphire co-doped with vanadium and titanium.

Figure 4. Fluorescence lifetime as a function of temperature for V,Ti:Sapphire. Circles are lifetimes measured with 514 nm excitation, diamonds for 590 nm excitation.

The fluorescence spectrum of the co-doped sample excited by the 514.5 nm line of an argon ion laser was identical to that of the Ti:Sapphire. Fluorescence lifetime measurements of V,Ti:Sapphire as a function of temperature were conducted using a pulsed dye laser tuned to 514 nm as the excitation source. Fluorescence decay was monitored with a photomultiplier mounted to a monochromator set to the peak of the fluorescence curve. See BYVIK and BUONCRISTIANI [3] for details. The temperature dependence of the fluorescence lifetime of Ti:Sapphire and V,Ti:Sapphire is shown in Figure 4. No significant differences in the temperature dependence of the fluorescence lifetimes for these samples were noted.

The quantum efficiency (QE) of samples of Ti:Sapphire and V,Ti:Sapphire was determined using a 20-inch integrating sphere (see ANDREWS [4] for experimental details). Samples of Ti:Sapphire, V,Ti:Sapphire and pure sapphire were mounted on thin steel rods coated with barium sulfate. The 514.5 nm line from an argon ion laser was used as the source of fluorescence excitation. The quantum efficiency was determined as the ratio of the luminescence intensity to the intensity of the absorbed radiation, that is,

$$QE = \frac{\text{Luminescent Intensity}}{\text{Absorbed Intensity}} = \Delta I_L \, G_L / (\Delta I_A \, G_A \, S_L),$$

where

$$S_L = \frac{\int \{S(\lambda)/S(\lambda_{max})\} \, L(\lambda) \, d\lambda}{\int L(\lambda) \, d\lambda},$$

where ΔI_L is the difference in the photomultiplier current between the luminescence of the doped sapphire and the pure sapphire. ΔI_A is the difference in the photomultiplier current monitoring the absorption of the pure and doped sapphire, G_L is the filter response function for a 540 nm bandpass filter, G_A is the filter response function for a 514.5 nm laser line filter, and S_L is the response of the photomultiplier $S(\lambda)$ integrated over the corrected fluorescence spectrum $L(\lambda)$ and normalized to the integrated fluorescence intensity ANDREWS [4].

The quantum efficiency for Ti:Sapphire and V,Ti:Sapphire as determined by this technique are given in Table 2. No significant difference in quantum efficiencies for these two doped samples were detected. This result is consistent with the results of the fluorescence lifetime experiments.

In conclusion, elemental analysis has indicated that transition metal ions are concentrated in the laser crystals during the growth process for both Czochralski and Heat Exchange growth methods. Purity of the starting materials and growth containers may affect improvements in

Table 2.

ROOM TEMPERATURE QUANTUM EFFICIENCY AND FLUORESCENCE LIFETIMES

CZOCHRALSKI CRYSTALS	QUANTUM EFFICIENCY	FLUORESCENCE LIFETIME (microsec)
TI:SAPPHIRE	0.7	3.07
V,TI:SAPPHIRE	0.7	(514nm) 3.18 (590nm) 2.8

laser quality of Ti:Sapphire and other solid-state lasers such as Nd:YAG, Cr:GSGG, and alexandrite. Experiments measuring the quantum efficiency and fluorescence lifetime indicate that the presence of unintentionally doped vanadium does not significantly effect the lasing efficiency when Ti:Sapphire is pumped by 514.5 nm radiation.

REFERENCES

1. P. F. Moulton: J. of the Optical Society of America B, vol. 3 p. 125 (1986).

2. C. E. Byvik, A. M. Buoncristiani and J. C. Barnes: Extended Abstracts of the Electrochemical Society, vol. 86-1, p. 630 (1985).

3. C. E. Byvik and A. M. Buoncristiani: J. of Quantum Electronics, vol. QE-21, no. 10, p. 92 (1985).

4. L. J. Andrews, A. Lempecki and B. C. McCollum: J. of Chem. Phys. vol. 74, p. 5526 (1981).

Injection-Controlled Titanium-Doped Sapphire Laser Using a Pulsed Dye Laser

C.H. Bair, P. Brockman, J.C. Barnes, R.V. Hess, and E.V. Browell

NASA Langley Research Center, Hampton, VA 23665, USA

Titanium-doped sapphire (Ti:sapphire) lasers have been extensively studied since their recent introduction [1-3]. Because of their wide tuning range (≈700 to 1000 nm), Ti:sapphire lasers are candidates for remote lidar measurements of H_2O vapor (≈720 and 940 nm) and pressure and temperature (≈760 nm). These measurements require efficient narrow bandwidth operation and accurate wavelength control.

Conventional interferometric spectral narrowing techniques are readily applicable to CW lasers. However, for short gain- or Q-switched pulses, the restricted number of passes through the narrowing elements makes spectral narrowing more difficult. Injection control of pulsed lasers with pulsed and CW lasers has been studied for line-tunable and wide bandwidth gas, dye, and solid-state lasers [4-11]. Injection control of a tunable Ti:sapphire laser using a narrow bandwidth pulsed dye laser operating at a wavelength removed from the peak of the Ti:sapphire laser gain curve is reported. Essentially complete energy extraction was achieved in a narrow bandwidth matching the injection source without use of conventional line-narrowing elements.

The frequency doubled Nd:YAG pump laser and narrow band dye laser used in this experiment are part of a flight-measurements system and are described in [12]. For the injection experiments, the dye laser used Exiton LDS-750 dye and was operated at 727 nm with a bandwidth of 2.5 pm.

The optical configuration for injection control is shown in Fig. 1. The 532 nm pump energy was split between the dye and Ti:sapphire lasers. Approximately 20 mJ of this energy was focussed through the dichroic rear Ti:sapphire cavity mirror and into the 0.5 cm laser rod whose c-axis was aligned parallel to the pump polarization. Approximately 6 mJ of the pump energy was absorbed by the rod. The Ti:sapphire laser cavity rear mirror

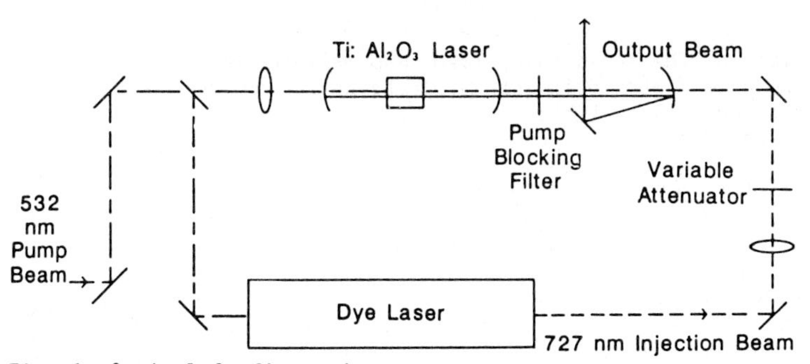

Fig. 1. Optical Configuration

was nearly 100% reflecting across the wavelength range of 620 to 780 nm and the output coupler had a reflectivity of 80% from 680 to 780 nm. The cavity length was 0.6 m and the nominal output energy was 0.5 mJ.

The major portion of the 532 nm pump energy was used to pump the dye laser. The dye laser output was focussed through a variable attenuator, a beamsplitter, and the output coupler of the Ti:sapphire laser cavity. For this experiment, no attempt was made to provide transverse or longitudinal mode matching of the injected laser to the Ti:sapphire laser cavity.

A series of measurements of the Ti:sapphire laser output were performed. A nearly 100% blocking filter at 532 nm was used to minimize residual pump energy in the measurement paths. Pulse timing and pulse shape were measured with a photo diode. Figure 2a shows the pump and Ti:sapphire laser pulses for a free-running condition (no dye laser injection energy). The 100 ns delay between the 532 nm pump pulse and Ti:sapphire pulse and the Ti:sapphire pulse shape were typical of the free-running laser. Figure 2b shows an injection-controlled condition. The dye laser pulse occurred approximately 10 ns after the pump pulse. This timing, set by the path lengths between the lasers, was used for all the data reported here. A comparison of Fig. 2a and 2b indicates that the injection-controlled Ti:sapphire laser pulses started earlier and persisted longer than the free-running pulse. Increasing the delay of the dye laser pulse to 35 ns after the pump pulse made injection control more difficult to achieve.

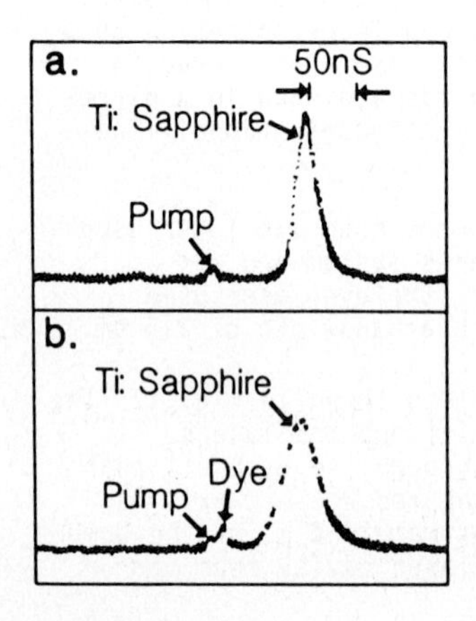

Fig. 2. Pump, Dye and Ti:sapphire Signals vs. Time
a. No Injection
b. With Injection

Spectral content of the Ti:sapphire laser was measured by passing the laser output through a fiber optic into a spectrograph and monitoring the signal with a photomultiplier tube. The PMT output was sampled with a gated integrator and recorded on an x-y plotter. The gate was opened immediately after the dye laser pulse and was kept open ≈500 ns to include the entire Ti:sapphire pulse. Sample spectra are shown in Fig. 3. The top curve, which shows the spectral output when no injection signal was introduced into the Ti:sapphire cavity, indicates that the free-running laser had broadband laser emission from 760 to 790 nm. Note that the injected wavelength, 727 nm, was well removed from the free-running laser wavelength range. The next three curves show the change in spectral character with injected energy at 727 nm as a parameter. The injected dye laser energy, indicated on the left side of each curve, was measured inside

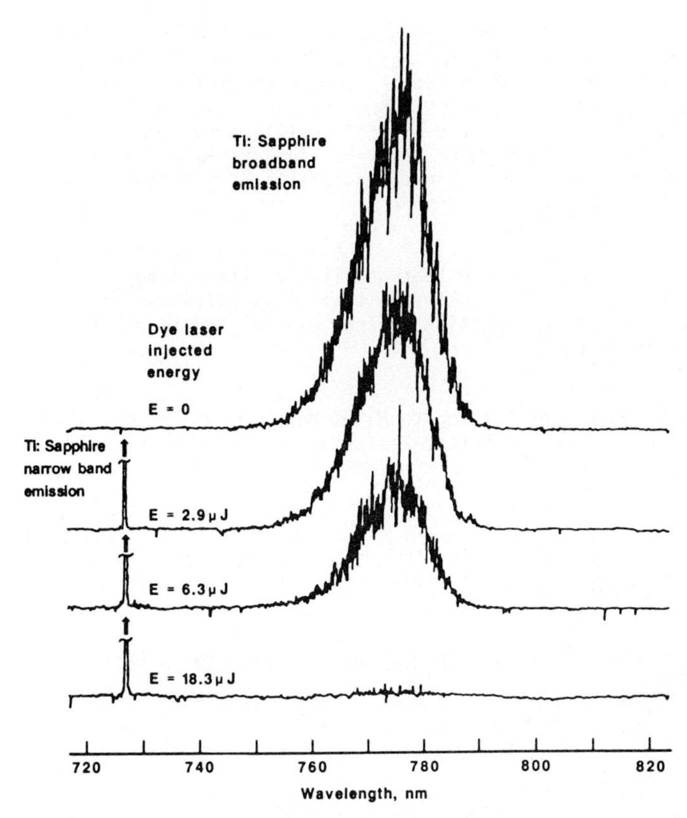

Fig. 3. Spectral Scans of Ti:Sapphire Output

the output mirror of the Ti:sapphire cavity. Onset of Ti:sapphire laser oscillation at the injected wavelength of 727 nm was observed to occur with approximately 0.07 μJ dye laser-injected intracavity energy. At the injected intracavity energies shown on Fig. 3, the injection-controlled Ti:sapphire laser output at 727 nm saturated the PMT so that the signal was off-scale. The bottom curve shows that, with approximately 18μJ of dye laser-injected intracavity energy, nearly all of the broadband emission was suppressed and nearly all the laser energy was extracted in a narrow band at the injection 727 nm wavelength. Oscilloscope measurements of the PMT output vs. time were performed with the spectrograph set both at the injected frequency and at the center of the broadband emission. For a partially injection-controlled condition, the Ti:sapphire laser output at the injected frequency occurred earlier than the broadband emission.

A Fabry-Perot etalon, described in [12], with a scanned array detector was used to measure the bandwidth of the Ti:sapphire laser. When the Ti:sapphire laser was operated without injected dye laser energy, the array detector exhibited a broadband output with no visible fringes. As the injection energy was increased, fringes superimposed on the broadband signal began to appear. With approximately 18 μJ of injection energy, well-defined fringes with a full width at half maximum of 2.5 pm and no noticeable broadband signal were observed. To assure that the observed fringes were due to Ti:sapphire and not reflected dye laser emission, the 532 nm pump laser input to the Ti:sapphire laser was blocked. At dye laser energies corresponding to those used in the injection experiments, the residual fringing was barely noticeable. To independently measure the dye laser bandwidth, the dye laser energy was significantly increased while keeping the 532 nm Ti:sapphire pump laser blocked. The dye laser bandwidth thus measured was also 2.5 pm. The etalon had a free spectral range of 50 pm and a resolution of ≈ 1.25 pm.

Measurements of the laser output energy for the injection controlled and free running laser conditions indicated that nearly identical energy extraction was being obtained. This indicates that Ti:sapphire is homogeneously broadened and that injection control can be used to obtain efficient narrow linewidth operation.

The authors express their appreciation to M. N. Mayo, W. J. McCabe, and E. A. Modlin for their excellent technical assistance throughout this research.

References

1. P. Moulton: Solid-State Res. Rep., Lincoln Lab., MIT, pp. 15-21, 1982.
2. G. F. Albrecht, J. M. Eggleston, J. J. Ewing: Optics Communications, 52, No. 6, pp. 401-404, 1985.
3. C. E. Byvik, A. M. Buoncristiani: IEEE J. Quantum Electron., QE-21, pp. 1619-1624, 1985.
4. A. Javan: Heterodyne Systems and Technology. NASA Conference Publication 2138, pp. 511-528, Williamsburg, VA, March 25-27, 1980.
5. R. T. Menzies, P. H. Flamant, M. J. Kavaya, E. N. Kuiper: Appl. Opt., 23, No. 21, pp. 3854-3861, 1984.
6. P. Juramy, P. Flamant, V. H. Meyer: IEEE J. Quantum Electron., QE-13, No. 10, pp. 855-865, 1977.
7. V. Ganiel, A. Hardy, D. Treves: IEEE J. Quantum Electron., QE-12, No. 11, pp. 704-716, 1976.
8. Y. K. Park, G. Giuliani, R. L. Byer: IEEE J. Quantum Electron., QE-20, No. 2, pp. 117-125, 1984.
9. D. J. Harter, J. J. Yeh, A. J. Heiney, D. R. Siebert: Topical Meeting on Tunable Solid-State Lasers, pp. FA2-1 to FA2-4, (Lasers and Electro-Optics Society of the Institute of Electrical and Electronics Engineers and Optical Society of America, Arlington, VA, May 16-17, 1985).
10. Y. K. Park, R. L. Byer: Optics Communications, 37, No. 6, pp. 411-416, 1981.
11. A. J. Berry, D. C. Hanna, and C. A. Sawyers: Optics Communications, 40, No. 1, pp. 54-58, 1981.
12. E. V. Browell, A. F. Carter, S. T. Shipley, R. J. Allen, C. F. Butter, M. N. Mayo, J. H. Siviter, Jr., W. M. Hall: Appl. Opt., 22, pp. 522-534, 1983.

Part VII

Color Center Lasers

New Color Center Lasers Based on Molecule-Doped Alkali Halides

D. Wandt [1]*, *W. Gellermann* [1], *F. Luty* [1], *and H. Welling* [2]

[1]University of Utah, Salt Lake City, UT 84112, USA

[2]Universität Hannover, D-3000 Hannover 1, Fed. Rep. of Germany

I. INTRODUCTION

Several types of F aggregate centers in alkali halide crystals have been developed in the last decade into active materials for tunable IR lasers. The ionic structures and electron distributions in ground and excited state of the more important centers, useful both for cw and mode-locked laser operation, are illustrated in Fig. 1. The centers permitting long-term stable laser operation (at cryogenic temperatures) can be categorized into three groups:

(1) F centers attached to either Li^+ impurities (F_A(II) centers) or Na^+ pairs (F_B(II) centers) in the hosts KCl and RbCl. Relaxing into an ionic saddlepoint configuration after optical excitation the centers have a strongly Stokes shifted emission with ~50% quantum efficiency. When pumped by ion lasers in the visible they provide tunable laser operation over the 2.2 to 3.4 μm region [1,2] with output power levels in the 10-

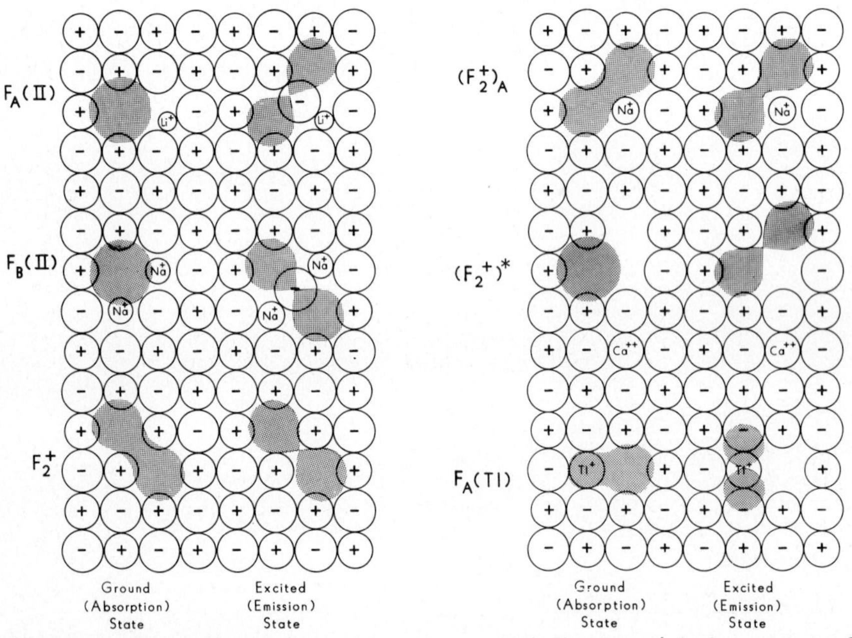

Fig. 1 Ionic configurations and electron distributions (shaded areas) of laser-active F-aggregate centers, shown for ground and relaxed excited states.

*On leave from Institut für Quantenoptik, Universität Hannover 3000 Hannover 1, FRG

100 mW range. The laser crystals are produced by simple additive coloration, can be handled at room temperature and show no output power fading effects up to ~2 W pump power levels.
(2) F centers attached to Tl^+ impurities ($F_A(Tl)$ or Tl°(1) centers), produced in several host lattices. Due to the large electron affinity of the Tl^+ impurity the F center electron is localized partly (~50%) at the Tl^+ site. The low-energy optical transitions of the centers in the IR produce a small Stokes shift and a nearly full quantum efficiency of the emission. Pumped by the 1.06 μm Nd:YAG laser line, $F_A(Tl)$ systems in the hosts KCl and KBr provide tunable laser operation over the 1.4 to 1.7 μm range [3]. In KCl cw output power levels in excess of 1 W have been achieved at ~5 W pump power. Laser crystal preparation of these systems requires coloration by e^--beam exposure and is thus restricted to facilities with a suitable e^--accelerator.
(3) F_2^+ centers attached to Li^+ or Na^+ cationic impurities ($(F_2^+)_A$ centers). These F_2^+-like systems retain the basic optical properties of the laser-active pure F_2^+ center (an electron shared by two <110> neighboring anion vacancies), i.e. they have essentially an (H_2^+-molecule-like) $1s\sigma_g \leftrightarrow 2p\sigma_u$ transition [4] in the infrared with small Stokes shift and full emission quantum efficiency. Color center lasers based on the pure F_2^+ center are not stable due to pump laser-induced center reorientations which lead to migration and aggregation into other defects. In the $(F_2^+)_A$ center systems the reorientations are still possible but are limited to sites near the "anchoring" cation impurity. Stable $(F_2^+)_A$ center laser operation has been realized in the 1.6 to 2.5 μm range [5], with output powers up to several hundred mW in one case, KCl:Li^+. The centers can be produced by simple additive coloration, however it is difficult to produce these centers in high concentrations in larger host ion lattices. Since $(F_2^+)_A$ centers are positively charged they require stable electron traps for their production. The latter are assumed to be provided by F_A centers [5]. In order to maintain a high concentration of $(F_2^+)_A$ centers during laser operation,continuous irradiation of the crystal with near UV light is necessary.

In order to extend the total tuning range of stable cw color center lasers to new wavelength regions, either new laser-active defects have to be developed or methods have to found which allow to produce existing laser-active defects in sufficiently high concentrations in new host lattices.

Recently Pinto et al. [6] discovered stable color center laser operation in additively colored NaCl crystals. They identified the new laser-active defect tentatively as an F_2^+ center attached to a neighboring K^+ impurity cation, thus forming an $(F_2^+)_A$ center. In this paper we show that the new defect is not related to alkali impurities but rather is formed in crystals doped with OH^- or O_2^- molecules. This leads to production of the centers in several hosts and to laser operation besides in NaCl so far also in KCl.

The center has F_2^+-like properties in terms of <110> symmetry, transition energies, Stokes shift and optical gain. It is likely to consist of F_2^+ centers attached in various configurations to an anionic impurity (O^{--}) replacing a halogen lattice ion. We will therefore refer to it as $(F_2^+)_H$ center.

II. CENTER FORMATION

Initial efforts to produce the new defects in K^+ doped NaCl were not successful. The crystal showed essentially the same coloration behavior as an undoped one. This observation led us to suspect that OH^- molecules, which are easily doped into alkali halides and therefore are likely candidates for accidental impurities play a major role in the center formation.

Indeed, when using an NaCl crystal which was free of K^+ impurites but doped intentionally with OH^- molecules, $(F_2^+)_H$ centers could be formed in high concentrations.

The $(F_2^+)_H$ center formation requires two main steps: additive coloration of the crystals in alkali metal vapor, and subsequent optically induced F center aggregation in the temperature range of anion vacancy mobility. By an added F band light exposure at low temperatures (77 K) F_2^+ center reorientations or configuration changes can be induced.

Absorption spectra (at 15 K) obtained after these production steps are shown in Fig. 2 for NaCl:OH^-. After additive coloration and quenching the dotted spectrum is obtained, corresponding mainly to F centers (strong absorption at 450 nm), F_2 centers (weaker absorption at 700 nm) and oxygen defects (band at 290 nm). The latter are formed together with U centers (negatively charged hydrogen anions with a characteristic IR local mode absorption at 562.5 cm^{-1}) by dissociation of the OH^- molecules and reaction with F centers during the additive coloration process. In addition to these defects a weak absorption band exists at 1060 nm, with an optical density depending on the duration of the quenching time. This band is attributed to the lowest energy transition of the $(F_2^+)_H$ center. After F light exposure in the temperature range of anion vacancy mobility, at 300 K, the most pronounced absorption change is a strong increase of the $(F_2^+)_H$ band (dashed spectrum). Low-temperature F light exposure produces a slight broadening and red shift (peak at 1090 nm) of the $(F_2^+)_H$ absorption (solid curve).

In order to determine which OH^- dissociation product (U center or oxygen defect) is responsible for the existence of the $(F_2^+)_H$ center, the coloration behavior of a NaCl crystal doped with Na_2O_2 was investigated. After crystal growth three absorption bands are obtained (at 460, 290 and 230 nm), shown in Fig. 3. The wavelength positions and relative strengths of these bands are nearly identical to the corresponding transition properties of O^{--}-vacancy

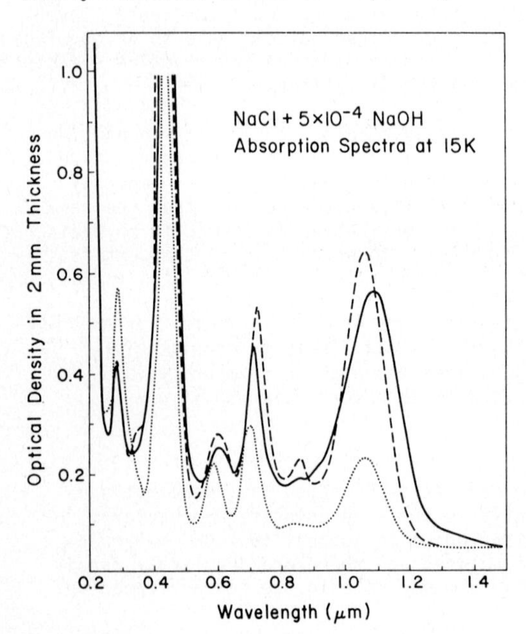

Fig. 2 Absorption spectra of NaCl:OH^-, measured at 77 K, showing absorptions after additive coloration (dotted curve), F center aggregation (dashed curve) and exposure to F band light at 77 K (solid curve).

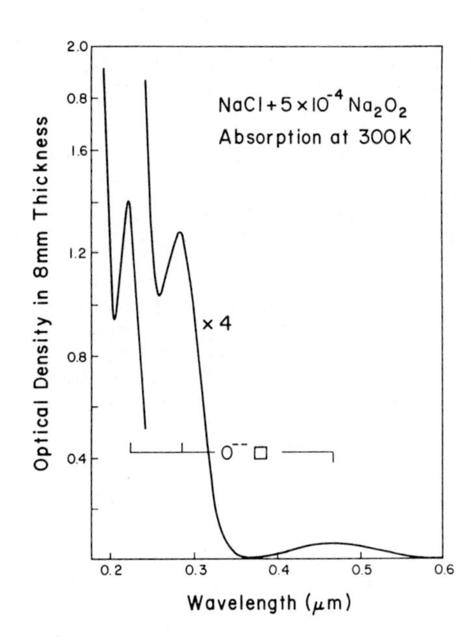

Fig. 3 Absorption spectra corresponding to transitions of oxygen-vacancy defect pairs (O^{--} - □) in Na_2O_2-doped NaCl, measured at 300 K after crystal growth.

defect pairs in the host KCl, investigated in detail by Fischer and co-workers [7]. Therefore we assign the absorption spectrum of Fig. 3 also to transitions from O^{--}-vacancy pairs.

When using the same center production steps for the oxygen-doped NaCl crystal as previously for the OH^- doped NaCl, the same absorption spectra were obtained as shown in Fig. 2. This behavior proves that $(F_2^+)_H$ centers are formed by the association of F centers and O^{--}-vacancy pairs.

Based on the identification of the defect constituents $(F_2^+)_H$ centers were produced besides in NaCl so far also in KCl, KBr and RbCl hosts. In all latter cases the crystals were doped with O_2^- molecules and the O^{--}-vacancy pairs, required for $(F_2^+)_H$ center formation, were produced during additive coloration according to following reducing reaction of F centers with O_2^- defects:

$$O_2^- + 3\text{ F centers} \rightarrow 2 \quad O^{--}\text{-vacancy pairs}$$

In Fig. 4 the absorption spectra corresponding to the main $(F_2^+)_H$ center production steps are shown for KCl and KBr containing optimized O_2^- doping-impurity levels. After F center aggregation (dashed curves) $(F_2^+)_H$ center bands are formed at 1.45 μm in KCl and at 1.53 μm in KBr. Subsequent F light irradiation leads to formation of F_2^+ centers in KCl (shift of the $(F_2^+)_H$ band to 1.38 μm, not shown in Fig. 4) while in KBr a small long wavelength shift of the band is observed (solid curve), similar to the NaCl case.

III. OPTICAL PROPERTIES

Excitation of the low-energy $(F_2^+)_H$ absorption transitions leads in all three host crystals to strong emissions with similar Stokes shifts, shown in Fig. 5. For each crystal a set of two slightly shifted transitions is obtained, corresponding to the absorption in the state of F center aggregation alone (dashed curve) or to the state of subsequent low-temperature (77 K) F light irradiation (solid curves). In KCl the shifted transitions obtained after

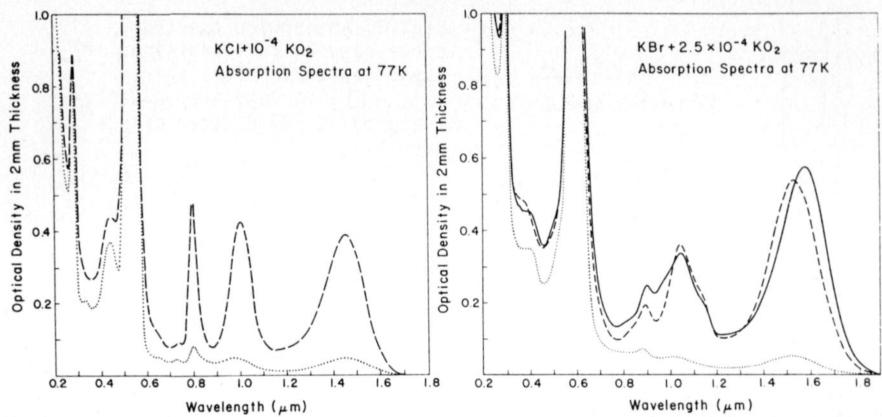

Fig. 4 Absorption spectra of O_2^- doped KCl and KBr measured at 77 K, showing absorptions after additive coloration (dotted curves), F center aggregation (dashed curves) and (for KBr) after exposure to F band light at 77 K (solid curve).

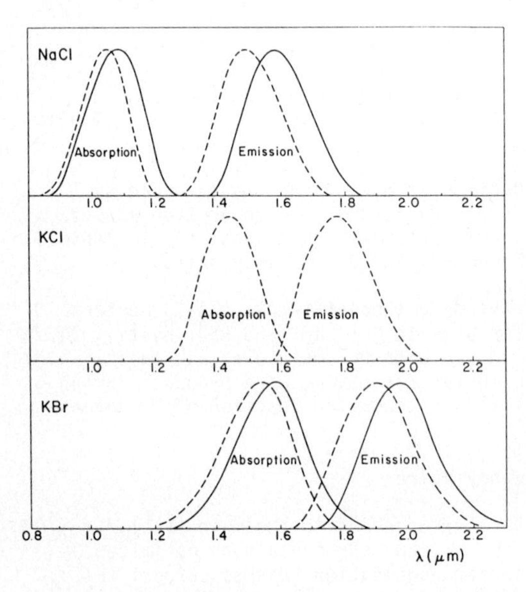

Fig. 5 Absorption and emission bands corresponding to the low energy optical transitions of $(F_2^+)_H$ centers in NaCl, KCl and KBr. For each crystal a set of two transitions is obtained, one set (dashed curves) obtained after F center aggregation and a second one (solid curve) obtained after subsequent low temperature F light exposure. In KCl the solid curves (not shown) coincide in wavelengths with the pure F_2^+ center transitions.

77 K F light exposure coincide with the pure F_2^+ center transitions; in NaCl and KBr the shifted transitions are positioned further away from the pure F_2^+ center transitions.

In all three hosts the wavelength shift obtained under 77 K light exposure can be reversed by heating of the crystals (in the dark) to 300 K.

Polarized excitation and emission measurements were performed for all $(F_2^+)_H$ systems in various geometries. They showed that the emission intensity analyzed parallel to a <110> polarized excitation of the low energy absorption is about twice as strong as the intensity obtained with crossed analyzer. This ratio is consistent with a dipole emission originating from the six possible orientations of a <110> symmetry defect. Besides the IR transitions of the $(F_2^+)_H$ center we identified in the hosts KCl and NaCl strong absorption transitions in the F band region, corresponding to transitions polarized perpendicular to the F_2^+ center axis.

The absorption spectra of Figs. 2 and 4 show clearly that the $(F_2^+)_H$ center is formed in all three hosts by F band light exposure in the temperature range of anion vacancy mobility. Since the crystals contain besides F centers large concentrations ($\sim 10^{17}$ cm^{-3}) of oxygen-vacancy defect pairs (identified by their characteristic UV absorption transitions) the latter must be the aggregation partners for the F centers. The close spectral proximity of the low-energy $(F_2^+)_H$ center transitions to those of the F_2^+ center, observed in all three hosts suggests that the aggregated F center and the charge-compensating empty anion vacancy (neighboring the O^{--} impurity) are combined on adjacent anion sites. In this way an F_2^+-like double well potential would be formed for the electron, distorted by the influence (electric field, ion size, ...) of the neighboring (double negatively charged) oxygen anion.

Depending on the lattice site of the associated anion vacancy with respect to the charge-compensating anion vacancy, four $(F_2^+)_H$ center configurations are possible in principle, as shown schematically in Fig. 6. Having different symmetries and O^{--} perturbations, different optical transition energies are expected for each configuration. Since the double negatively charged oxygen impurity should have a strong trapping efficiency for (positive) anion vacancies, the second F center vacancy will likely be trapped next to it. This would form an $(F_2^+)_H$ defect configuration, in which the oxygen impurity and the two vacancies are located at the corners of an equilateral triangle (C_{2V} symmetry, F center vacancy in lattice site 1 in Fig. 6). Assuming that in this configuration the perturbation of the O^{--} impurity on the optical transitions of the pure F_2^+ center is strongest, we tentatively assign it to the $(F_2^+)_H$ center transitions with largest shift from the pure F_2^+ center transitions.

Obviously the formation conditions for the various $(F_2^+)_H$ center configurations and the response of the latter to low-temperature F light exposure are both similar for NaCl and KBr but differ sharply for KCl.

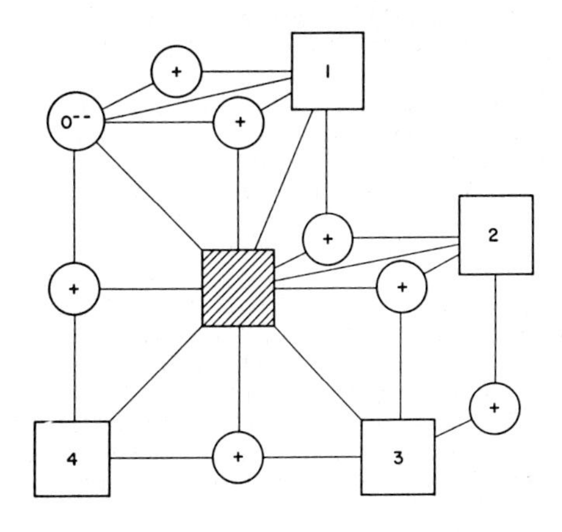

Fig. 6 Ionic configurations of the $(F_2^+)_H$ center obtained by association of an F center with an oxygen-vacancy defect pair. Four defect configurations with different symmetries are possible for closest (110) neighboring positions of the F center vacancy site (labelled 1 to 4) with respect to the charge-compensating vacancy (shown as hatched box).

In NaCl and KBr F center aggregation produces the $(F_2^+)_H$ center configurations with only slightly shifted transitions (with respect to the pure F_2^+ center). Inspite of having a high thermal stability these configurations are not stable under F light exposure. Instead, they are transformed, via F light-induced lattice site changes of the anion vacancies, into the $(F_2^+)_H$ configuration with large wavelength shift. Once in this configuration, the center has complete "optical" stability under further F light exposure. Lattice-site changes of the anion vacancies are still possible but they are limited to the immediate neighborhood of the anchoring O^{--} impurity and can lead only to reorientations of the same configuration.

In KCl the $(F_2^+)_H$ center configuration with large shift of the transitions is formed already during F center aggregation, indicating that it has a high thermal stability in this host. Surprisingly, its optical stability is very poor, however, as seen by its destruction under low temperature F light exposure. Apparently the binding effect of the O^{--} impurity on the neighboring F_2^+ center becomes very weak under high-energy optical excitation in this host and the F_2^+ center can migrate away from it in a sequence of reorientations.

IV. LASER EXPERIMENTS

Laser operation of $(F_2^+)_H$ centers has so far been realized with NaCl and KCl hosts. For each host the laser properties were investigated before and after F light exposure. In both cases stable laser performance was obtained with the long wavelength emission bands. This required F light exposure for NaCl while for KCl any F band excitation had to be avoided.

The NaCl $(F_2^+)_H$ center laser was found to oscillate, prior to F light exposure, simultaneously at 1.49 μm and 1.58 μm (using broad band reflectors), with ~90% of the total output on the 1.58 μm line. When adding the auxiliary F light exposure to the pump the laser operated stable and with strongly increased power only on the long wavelength transition.

The KCl $(F_2^+)_H$ center laser operated at 1.78 μm. When F light exposure was avoided the output power on this line was stable (following an initial 50% reduction) over a tested period of six hours. When exposing the crystal to F light (producing $(F_2^+)_H \rightarrow F_2^+$ conversion) the laser operated at the (pure) F_2^+ center emission peak near 1.70 μm and faded in a time period of ~30 min due to irreversible center destruction, well known for pure F_2^+ centers.

The laser performance of the crystals is summarized in terms of tuning behavior and dependence of output power versus input power in Figs. 7 and 8.

The NaCl $(F_2^+)_H$ center laser (Fig. 7) was tunable from 1.45 to 1.74 μm using a grating in Littrow configuration as tuning element. At 10 W pump power the output in the peak of the tuning curve at 1.6 μm was ~1 W continuous-wave. Using a 22% output coupling of the grating the laser was still undercoupled. When replacing the grating with a 40% output coupling mirror a maximum cw output power of ~1.3 W could be obtained. A still higher output power should be obtainable with an intermediate optimum output coupling of ~30%. The plot of output versus input power in Fig. 7 shows that the output power increases linearly with increasing pump power up to the highest available level of 10 W. No saturation or temperature effects were observed in this host. The measured slope efficiency is about 60% taking only the absorbed pump power into account. This is close to the limit of 66% expected for a center emission with full quantum efficienty and the given Stokes loss. In order to maintain high-power laser operation the crystal had to be irra-

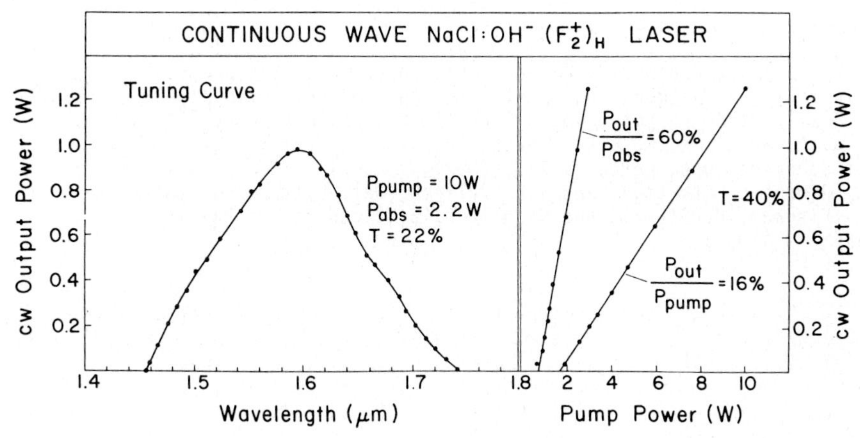

Fig. 7 Tuning curve and output versus input power characteristics of the NaCl $(F_2^+)_H$ center laser.

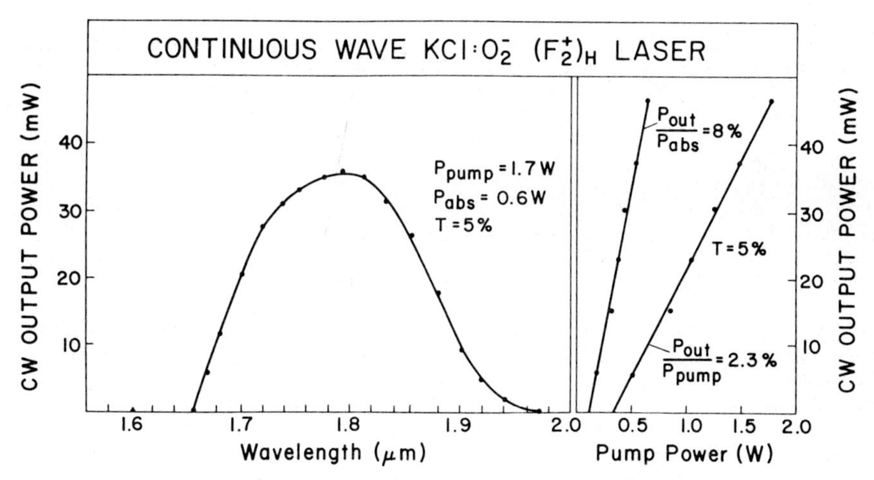

Fig. 8 Tuning curve and output versus input power characteristics of the KCl $(F_2^+)_H$ center laser.

diated with auxiliary green argon laser light exciting the higher energy transition of the center in the visible. Without the auxilliary light the output power dropped to a much reduced but still stable level of ~400 mW.

The KCl $(F_2^+)_H$ center laser (Fig. 8) tuned from 1.66 to 1.97 μm with about 40 mW peak output power at 1.8 μm using a prism and 10% output coupler. The output power increased linearly with increasing pump power up to 1.8 W, which was the upper pump power level at the time of the experiments. The best slope efficiency obtained so far with this not yet optimized system was 8%.

ACKNOWLEDGEMENTS
This work was supported by grants from the Deutsche Forschungsgemeinschaft and National Science Foundation (DMR 81-05332 and DMR 82-11857).

References

1. L.F. Mollenauer and D.H. Olson, Appl. Phys. Lett. 24, 386 (1976).
2. G. Litfin, R. Beigang and H. Welling, Appl. Phys. Lett. 31, 382 (1977).
3. W. Gellermann, F. Luty, and C.R. Pollock, Opt. Comm. 39, 391 (1981).
4. M.A. Aegerter and F. Luty, Phys. Status Solidi (b) 43, 245 (1971).
5. I. Schneider Opt. Lett. 7, 271 (1982).
6. J. Pinto, L. Stratton, and C.R. Pollock, Opt. Letters 10, 384 (1985).
7. F. Fischer, H. Gründig and R. Hilsch, Zeitschrift für Physik 189, 79 (1966).

Stable Color Center Laser in OH-Doped NaCl Operating in the 1.41–1.81 μm Region

J.F. Pinto, E. Georgiou, and C.R. Pollock

School of Electrical Engineering, and Material Science Center,
Cornell University, Ithaca, NY 14853

I. Introduction

Over the past year we have been investigating a new color center laser in NaCl [1] . We tentatively attribute the laser action to a new type of color center, namely the F_2^+ center stabilized beside an O^{--} ion. Optimum laser performance to date displays stable cw operation with tuning from 1.41 to 1.81 μm with a maximum cw output power of 1.25 W when pumped by a 9 W 1.06 μm Nd:YAG laser. The laser crystals can be stored for weeks at room temperature with no apparent degradation, however efficient operation requires that the crystal be cooled to 77 K during laser operation. The crystals are additively colored in a heat pipe, which partially explains the storage and operational stability of the center. In addition, modelocked pulses of 5 psec duration have been generated in a synchronously pumped arrangement, tunable from 1.47 to 1.73 μm with an average power of 450 mw at the peak of the tuning curve.

II. Laser Properties of the NaCl:OH Color Center Crystal

NaCl crystals doped with various amounts of OH^- were obtained from the Crystal Growing Facility at Cornell University and from the Crystal Growth Laboratory of the University of Utah. OH^- was incorporated into the crystal by addition of appropriate amounts of NaOH to the melt. Optimum concentrations of incorporated OH^- in the crystal are in the 10-70 ppm range as determined from FTIR measurements of the OH^- vibrational band at 3654 cm^{-1}. Higher concentrations were found to lead to the formation of (OH)-dimers in the lattice, with a reduction in laser performance.

Cleaved sections approximately 4 mm thick were additively colored in a heat pipe [2] at various pressures for 30 minutes. Twenty torr pressure at 700°C provided an optimum coloration density, corresponding to an F-center concentration of approximately 7×10^{17} cm^{-3}. Colored crystals could be stored for weeks at room temperature with no apparent degradation in their laser performance. No further annealing was required prior to laser use for properly colored crystals.

A colored crystal was typically sanded to 3 x 5 x 10 mm dimensions, and polished in normal room light with 0.05 μm polishing compound dampened with pure methanol. The polished crystal was mounted at Brewster's angle in the laser cavity and exposed *at room temperature* for one hour to 365 nm light obtained from an 100 watt Hg lamp filtered through a 10 cm cell filled with $CuSO_4$ solution. This exposure caused a photo-aggregation between color centers and defects, producing a broad absorption band centered at 1.04 μm with a 60% small signal absorption in a 3 mm thick crystal. Fig. 1 shows this absorption profile, with its associated emission band. Longer exposure times did not increase the magnitude of the absorption; in fact, the absorption band reached a saturation level after about 1 hour, and slowly began to decrease with prolonged exposure exceeding one hour. If the photoaggregation was done at reduced temperatures (<-10° C), the 1.04 μm band did not form to its full depth. Following this room-temperature photoaggregation, the crystal was cooled to 77° K and exposed again to 365 nm light. This second exposure caused the absorption band to shift to 1.09 μm within a matter of seconds, forming the laser active center. The absorption and emission bands for this new center are also shown in Fig. 1. This center proved to be the laser-active center.

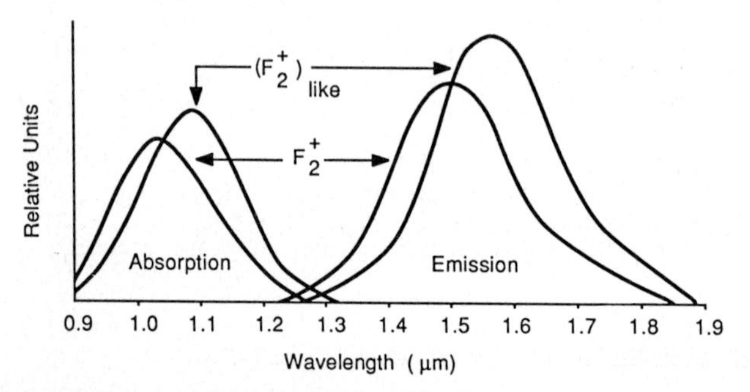

Fig.1. Relative absorption and emission bands of the laser-active color center.

The laser cavity used for the following data is typical of most color center lasers [1]. The crystal is placed at Brewster's angle at the 30 μm beam waist of the cavity. Tuning is accomplished with a single quartz or sapphire Brewster prism. Power tuning curves of the laser are shown in Fig. 2 for both 5% and 10% output coupling. The differences in tuning range for the two output couplers is in part due to the rapid decrease in the reflectivities of the couplers near the tuning extremes. The flat power profile of the 5% coupler indicates undercoupling of the laser. Output couplers with transmission values of 20 and 35% were also tested, however these led to lower output power and reduced tuning ranges. In almost all of the crystals tested, 10% output coupling appeared to be nearly optimum.

During lasing, 365 nm light from the filtered Hg lamp is focussed onto the crystal through a window on the front of the laser Dewar. The action of the UV light is not fully understood. Without auxilliary light exposure, the laser output

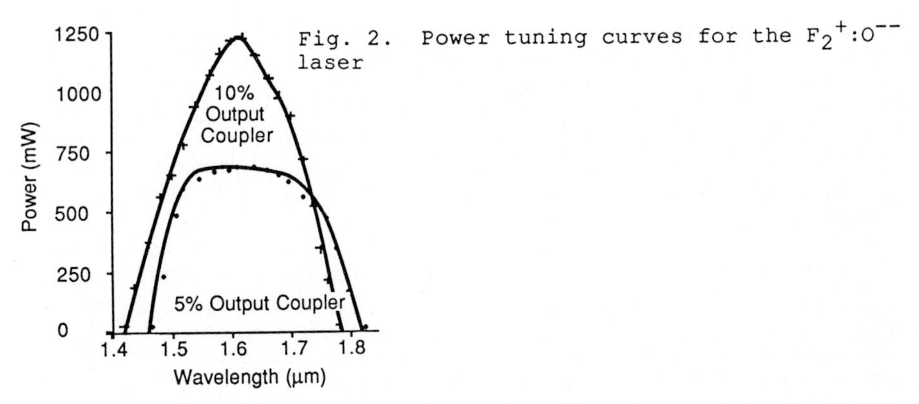

Fig. 2. Power tuning curves for the F_2^+:O^{--} laser

power is reduced to about 30% of full power. Output power increases logarithmically with UV exposure, which is similar to the behavior of the $(F_2^+)_A$ laser in KCl:Li [3].

Spontaneous emission lifetime measurements of this color center indicate a radiative decay time of approximately 150 nsec, corresponding to a gain cross-section of $\sigma = 8.5 \times 10^{-17}$ cm^2. The measured lifetime appears to be temperature independent for temperatures ranging from 85 °K to 330 °K, suggesting the absence of phonon-assisted nonradiative decay processes from the upper laser level. In view of this, it may prove possible to operate this laser at elevated temperatures. Unfortunately, as the temperature rises above 100° K, the absorption and emission bands become wider and shallower, resulting in poorer pumping efficiency and reduced gain. The cw laser power drops linearly with increasing temperature, until reaching threshold around 200° K.

The NaCl:OH^- laser can be modelocked using synchronous pumping techniques with a modelocked Nd:YAG laser. We have obtained stable pulse trains with pulse durations of 4.0-6.5 psec over the tuning range from 1.47 to 1.73 μm when using pump pulses of approximately 120 psec duration. A dispersing prism is required in the cavity to obtain modelocked pulses with no excess bandwidth. Average power at the peak of the tuning curve exceeds 450 mw. The pulse widths do not vary appreciably with tuning or pump power.

III. Center Identification

We believe that this laser active color center is a perturbed F_2^+ center: an F_2^+ center adjacent to an O^{--} ion. Polarized absorption spectra of the 1.09 μm band indicate that the color center defect is aligned along a <110> direction in the crystal and can be reorientated with UV light. As shown in Fig. 1, the absorption and emission bands of the center are shifted to slightly lower energy as compared to the NaCl F_2^+ center. This shift is consistent with the perturbation effect observed with alkali impurities on the spectra of the $(F_2^+)_A$ center [4]. However, chemical analysis of the laser-active crystals indicates that alkali impurities, specifically K or Li, have concentrations well below 2 ppm. Thus the active center could not be an $(F_2^+)_A$ center as was previously suggested [1]. It should be stressed that NaCl crystals which contained no OH^- did not form either the 1.04 or 1.09 μm absorption band upon coloration. Optical and thermal properties of the laser-active color center show a great similarity to other perturbed F_2^+ systems observed in OH-doped NaF (the $(F_2^+)^{**}$ center [5]) and OH-doped LiF [6].

There is evidence that the center consists of an F_2^+ center adjacent to an O^{2-} ion, which is substitutionally located at an anion location of the crystal. Excited state spectra taken of the 1.09 μm band show a spectrum which is similar in form to the measured and theoretical spectrum of the F_2^+ center [7]. A chopped cw Nd:YAG laser was used to pump the 1.09 μm band of a processed laser crystal mounted inside a spectrometer. The chopped pumping caused the populations of the ground and first excited state to vary in time, in-phase and out-of-phase with the pump modulation, respectively. Observation of the absorption spectrum using a lock-in amplifier at the chopping frequency revealed three ground-state transitions and five excited-state transitions. The observed transition energies are listed in Table 1 along with data taken for the F_2^+ center in pure NaCl [7]. The general form of the data indicates that the laser-active center is in fact a perturbed F_2^+ center.

The presence of OH^- in the NaCl samples was confirmed by observing the strong electronic transition of oxygen in the UV region. For uncolored NaCl:OH^- samples, with OH^- concentrations less than approximately 100 ppm, only the tail of an absorption peak at 185 nm was observed corresponding to the OH^- ion. Following additive coloration, the OH^- ion decomposed [8], creating UV absorptions at 190 nm and 228 nm which corresponded to U-centers and O^{2-}-vacancy dipoles, respectively [9].

The proposed F_2^+:O^{2-} model assumes that exposure of these crystals to UV light at room temperature causes individual F-centers to aggregate with O^{2-}-vacancy dipoles. Figure 3 shows a possible structure for the color center. Such a color

Table 1. Comparison of the excited-state transition energy between the F_2^+ and the perturbed F_2^+ center in NaCl

Transition	F_2^+	$(F_2^+)_{OH}$
3Ps	-	4.00 eV
4Ds	-	3.29
4Ss	-	2.95
2Pπ	2.92 eV	2.69
3Dπ	2.48	2.41
2Ss	2.01	1.95
3Ds	1.76	1.68
2Ps	1.20	1.14

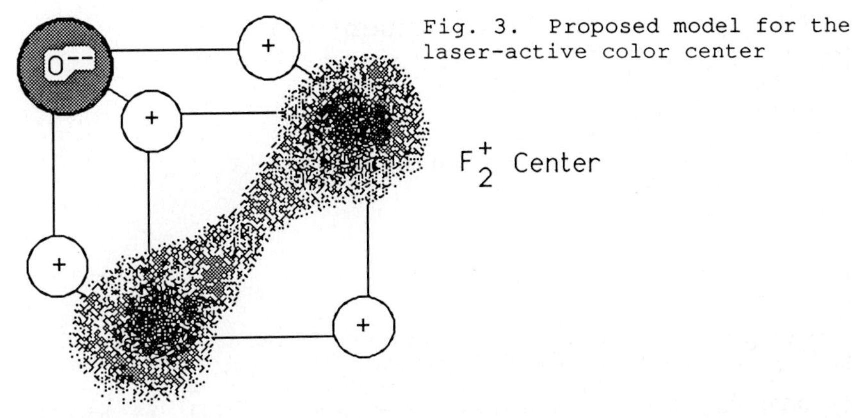

Fig. 3. Proposed model for the laser-active color center

center should be extremely stable, since the products of the OH^- dissociation form stable electron traps, which enhance the production of electron-deficient centers. Furthermore, the O^{2-} ion serves as a strong electro- negative point defect in the lattice which is probably capable of trapping the F_2^+ center beside it, thus preventing destructive migration effects.

IV. Conclusion

The NaCl:OH^- laser represents a significant improvement in output power and tuning range over presently available color center lasers in the 1.6 μm region. Perhaps the most important characteristic of this new color center laser is its dramatic stability as compared to other presently known alkali--halide-based color center lasers. Properly colored laser crystals can be stored at room temperature for extended periods in the dark (months) with no observable degradation of laser performance. Operationally, no fading of the output power has been observed after 200 hours of active use. Future work will concentrate on making a positive identification of this center and optimization of the laser performance.

Acknowledgements: We would like to acknowledge the help and advice of Gerhardt Schmidt concerning crystal growth. This work was partially supported by ITT and by NSF grant ECS-8352217.

1. J. F. Pinto, L. W. Stratton, and C. R. Pollock, Opt. Lett. 10, 384 (1985)
2. L. F. Mollenauer, Rev. Sci. Instrum. 49, 809 (1978)
3. I. Schneider and C. R. Pollock, Jour. App. Phys. 54, 6193 (1983)
4. I. Schneider, Opt. Lett. 6, 157 (1981)
5. L. F. Mollenauer, Opt. Lett. 6, 342 (1981)
6. V. M. Khulugurov and B. D. Lobanov, Pis'ma v ZhTF 4, 1471 (1978)
7. L. F. Mollenauer, Phys. Rev. Lett. 43, 1524 (1979)
8. S. P. Morato and F. Luty, Phys. Rev. B 22, 4980 (1980)
9. G. Gummer, Z. Phys. 215, 256 (1968)

Recent Progress in the Development of $(F_2{}^+)_A$ Color Center Lasers

D.R. Foster and I. Schneider*

Naval Research Laboratory, Washington, DC 20375, USA

We review our recent work on the development of color-center lasers based on lithium-$(F_2{}^+)_A$ centers in KI and RbI crystals. These laser materials have received considerable attention in recent years since they should provide broadly-tunable laser action at wavelengths longer than 3 μm [1,2]. We describe three significant advances that include: (1) the attainment of tunable laser action using electron-beam colored RbI, (2) the production of $(F_2{}^+)_A$ centers in stable, additively-colored KI:Li crystals and (3) the application of a two-step photoionization technique to achieve high concentrations of $(F_2{}^+)_A$ centers in both materials.

I. Laser Action in RbI:Li

The $(F_2{}^+)_A$ center is a stable aggregate consisting of an intrinsic $F_2{}^+$ center (an electron bound to two adjacent anion vacancies) and neighboring substitutional Li^+ impurity. A particularly important property of these centers in KI:Li and RbI:LI is that their broad emission bands peak at 3.0 μm and 3.2 μm, respectively [2], which are the longest-wavelength vibronic transitions for color centers. The $(F_2{}^+)_A$ KI:Li laser is tunable from 2.38 to 3.99 μm [1]. Since the center's emission in RbI:Li shows a larger Stokes shift than in KI:Li, tuning out to about 4.2 μm should be possible using a $(F_2{}^+)_A$ RbI:Li laser.

The RbI crystals being used in our current studies are grown with 1 mole% LiI and 100 ppm Pd^{2+} added to the melt. $(F_2{}^+)_A$ centers are produced in crystals initially colored at 77 K with 2 MeV electrons by a series of thermal and optical treatments similar to those used for the KI:Li laser [1]. The irradiated crystals are first annealed at 230 - 250 K to form F_2 , $(F_2)_A$, and $F_2{}^+$ centers, and are then exposed to 514.5 nm light from an Ar^+ laser for about 5 hr at 150 K to convert these defects to $(F_2{}^+)_A$ centers. The optical configuration of the laser is also similar to the KI:Li laser [1]. The crystal is located at the beam waist of a linear three-element cavity purged with nitrogen gas. A Littrow-mounted diffraction grating is used as the output coupler and tuning element. The crystal is pumped coaxially through the input mirror with the 1.73 μm line of an Er:YLF laser operating in a free-running mode.

Figure 1 shows that the crystal containing $(F_2{}^+)_A$ centers lases with a tunability that extends from 2.84 to 3.68 μm. The threshold energy and slope efficiency were 0.47 mJ/pulse and 1%, respectively. The tuning curve was measured with an input energy of 11 mJ/pulse while the crystal was exposed to an auxiliary NIR light, i.e., a tungsten source and Corning 7-56 filter. The auxiliary light increases the output energy by a factor of two which may be attributed to an increase in $(F_2{}^+)_A$ concentration by a two-step ionization process (see section III, below). As yet, the $(F_2{}^+)_A$ concentration is insufficient to attain the full tuning range of the RbI:Li laser.

---

* Sachs/Freeman Associates, Landover, Md. 20785

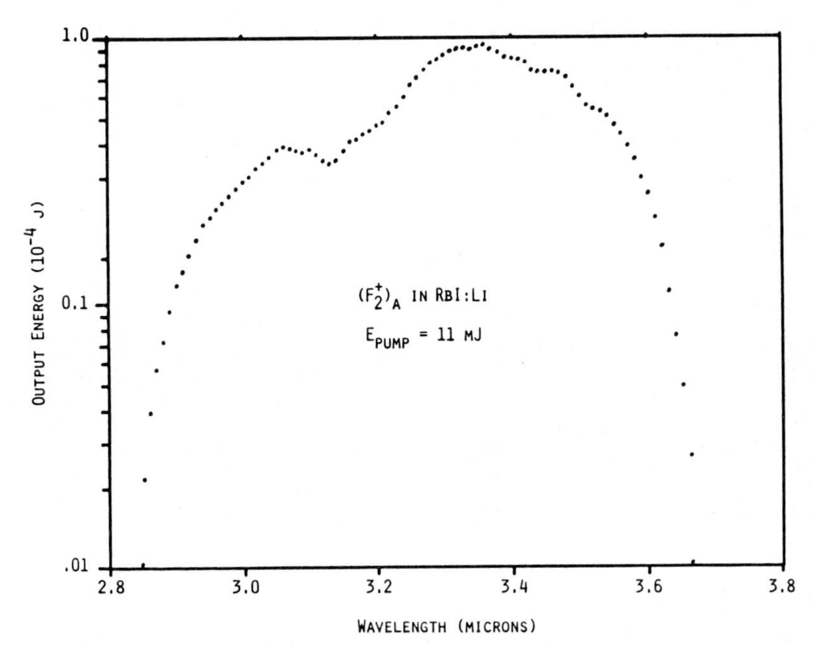

Fig. 1. Tuning curve for the $(F_2^+)_A$ RbI:Li laser.

II. Production of $(F_2^+)_A$ Centers in Additively-Colored KI:Li

Large concentrations of laser-active centers can often be produced in crystals colored with high-energy electrons, as observed for RbI:Li and KI:Li [2], but the crystals require constant cryogenic storage to avoid fading. On the other hand, the coloration in additively-colored crystals, such as KCl:Li [3], is more stable near room temperature. We can now produce $(F_2^+)_A$ centers in additively-colored KI:Li [4] with which a practical laser system may be constructed with tunability out to 4 μm.

The essential requirement for producing $(F_2^+)_A$ in additively-colored KI:Li is the elimination of nitrogen during each step of crystal preparation. We found that nitrogen diffuses into the KI lattice at elevated temperatures, and reacts with the lithium dopant. This reaction depletes the concentration of isolated Li^+ ions, thereby inhibiting $(F_2^+)_A$ center formation. Consequently, KI crystal boules were grown in an argon atmosphere doped with 1 mole% LiI in the melt. Samples cleaved from these boules were additively colored with potassium metal at 873 K for 8 hr in stainless-steel tubes which were sealed after being evacuated and then backfilled with argon. Before use, each colored crystal was annealed for several minutes at 848 K to disperse colloids that form during additive coloration. The dashed curve in Fig. 2 shows the absorption spectrum of a freshly annealed crystal containing mostly F and F_2 centers.

$(F_2^+)_A$ centers were produced by first converting F centers to F_2, F_A, and $(F_2)_A$ centers by exposing the crystal to blue-green light for 5 min at 240 - 250 K. F_2 centers were then converted to $(F_2)_A$ centers by exposing the crystal to the same light at 200 K (~1 hr) and at 150 K (~4 hr). The solid curve in Fig. 2 shows that the crystal now contains a strong $(F_2)_A$ absorption at 1.10 μm and a relatively weak F_2 absorption at 1.01 μm. F_2^+ and $(F_2^+)_A$ centers were then

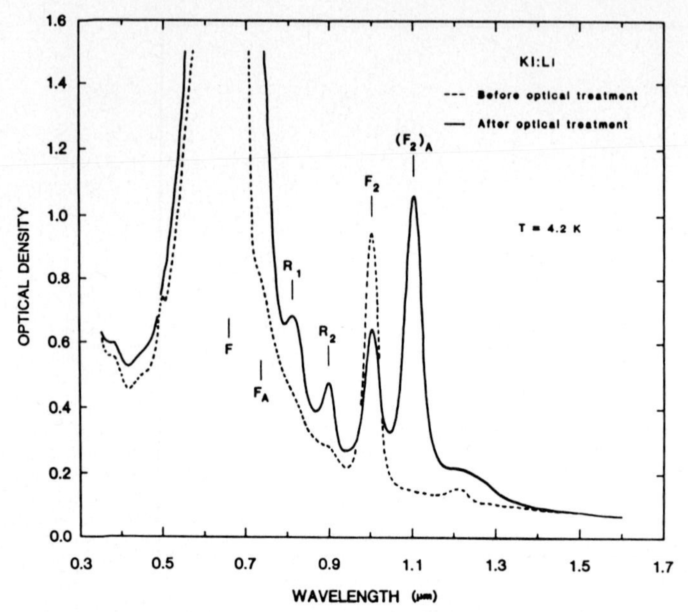

Fig. 2. Absorption spectra measured at 4 K of an additively-colored KI:Li crystal, 1 mm thick.

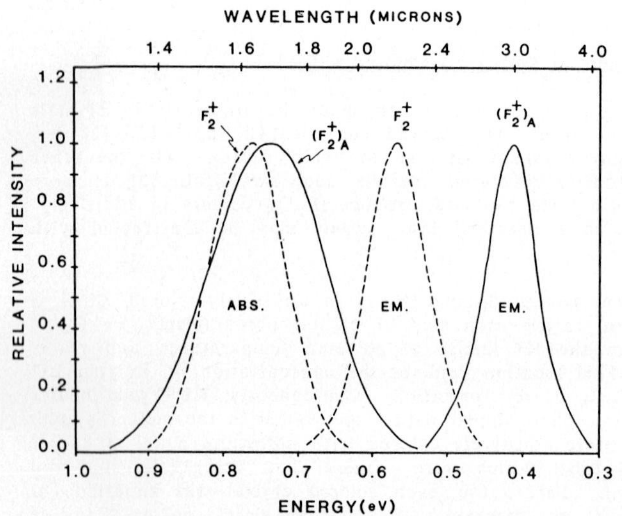

Fig. 3. Normalized absorption and emission spectra measured at 77 K of F_2^+ and lithium-$(F_2^+)_A$ centers in KI.

formed by ionizing F_2 and $(F_2)_A$ centers with blue-green light at 77 K. Absorption and emission spectra for these ionized centers are shown in Fig. 3. These spectra are similar to those obtained for crystals colored with high-energy electrons [2].

III. Two-Step Photoionization of $(F_2)_A$ Centers

In additively-colored crystals, $(F_2^+)_A$ centers are generated through a photodynamic equilibrium in which UV [5] or visible [4] excitation partially converts $(F_2)_A$ centers to $(F_2^+)_A$. This single-step process produces enough centers to sustain intense laser action in KCl:Li [3,6]. However, $(F_2)_A$ centers can not be completely converted to $(F_2^+)_A$ because the excitation also discharges electrons from other centers, such as the F, F_2, F^-, and F_A^- [5]. This constitutes an efficient reverse reaction where electrons transfer back to $(F_2^+)_A$ centers. Therefore, we are investigating other methods that can be used to increase the conversion to $(F_2^+)_A$ centers.

A particularly attractive technique was discovered by Mollenauer and Bloom [7]. They used a two-step photoionization process to achieve almost complete conversion of F_2 to F_2^+ centers in electron-beam-colored NaF and KF crystals doped with certain divalent-transition metal ions. We found that this two-step method produces a high concentration of $(F_2^+)_A$ centers in additively-colored KI:Li, and may explain the increased output of the $(F_2^+)_A$ RbI:Li laser described in section I. The two-step ionization involves the sequential absorption of two photons by the $(F_2)_A$ center. This process provides a highly selective way to ionize these centers since the energy required to excite the lowest-energy transition of the $(F_2)_A$ center is not enough to excite other centers, such as F centers which are a major source of free electrons.

In one experiment, the 1.06 µm line of a Q-switched Nd:YAG laser was used to ionize $(F_2)_A$ centers in an additively-colored KI:Li crystal. Neutral $(F_2)_A$ and F_2 centers were created by the procedure described in section II. F and F_A centers acted as intrinsic electron traps. The solid curve in Fig. 4 shows the absorption spectrum of a crystal containing $(F_2)_A$ and F_2 centers. Exposure of the crystal first to a 1.06 µm cw laser did not change its absorption spectrum. However, as shown by the dashed curve in Fig. 4, the pulsed 1.06 µm laser produced an absorption peaking at 1.65 µm, due mostly to $(F_2^+)_A$ centers, and

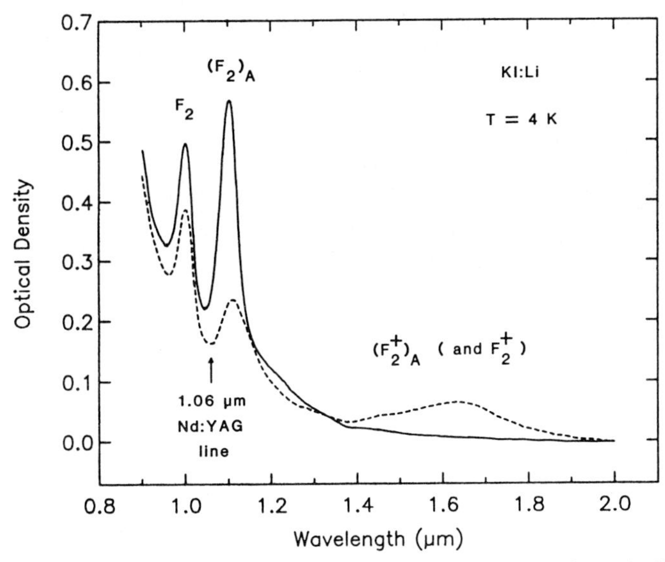

Fig. 4. Absorption spectra measured at 4 K before (solid curve) and after (dashed curve) a 10 min exposure to a 1.06 µm Q-switched Nd:YAG laser.

reduced the $(F_2)_A$ absorption by about 75%. Another exposure to the cw laser neutralized the charged centers (see solid curve in Fig. 4). By comparison, blue-green excitation produces about a 10-15% steady-state conversion to $(F_2^+)_A$ centers.

The absorption spectra in Fig. 4 clearly show that the two-step ionization by pulsed laser excitation produces a rather large conversion of $(F_2)_A$ to $(F_2^+)_A$ centers. Complete conversion, however, does not occur because F^- and F_A^- centers also absorb the 1.06 μm light-liberating electrons that transfer back to $(F_2^+)_A$ centers. The inability of the cw laser to ionize $(F_2)_A$ centers indicates that the two-step photoionization is able to effectively compete with this reverse reaction only at high laser intensities.

As we noted in section I, the use of conventional NIR light as an auxiliary source was quite effective in increasing the output of the $(F_2^+)_A$ RbI:LI. This approach was based on a hint provided by the work of Mollenauer and Bloom [7]. They suggested that conventional light sources could possibly produce a two-step ionization as long as (1) the light is filtered to avoid excitation of F centers and (2) there are a sufficient number of optically-stable electron traps. Accordingly, we used this technique to enhance the formation of laser-active $(F_2^+)_A$ centers in electron-beam-colored RbI:Li crystals co-doped with Pd^{2+}. In this case, the excitation light apparently does not discharge the Pd traps.

IV. Conclusion

We have reported three recent advances in the development of tunable infrared lasers using $(F_2^+)_A$ centers in KI:Li and RbI:Li. Preliminary results show that $(F_2^+)_A$ centers lase between 2.84 and 3.68 μm in electron-beam colored RbI:Li. With further improvements, we can possibly extend this range out to about 4.2 μm. Secondly, a procedure is now available to produce $(F_2^+)_A$ centers in additively-colored KI:Li. This stable material makes it possible to build a practical laser that is continuously tunable in the 3 to 4 μm region. Finally, a two-step photoionization technique is shown to be effective in generating a high concentration of $(F_2^+)_A$ centers in additively-colored KI:Li. This method was also applied to increase the output of a $(F_2^+)_A$ RbI:Li laser system.

REFERENCES

1. I. Schneider, Opt. Lett. 7, 271 (1982).
2. I. Schneider, Opt. Lett. 6, 157 (1981).
3. I. Schneider and C.L. Marquardt, Opt. Lett. 5, 214 (1980).
4. D.R. Foster and I. Schneider, Opt. Lett. 11, 213 (1986).
5. I. Schneider and C.R. Pollock, J. Appl. Phys. 54, 6193 (1983).
6. C.R. Pollock and D. Jennings, Appl. Phys. B28, 308 (1982).
7. L.F. Mollenauer and D.M. Bloom, Appl. Phys. Lett. 33, 506 (1978).

Electric Field Modulation of a Colour Centre Laser

G. Baldacchini[1], *U.M. Grassano*[2], *M. Meucci*[3], *P. Minguzzi*[3], *and M. Tonelli*[3]

[1]ENEA, TIB-FIS, Centro Ricerche Energia, C.P. 65, I-00044 Frascati, Italy
[2]Dipartimento di Fisica, Seconda Università di Roma, I-00173 Roma, Italy
[3]Dipartimento di Fisica, Piazza Torricelli 2, I-56100 Pisa, Italy

1. Introduction

Since the first discovery of the laser effect in F_A(II) colour centres of a KCl:Li crystal /1/ and since the first development of a tunable CW colour centre laser /2/, these sources have been the subject of intense investigation to improve their performances. Different crystals, various types of colour centres and different optical configurations have been studied with the purpose of extending the spectral coverage and increasing the operation reliability. Recently, in our laboratories, a new method has been designed to increase the efficiency of the coupling between the pump laser and the active medium /3/.

In this work we describe the first results of an experiment aimed to study the emission of a colour centre laser when an intense electric field is applied to the active crystal. We obtain indeed the first evidence for an electric field dependence of the output power.

The Stark effect of the F centres in KCl has been first observed in absorption by CHIAROTTI et al. /4/; the effect on the F_A centres has been studied by ROSENBERGER and LUTY /5/, who measured how the absorption coefficient depends on the field direction, on the direction and polarization of the incident light and on the temperature of the crystal. The effects on the luminescence of the F centres have been investigated by KUHNERT /6/ in KCl and by BOGAN and FITCHEN /7/ in several alkali halides. They observed an induced polarization of the emission parallel to the electric field, a red-shift of the bands, and in certain cases also a broadening. They analyzed these effects at different temperatures and for a wide range of field strengths. The vibronic theory of the Stark effect on the relaxed excited state of the F centre has been developed by HAM and GREVSMUHL /8/ , by IMANAKA et al. /9/, and recently by L. MARTINELLI et al. /10/. As far as we know, the effect of an electric field on the luminescence of F_A centres has never been studied.

2. Experimental Apparatus

Our experimental apparatus consists of a conventional KCl:Li F_A(II) colour centre laser /11/. It features an astigmatically compensated resonator consisting of a folded configuration with two concave mirrors (curvatures 8 and 5 cm); the whole structure is supported by a super-invar plate to improve the mechanical and thermal stability. The pump radiation is the 647 nm line of a Kr-ion laser (maximum power 2 watt) and it is injected into the infrared cavity through a coated ZnSe beam splitter. The crystal is placed at Brewster angle and is oriented with vertical <110> direction. The polarization of the pump beam can be chosen either parallel or perpendicular to this direction, the reflectivity of the beam-splitter for the pump light being 98% and 95% respectively. The perpendicular polarization yields a larger operation efficiency in our experimental configuration, as discussed in Ref. /3/. The crystal dimensions are 7x8 mm with a thickness of 2 mm, and the density of the $F_A(II)$ centres is about 4×10^{16} cm^{-3}. The laser can be operated either with a mirror output coupler (95% reflectivity) or in a frequency - selective configuration. In the latter case the dispersing element is a grating of 420 g/mm, blazed at 2.15 um which is mounted in a Littrow configuration and allows the tuning from 2.5 to 2.9 um. The output power is obtained from the zeroth-order of the grating.

To allow the application of the electric field we placed two metal meshes with a period of 1 mm on the crystal surfaces. The square holes between the wires are much wider than the waist of the infrared beam, thus the meshes do not appreciably affect the optical propagation. A schematic drawing of the crystal assembly is shown in Fig. 1; the electric field is along the direction <001>.

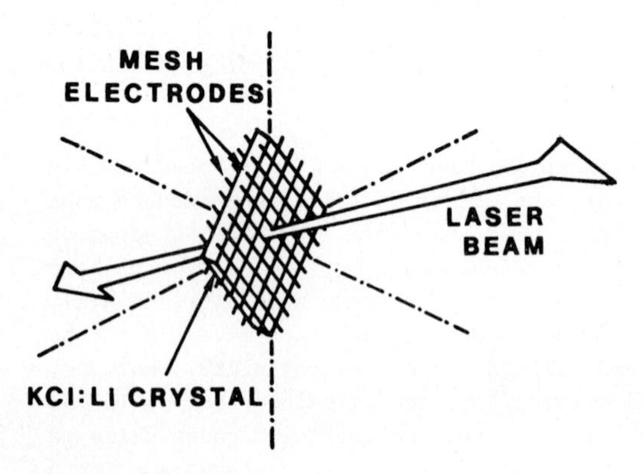

Fig. 1. Schematic diagram of the crystal assembly with Stark electrodes

A high-voltage square-wave modulation is generated by a simple amplifier employing a power tetrode in the final stage, and it is AC-coupled to the mesh electrodes. The maximum available intensity for the electric field is 7.5 kV/cm (zero-to-peak). The laser output is observed by a room-temperature PbS detector and the signal is processed by a lock-in amplifier tuned to twice the frequency of the square wave (1-2 kHz). In Fig. 2 we show a simplified diagram of the apparatus used for detecting the laser power-modulation.

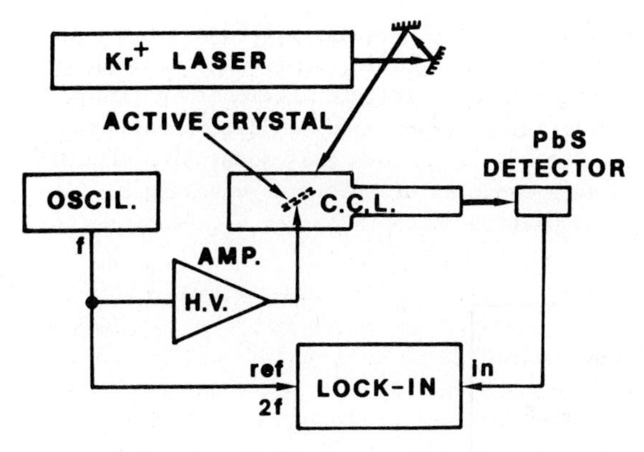

Fig. 2. Simplified diagram of the experimental apparatus

3. Results and Discussion

Figure 3 presents the results obtained at liquid nitrogen temperature, with a mirror output coupler and a perpendicular polarization of the pump: the power modulation increases quadratically with the applied electric field all over the investigated range. The maximum field in our measurements is presently limited only by the available power supply and not by possible crystal damage or electric breakdown. The arbitrary units for the vertical axis denote a scale which is obtained by dividing the modulation amplitude by the total output power, and they correspond approximately to parts per thousand.

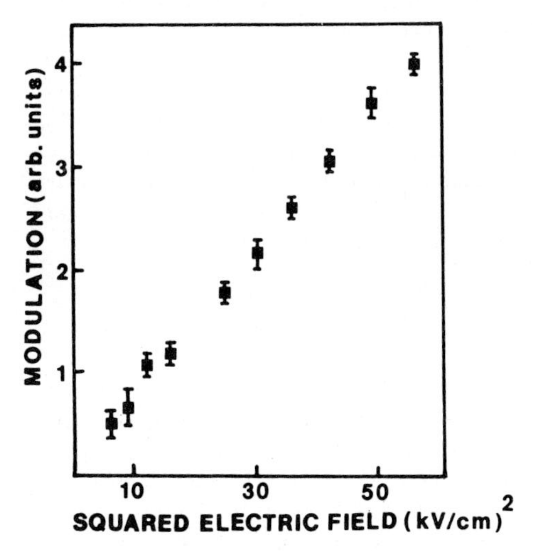

Fig. 3. Plot of the modulation amplitude as a function of the square of the field intensity

We repeated the measurements with a parallel pump polarization and we obtained a quite similar behaviour. With the grating replacing the output mirror we investigated the effect of the electric field at different wavelengths and we observed a quadratically-dependent modulation all over the tuning range of the laser. In Fig. 4 we show the modulation depth obtained at a field of 7.5 kV/cm as a function of the laser wavelength. In all cases we noticed that the modulation depth is larger if the laser is operated near the threshold.

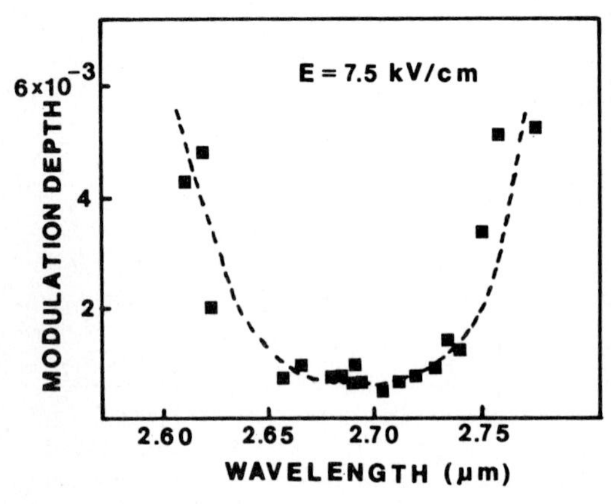

Fig. 4. Plot of the modulation depth as a function of the laser wavelength

From the studies of Ref. /6/ it appears desirable to cool the crystal at temperatures lower than 77 K since in this case the overall effects of the electric field may be stronger. Unfortunately there are indications that the mesh assembly slightly degrades the thermal coupling of the crystal to the cryostat, thus we expect that a device with improved cooling will produce a larger effect. The modulation signal is typically a few thousandths smaller than the total output level of the laser; however this order of magnitude should be considered as a rather rough estimate and a first achievement. Remarkable improvements are expected in the future by use of larger field strengths, lower crystal temperatures and, possibly, optimized techniques of electric field application.

At this stage of the investigation we believe that both the electric field dependence of the absorption coefficient and the polarization induced by the electric field on the F_A emission (analogous to that found in F centres) could be the origin of the modulation in the output power of the laser. However we cannot exclude an influence of the possible wavelength shift of the emission band. Experiments are in progress to clarify the nature of the amplitude modulation and to detect the existence, if any, of a frequency modulation. This work proves the feasibility of the electrical control of colour centre lasers and opens new possibilities both for the many applications of these devices and for the study of the fundamental physics of the centres in the crystals.

References

1. B. Fritz and Menke: Solid State Commun. 3, 61 (1965)
2. L.F. Mollenauer and D.H. Olson: Appl. Phys. Lett. 24, 386 (1974)
3. M. Meucci, M. Tonelli, G. Baldacchini, U.M. Grassano, A. Scacco, F. Somma: Opt. Commun. 51, 33 (1984)
4. G. Chiarotti, U.M. Grassano, R. Rosei: Phys. Rev. Lett. 17, 1043 (1966)
5. F.E. Rosenberger and F. Luty: Solid State Commun. 7, 983 (1969)
6. H. Kuhnert: Phys. Status Solidi 21, K171 (1967)
7. L.D. Bogan and D.B. Fitchen: Phys. Rev. B 1, 4122 (1970)
8. F.S. Ham and U. Grevsmuhl: Phys. Rev. B 8, 2945 (1973)
9. K. Imanaka, T. Iida, and H. Ohkura: J. Phys. Soc. Japan 43, 519 (1977)
10. L. Martinelli, G. Pastori Parravicini, and P.L. Soriani: Phys. Rev. B 32, 4106 (1985)
11. G. Baldacchini, G.P. Gallerano, D. Censi, M. Tonelli, P. Violino, U.M. Grassano, M. Meucci, and A. Scacco: Revue Phys. Appl. 18, 301 (1983)

Synthetic Diamond for Color Center Lasers

S.C. Rand

Hughes Research Laboratories, 3011 Malibu Canyon Road,
Malibu, CA 90265, USA

1. Introduction

Color centers which are substitutional in character, like many of the nitrogen-related defects in diamond, are well suited to the achievement of tunable laser action at room temperature. In particular, high densities of H3 centers permit efficient, stable generation of coherent visible radiation /1/ in the spectral range 500-600 nm. The optical, thermal and mechanical properties of diamond are also excellent, but the crystal is difficult and expensive to grow. Here we present experimental results which demonstrate that the prospects for a family of color center lasers spanning the visible and near-infrared in synthetic diamond are nevertheless very good.

2. Experimental Results

Most natural diamonds contain nitrogen, but it is rarely present as predominantly single substitutional atoms (Type Ib) which can serve as the precursors for a wide variety of color centers. Most synthetic diamonds on the other hand, do contain this form of nitrogen impurity. Moreover, single crystal, type Ib synthetic diamond plates of high optical quality are available commercially at low cost compared to other solid-state laser materials, as the result of continuing research in high-pressure, high-temperature synthesis techniques/2/. The light yellow coloration of these synthetics is due to the ultraviolet absorption band of the singly-substituted nitrogen impurities and nitrogen complexes in the covalent carbon matrix (Fig. 1a). In the infrared, the presence of high densities of A and B nitrogen aggregates can also be inferred from the absorption spectrum (Fig. 1b). The visible, laser-induced fluorescence spectrum shown in Fig. 2 is in fact dominated by emission from B aggregates associated with single vacancies (H4 centers) in the 500-600 nm region in "as grown" material. Since irradiation and annealing procedures produce high yields of several color centers from this type of diamond, we have made spectroscopic measurements to evaluate the laser potential of those which are simple in structure, thermally and photo-stable as well as easy to fabricate in this material. Type Ia diamond, which also contains useful precursors of H3 centers, but is both more difficult and more costly to produce as homogeneous, single crystal material, is not considered in this paper.

In Table I, calculated gain coefficients are given for a family of nitrogen-related centers which have been produced from Type Ib synthetic diamond. For comparison, spectroscopic parameters and gain coefficient of F_2^+ centers in KCl (at 77K) are also given. Fluorescent decays and emission spectra measured in this work were obtained using a Nd:YAG-pumped dye laser system with pulsewidth τ_p=7 ns and cw Ar^+ laser excitation respectively.

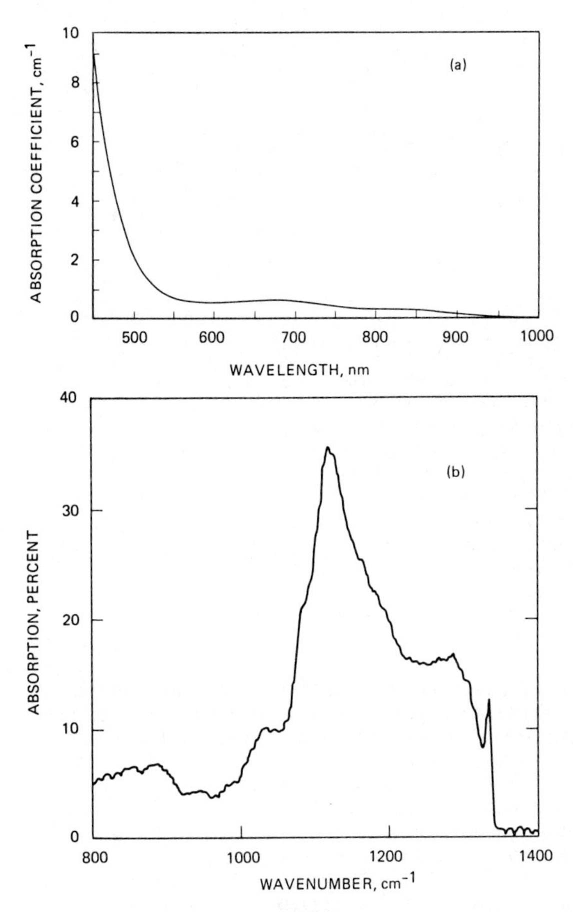

Fig. 1a. Absorption coefficient of Type Ib synthetic diamond host versus wavelength in the spectral regions covered by new candidate laser centers in diamond.
Fig. 1b. Absorption spectrum of Type Ib synthetic diamond showing a composite band near 7.8 μm due to single nitrogen as well as A and B aggregates. GE sample thickness was 2.5 mm.

Table 1. Comparison of diamond and F_2^+:KCl color-center gain coefficients γ for constant inversion density $N=10^{16}$ cm^{-3}.

Center	n	T(K)	τ_{f1}(ns)	η	$\Delta\nu(10^{13}s^{-1})$	$\lambda_0(\mu m)$	$\gamma(cm^1)$
F_2^*	1.4841[a]	77	$\tau_{f1}/\eta=80$[a]		1.69[a]	1.680[a]	3.52
GR1	2.3919[b]	295	~3	~0.95[h]	8.57[c]	0.835[c]	1.684
N-V	2.4065[b]	295	13[c]	~0.95[hc]	6.53[c]	0.697[c]	0.351
H3	2.4262[b]	295	16[g]	0.95[d]	5.27[g]	0.531[g]	0.201
H4	2.4250[b]	295	19[g]	~0.95[he]	3.76[g]	0.538[g]	0.245
N3	2.4468[b]	295	41[g]	0.29[f]	8.40[g]	0.445[g]	0.010

a Ref.4 c Ref.6 e Ref.8 g This work
b Ref.5 d Ref.7 f Ref.3 h Assumed value

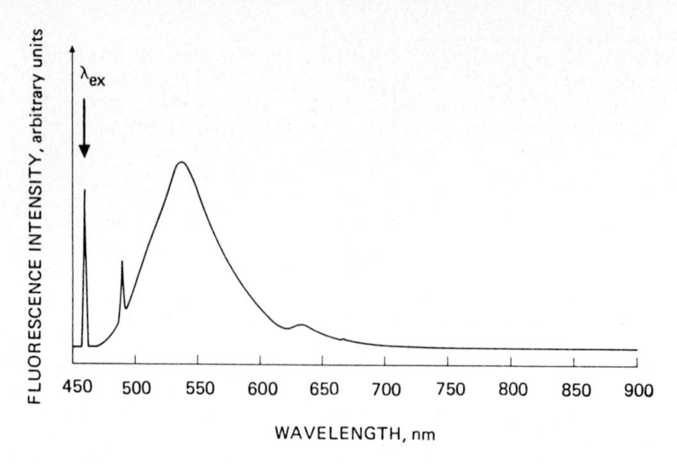

Fig. 2. Fluorescence spectrum of Type Ib synthetic diamond at T=295K resulting from Ar^+ laser excitation at λ_{ex}= 457.9 nm. The sharp peak at 494 nm is the zero phonon line of the H4 center and the broad peak its vibronic sideband. The small broad peak at 635 nm is probably due to the 594 nm zero phonon center.

3. Discussion

Before turning to a discussion of color center synthesis, we note on the basis of the estimated gain coefficients that the GR1, N-V and H4 centers compare very favorably with the H3 laser center in diamond. The GR1 has the highest calculated gain, but would be the most difficult to pump due to its short lifetime. The N-V center on the other hand exhibits a longer lifetime with somewhat reduced gain. An additional factor, possibly not taken into account in the table above, may be important in this case however. The N-V center is thought to have metastable triplet levels below the first singlet resonance/6/ which could reduce the quantum efficiency. The measured fluorescence lifetime of this center is independent of temperature however in the range 77-700 K /6/, suggesting that the decay is primarily radiative and that the intersystem crossing rate may be very low. The existence of slow cross-over should not significantly affect pulsed operation of an N-V laser, but could still strongly influence continuous-wave operation. We have previously used a double resonance technique /1/ to show that the high cross-over rate to a similar metastable state in the N3 center renders cw operation unlikely for that structure. Further study of the cross-over rates is clearly important to ascertain the full laser potential of several diamond color centers.

With the exception of the GR1 center, all the listings in Table 1 refer to aggregated defects which require migration of nitrogen atoms or vacancies or both within the diamond matrix. The high cohesive energy of carbon atoms in diamond and the high-pressure stabilizing environments used in many previous annealing experiments /9/ at first suggest that conditions for the formation of these defects could be extreme. Hence it is worth considering in detail whether high densities of the candidate centers can be prepared in an economical fashion in each case. Secondly, it is important to know whether secondary defects with undesired absorptions or energy coupling to the primary laser centers are formed during the preparation steps.

The GR1 center is the primary product of electron irradiation in diamond and is believed to be a neutral vacancy left by a single displaced carbon. As such, its recipe is the simplest and consists /10/ of room-temperature irradiation with 0.6-3 MeV electrons to a total dosage in the range of $10^{18}cm^{-2}$ for absorption coefficients of roughly 1 mm^{-1}. No strong absorptions are induced by this procedure in the GR1 emission range.

The N-V center is the next most complicated structure, in which a single substitutional nitrogen is associated with a vacancy. Earlier work /11/ in samples containing $4x10^{18}$ cm^{-3} isolated nitrogen atoms showed that N-V formation goes to completion during a two-hour vacuum anneal at 900C after 2 MeV electron irradiation to a dose of $5x10^{17}$ cm^{-2}. Again, absorption coefficients of approximately 1 mm^{-1} at the peak of the vibronic sideband result from this procedure. Emission intensity maximizes at 697 nm, which is in a region of low background absorption in the "as grown" crystal. It is important though to eliminate residual single vacancies, since these have strong absorption in the N-V emission region.

The H3 structure calls for two nitrogen atoms separated by an intervening vacancy. This center may be produced from Type Ib diamond without external pressure in three steps. First, single nitrogens are aggregated to form A precursor centers by annealing at 2300C for several hours /12/. This reaction goes to over 95% completion. Alternatively this step can be accomplished at the reduced temperature of 1500C if annealing is preceded by mild electron irradiation ($10^{16}cm^{-2}$, 2 MeV) as in Ref./13/. Next, the sample is heavily irradiated with 2 MeV electrons to a dosage of approximately $5x10^{18}$ cm^{-3} to ensure a high vacancy population. Finally, a vacuum anneal at 900C for two hours or more is used to cause migration of vacancies to A aggregates and formation of the final product. From Fig. 1a it might be expected that B aggregates could cause absorptive losses at the short wavelength end of the broad H3 emission region. However these B losses are greatly reduced during H3 center preparation by concurrent formation of H4 centers from the B aggregates, as we discuss next.

The H4 structure again requires only a few steps but is potentially more troublesome than H3 formation. In this case B precursor centers are initially formed from Type Ib diamond at a temperature of 2500 C /14/. It may also be possible to form B aggregates at lower temperatures after electron irradiation, just as COLLINS /13/ showed for A aggregates, but this has not been demonstrated to our knowledge. At the higher temperature, samples experience "platelet" formation /15/ which results in enhanced scattering and a general deterioration of crystal quality. Although re-irradiation and a low-temperature anneal at this stage furnish H4 centers in a manner similar to the H3 recipe above, the scattering problems suggests that an alternative procedure might be preferable here. The alternative is to begin with the sizeable B aggregate population already present in "as grown" synthetics (and evident from the absorption spectrum Fig. 1) and to perform a single irradiation followed by the low-temperature annealing step. Provided the population of B aggregates in the starting material is chosen to be high enough, this procedure achieves high H4 concentrations without unnecessary platelet formation and is in fact simpler than H3 preparation.

The N3 center exhibits a potential gain coefficient one order of magnitude lower than the other centers. In addition, non-radiative decay and re-absorption of light throughout the emission band from metastable levels /1/ makes this an unlikely laser candidate. Furthermore, we are not aware of any selective procedure for the preparation of high densities of this center at normal temperature and pressures. Hence the N3 is clearly a much poorer laser prospect than the other color centers considered here.

4. Conclusions

From the information in Table I, it is readily apparent that the characteristics of the H4 center are very similar to those of the H3. However, in terms of ease of synthesis and reduction of passive losses in Type Ib synthetic starting material, the H4 may provide an attractive alternative to the H3 laser in the same visible spectral region. More generally, the potential of the GR1, N-V, H3 and H4 members of the family of nitrogen-related centers in diamond for room-temperature, tunable lasers with emission bands spanning the visible and near-IR spectral regions appears to be good. More detailed spectroscopic studies should confirm this picture. In particular, four-wave mixing experiments with phase-correlated laser fields /16/, currently underway in this laboratory, are expected to make direct measurement of the cross-over rate in both N3 and N-V centers. Results of these studies will bear directly on the prospects for continuous-wave operation from new color centers in diamond.

The author wishes to thank R.G. DeVries and F.N. Mazandarany of the General Electric Company for the generous loan of a synthetic diamond sample as well as D.V. Manson and J. Shigley of the Gemological Institute of America who provided several natural diamond specimens. Other synthetic crystals (3.8mmx3.8mmx1.8mm) were purchased from Sumiden Carbide America, Inc. in Ohio.

References

1. S.C. Rand and L.G. DeShazer, Optics Letters 10, 481(1986).
2. H.T. Hall, U.S. patent No. 2,941,248, 1960; T. Kazuo, patent No. 4544540, 1985; for a recent review, see K. Nassau and J. Nassau, J. Cryst. Growth 46, 157(1979).
3. G. Davies, in The Properties of Diamond, ed. J.E. Field, Academic Press (1979), p. 165 and references therein.
4. L. Mollenauer, Methods of Experimental Physics (Academic Press, New York, 1979), Vol. 15B, Part 6.
5. F. Peter, Z. Phys. 15, 358(1923).
6. A.T. Collins, M.F. Thomaz and M.I. Jorge, J. Phys. C:Sol. St. Phys. 16, 2177(1983); C.D. Clark and C.A. Norris, J. Phys. C: Sol. St. Phys. 4, 2223(1971).
7. G. Davies, Diamond Research, 15-24(1977; M.D. Crossfield, G. Davies, A.T. Collins and E.C. Lightowlers, J. Phys. C7, 1909 (1974).
8. A.T. Collins, M.F. Thomaz and M.I. Jorge, J. Phys. C:Sol. St. Phys. 16, 5417(1983).
9. R.M. Chrenko, R.E. Tuft and H.M. Strong, Nature (London) 270, 141-144(1977).
10. H.B. Dyer and P. Fernando, Br. J. of Appl. Phys. 17, 419(1966).
11. G. Davies and M.F. Hamer, Proc. Roy. Soc (London) A348, 285-298(1976).
12. B.P. Allen and T. Evans, Proc. Roy. Soc. (London) A375, 93(1981); A.T. Collins, Inst. Physics Conf. Ser. No. 59, Defects and Radiation Effects in Semiconductors, Institute of Physics, Bristol, 247-252(1981).
13. A.T. Collins, J. Phys. C13, 2641(1980).
14. T. Evans and Zengdu Qi, Proc. Roy. Soc. A381, 159(1982).
15. L.A. Bursill and R.W. Glaisher, American Mineralogist 70, 608-618(1985).
16. S.C. Rand, Opt. Lett. 11, 135(1986).

Part VIII

Rare Earth Lasers

Laser Action of Cr, Nd, Tm, Ho-Doped Garnets

E.W. Duczynski, G. Huber, and P. Mitzscherlich

Institut für Angewandte Physik, Universität Hamburg, Jungiusstr. 11, D-2000 Hamburg 36, Fed. Rep. of Germany

We report on fundamental laser properties of Cr-doped as well as Cr-sensitized Nd,Tm,Ho-garnets as GSGG, GSAG, YAG, YSGG, and YSAG at following wavelengths: 740 to 840 nm (Cr), 1060 nm (Nd), 1862 nm (Tm), 1924 nm, 1944 nm, 2086 nm, 2091 nm, and 2095 nm (Ho).

1. Introduction

In Cr-doped garnets, with a low crystal field for the octahedral site, Cr^{3+} can be used as a tunable four-level laser ion [1,2] as well as an efficient sensitizer ion for Nd^{3+} [3]. Due to the broad band emission of Cr^{3+}, the sensitizer effect is not only restricted to the acceptor Nd^{3+}. For instance, Tm^{3+} can be pumped via Cr^{3+} with a quantum efficiency near unity. Even Cr^{3+} in YAG with a rather narrow band emission at a strong crystal field site can be used to sensitize Tm^{3+} because of the overlap between the Cr^{3+}-emission and Tm^{3+}-absorption in YAG.

2. Tunable Cr-lasers

In general, 3d-ion lasers possess low gain (effective cross-sections of the order of 10^{-20} cm^2) and relatively long fluorescence lifetimes τ of the order of 100 μs, thus showing large energy-storage capabilities and high possible peak power operation.

The incorporation of Sc and/or Ga in garnets increases the lattice constant, therefore decreases the crystal field, and leads to tunable Cr lasers in a number of Ga-garnets: $Y_3Ga_5O_{12}$(YGG), $Y_3(Ga,Sc)_2Ga_3O_{12}$(YSGG), $Gd_3(Sc,Ga)_2Ga_3O_{12}$(GSGG), $Gd_3Ga_5O_{12}$(GGG), $(La,Lu)_3(Lu,Ga)_2Ga_3O_{12}$(LLGG). Considering crystal growth aspects, Ga_2O_3 has a rather high vapour pressure compared to Al_2O_3 leading to vaporization of the melt. In this regard, Al-garnets are a more favourable composition. The weak crystal field can also be realized in Al-garnets YSAG and GSAG by substitution of the octahedrally coordinated Al by Sc [4]. However, Ga-garnets have the advantage of stable flat interface growth which yields crystals without a core.

At 4K, practically all garnets show inhomogeneous Cr^{3+}-fluorescence lines and nonexponential fluorescence decays due to slightly inequivalent Cr-sites. At 300 K, however, the electron-phonon coupling creates homogeneous lines and exponential fluorescence decays of the coupled metastable $^2E,^4T_2$-states [5]. In different garnets, the transition probabilities of the ^2E-$^4A_2$ transitions correlate with the energetic position of the $^4T_2$ level, which is admixed into the ^2E level via spin-orbit coupling and breaks the spin selection rule of the ^2E-$^4A_2$ transition [5].

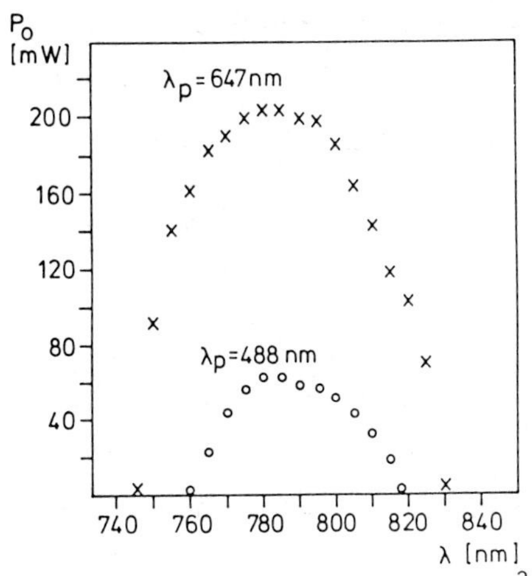

Fig. 1: Tunable output P_0 of the Cr^{3+}:GSGG laser under laser pumping at 488 nm and 647 nm with absorbed pump powers of 1.7 and 1.1 W, respectively [2].

Under laser pumping, power slope efficiencies up to 28 %, tunable cw output power of 200 mW, and a tuning range from 742 nm to 842 nm were achieved (see Fig. 1).

The smaller quantum efficiency of the 488 nm pump curve is attributed to the higher excited state absorption of the blue pump light in comparison to the red pump light. Such effects, as well as the creation of colour centers, have to be taken into account when Cr-garnets are operated with flash lamps [6].

3. Nd^{3+} sensitized by Cr^{3+}

Cross pumping of Nd^{3+} via Cr^{3+} is nearly as efficient as direct Nd^{3+} pumping in Cr,Nd:GSGG [3]. This is mainly due to an excellent spectral overlap of the broad band $^4T_2$-$^4A_2$ Cr-emission and the Nd^{3+} absorption. Dipole-dipole coupling is responsible for the nonradiative transfer. In spite of the statics of the Cr-Nd spacings, Cr-concentrations of $1 \cdot 10^{20}$ cm^{-3} and Nd-concentrations of $2 \cdot 10^{20}$ cm^{-3} insure an efficient dipole-dipole transfer with a quantum efficiency of 86 % [3]. Slightly different Cr- and Nd- concentrations may be used to match special geometries of the laser crystal and pump arrangement.

Under laser pumping Nd-laser slope efficiencies up to 41 % have been observed under Cr^{3+} cross pumping [3]. Consequently, broad band pumping of Cr,Nd:GSGG or Cr,Nd:GSAG is more efficient than broad band pumping of Nd:YAG (see Fig. 2).

The remaining fundamental problem of these crystals is the relative strong thermal lensing at high pump powers when compared to Nd:YAG. We emphasize, however, that thermal lensing must be compared under optimum conditions and identical output power (and not input power) for the different materials. The problems of additional impurities (such as Ca^{2+} in Sc_2O_3, which create Cr-coordinated infrared absorptions) are not of fundamental nature and can be solved (see Fig. 3).

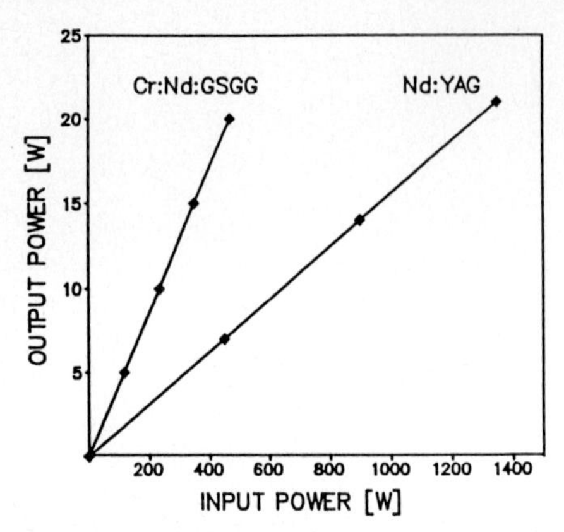

Fig. 2: Comparison of the efficiencies of Cr,Nd:GSGG and Nd:YAG at average output powers up to 20 W [7].

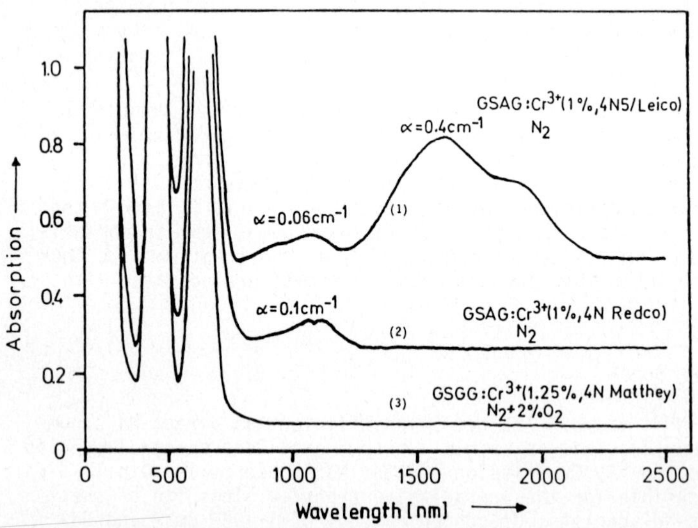

Fig. 3: Additional Cr-impurity center absorptions in Cr-doped garnets in the spectral region 800 nm to 2200 nm. Curves (1) to (3) differ mainly in the sources of the starting materials. Curve (3) represents an optimum laser grade Cr-garnet or Cr,Nd-garnet crystal. The spectral region below 750 nm shows the characteristic Cr^{3+}-bands.

4. Tm^{3+} sensitized by Cr^{3+}

In addition to Nd^{3+}, Er^{3+}, Tm^{3+}, and Ho^{3+} can be sensitized by Cr^{3+} [8,9]. The Cr-Ho and Cr-Tm interactions are also of dipole-dipole type [10] with the microscopic interaction parameters:

$$C_{DA}\ (Cr^{3+} \rightarrow Tm^{3+}) = 4.2 \times 10^{-39}\ cm^6 s^{-1}$$

$$C_{DA}\ (Cr^{3+} \rightarrow Ho^{3+}) = 2.8 \times 10^{-40}\ cm^6 s^{-1}$$

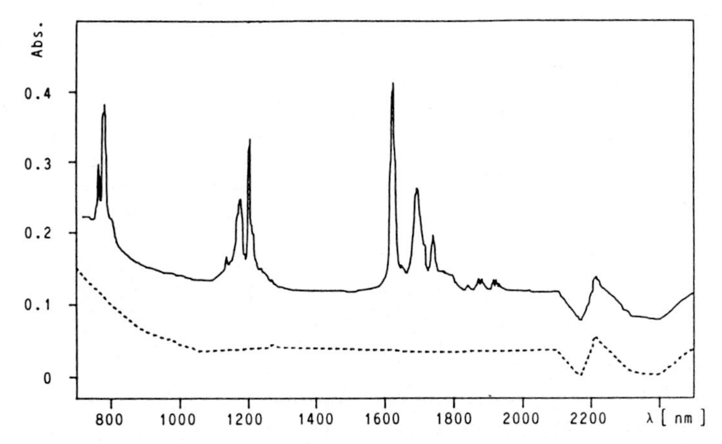

Fig. 4: Absorption spectrum of Cr,Tm,Ho:YSAG with $n(Cr)=2.5\cdot10^{20}$ cm^{-3}, $n(Tm)=8\cdot10^{20}$ cm^{-3}, $n(Ho)=5\cdot10^{19}$ cm^{-3}, and crystal length l = 2.1 mm. The dashed curve indicates zero absorption.

With these values direct $Cr^{3+}\rightarrow Tm^{3+}$ and $Cr^{3+}\rightarrow Ho^{3+}$ energy transfer at fast rates and high quantum efficiencies can be realized. Suitable concentrations for these processes in garnets are n(Tm) = $4\cdot10^{20}$ cm^{-3} and n(Ho) = $8\cdot10^{20}$ cm^{-3}. Figure 4 shows an absorption spectrum of Cr, Tm, Ho:YSAG.

The pump mechanism which was used to invert the Tm^{3+} level system is shown in Fig. 5. An efficient resonant $Cr^{3+}\rightarrow Tm^{3+}$ transfer system is realized by the excellent overlap between the $Cr^{3+}(^4T_2\rightarrow{}^4A_2)$ emission and the $Tm^{3+}(^3H_6\rightarrow{}^3F_4)$ absorption in the spectral region from 700 nm to 800 nm. At relative high Tm^{3+} concentrations ($4\cdot10^{20}$ cm^{-3}), the Tm^{3+} ions convert the energy down to the IR-region with a quantum efficiency of nearly 2. This is realized by the cross-relaxation process $Tm^{3+}(^3F_4\rightarrow{}^3H_4)$, $Tm^{3+}(^3H_6\rightarrow{}^3H_4)$ between adjacent Tm^{3+} ions [11]. This process is very important for a laser in the infrared region working at room temperature, because the thermal power dissipation inside the crystal is drastically decreased.

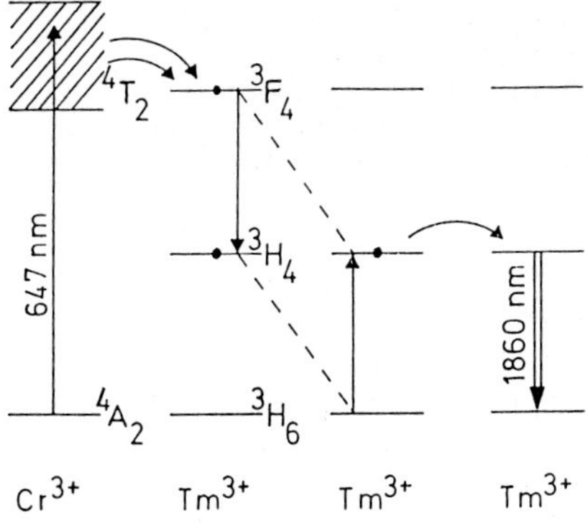

Fig. 5: Pump scheme (double cross-pumping) of the 1.86 μm Tm^{3+} laser with Cr→Tm transfer, Tm-Tm cross-relaxation (down conversion), and Tm-Tm energy migration.

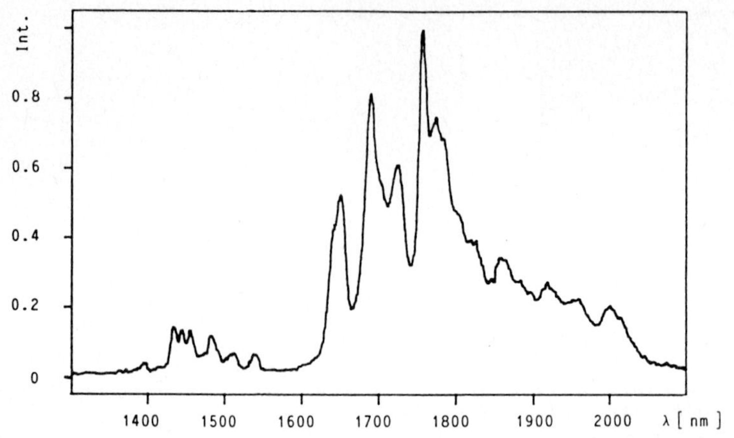

Fig. 6: Fluorescence spectrum of Cr,Tm:YSGG between 1.3 and 2.1 μm ($n(Cr)=2.5\cdot10^{20}$ cm^{-3}, $n(Tm)=4\cdot10^{20}$ cm^{-3}).

The crystals tested for fluorescence and lasing were grown from the melt with a standard Czochralski technique. The yttrium scandium gallium garnet (YSGG) has been doped with $Cr^{3+}(2.5\cdot10^{20}$ $cm^{-3})$ and $Tm^{3+}(4\cdot10^{20}$ $cm^{-3})$. Scandium was chosen to create the low crystal field strength for the $^4T_2$-$^4A_2$ transition of Cr^{3+}. Scandium and yttrium match well for the ionic radii of Cr^{3+} and Tm^{3+}, so that reasonable values of the distribution coefficients (not far from unity) can be expected.

Figure 6 shows the fluorescence spectrum of Cr,Tm:YSGG when pumped with the 647.1 nm line of a krypton laser, which excites the $Cr^{3+}(^4T_2)$ level. The coupled lifetime of the $Cr^{3+}(^2E,^4T_2)$ level is decreased from 140 μs of Cr:YSGG to the μs range in Cr,Tm:YSGG. The lifetime of the $Tm^{3+}(^3H_4)$ level is about 10 ms. This value proves a high storage capacity of the material.

The crystals (l = 1.1 mm) were placed into a concentric cavity formed by two 5 cm radius mirrors with reflectivities of more than 99.7 % at 1.86 μm. The crystals were pumped longitudinally with a krypton laser at 647.1 nm. True cw laser operation was obtained at 300 K. A threshold less than 40 mW of absorbed pump power was observed. Figure 7 shows a first input/output curve of Cr,Tm:YSGG under quasi-cw operation which leads typically to better results of about a factor of 2 than real cw operation. The pulse length was 10 ms at a duty cycle of 1:1. The power slope efficiency is 0.8 % and the laser wavelength is 1862 μm.

5. Ho^{3+} sensitized by Tm^{3+} and Cr^{3+}

A room-temperature laser around 2 μm in garnet crystals can be realized by sensitizing Ho^{3+} with Cr^{3+} and Tm^{3+} [12,13,14]. The pump mechanism which was used to invert the Ho^{3+} level system is shown in Fig. 8.

The mechanism to transfer the energy to the Tm^{3+} system is the same as for the Tm^{3+} laser at 1.8 μm. At high Tm^{3+} concentrations ($8\cdot10^{20}$ cm^{-3}) the Tm^{3+} ions converts the quantum energy down to the infrared region with a quantum efficiency of nearly 2. In order to operate the 2 μm Ho^{3+} at 300 K the Ho^{3+} concentration must be chosen low ($5\cdot10^{19}$ cm^{-3}) to avoid too big reabsorption losses.

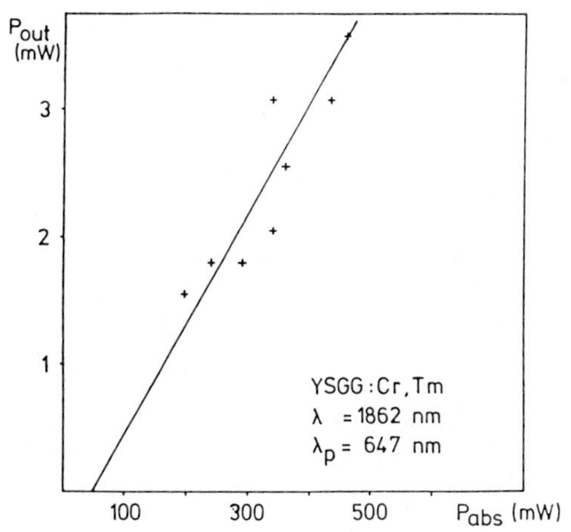

Fig. 7: Quasi-cw output power versus absorbed pump power in Cr,Tm:YSGG at a duty cycle of 1:1 and a pulse length of 10 ms.

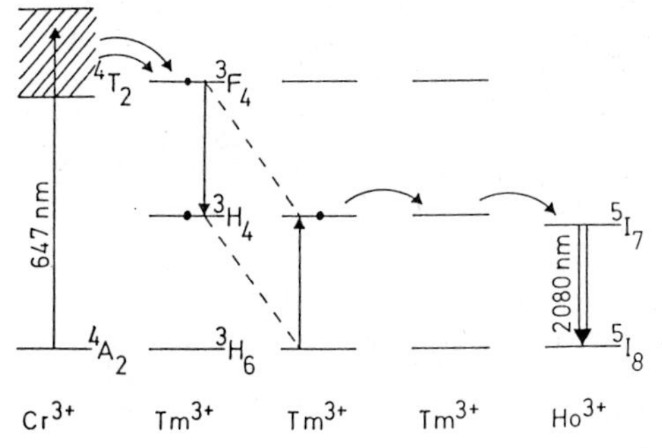

Fig. 8: Pump scheme (double cross-pumping) of the 2 μm Ho^{3+} laser with Cr→Tm transfer, Tm-Tm cross-relaxation (down conversion), Tm-Tm energy migration, and Tm→Ho energy transfer.

The crystals tested for fluorescence and lasing were grown from the melt with a standard Czochralski technique, too. All three garnets YSGG, YSAG, and YAG have been doped with Cr^{3+}($2.5 \cdot 10^{20}$ cm^{-3}), Tm^{3+}($8 \cdot 10^{20}$ cm^{-3}), and Ho^{3+}($5 \cdot 10^{19}$ cm^{-3}). Because also Ho^{3+} matches well for the ionic radius of yttrium all distribution coefficients are not far from unity.

Figure 9-11 show the fluorescence spectra of Cr,Tm,Ho:YSGG, YSAG, and YAG when pumped with the 647.1 nm line of a krypton laser. The major part of the fluorescence is channeled into the $^5I_7$-$^5I_8$ transition of Ho^{3+} in the spectral region between 1850 and 2100 nm.

Due to the low Ho^{3+} concentration the resonant reabsorption losses for the Ho($^5I_7$-$^5I_8$)-laser transition to the ground-state multiplet $^5I_8$ are below 0.1 cm^{-1} (see absorption spectrum of Fig. 4). We can roughly estimate an overall efficiency of more than 50 % for the Cr-Tm-Ho energy transfer. The lifetime of the upper Ho^{3+}($^5I_7$) laser level of approximately 8 ms proves the high storage capacity of the materials.

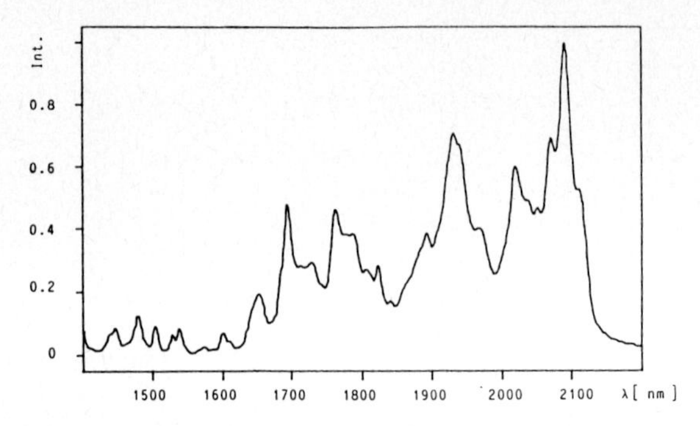

Fig. 9: Fluorescence spectrum of Cr,Tm,Ho:YSGG between 1.4 and 2.2 μm at concentrations $n(Cr)=2.5\cdot10^{20}$ cm^{-3}, $n(Tm)=8\cdot10^{20}$ cm^{-3}, $n(Ho)=5\cdot10^{19}$ cm^{-3}.

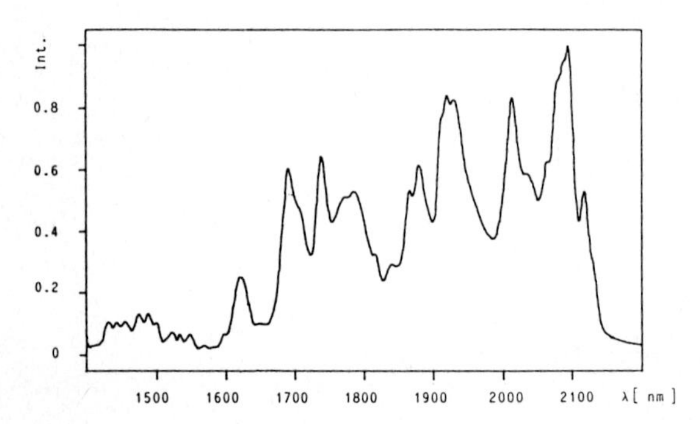

Fig. 10: Fluorescence spectrum of Cr,Tm,Ho:YSAG between 1.4 and 2.2 μm at concentrations $n(Cr)=2.5\cdot10^{20}$ cm^{-3}, $n(Tm)=8\cdot10^{20}$ cm^{-3}, $n(Ho)=5\cdot10^{19}$ cm^{-3}.

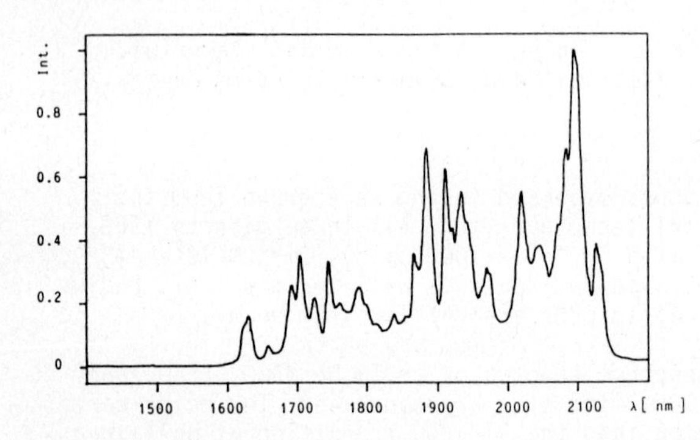

Fig. 11: Fluorescence spectrum of Cr,Tm,Ho:YAG between 1.4 and 2.2 μm at concentrations $n(Cr)=2.5\cdot10^{20}$ cm^{-3}, $n(Tm)=8\cdot10^{20}$ cm^{-3}, $n(Ho)=5\cdot10^{19}$ cm^{-3}.

The crystals were placed into a concentric cavity formed by two 5 cm radius mirrors having reflectivities of 98 % at 2.09 μm and 99.5 % at 1.94 μm. The crystals were pumped longitudinally via the $Cr^{3+}(^4T_2)$ absorption with a krypton laser at 647.1 nm. For all three crystals Cr,Tm,Ho:YSGG, YSAG, and YAG true cw laser operation was obtained at 300 K. Thresholds of less than 25 mW absorbed pump power were observed.

Figure 12 shows a typical input/output curve for Cr,Tm,Ho:YSGG under quasi-cw operation. The pulse length was 9 ms at a duty cycle of 1:1. The power-slope efficiency of 13 % demonstrates the efficient pumping mechanism described above.

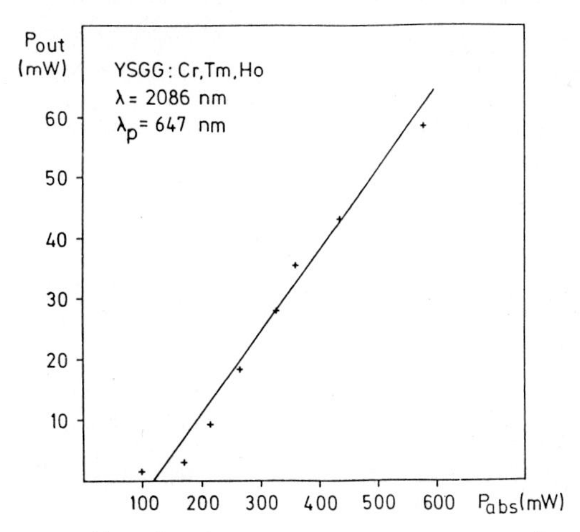

Fig. 12: Quasi-cw output power versus absorbed pump power in Cr,Tm,Ho:YSGG at a duty cycle of 1:1 and a pulse length of 9 ms.

The observed cw laser wavelengths at 300 K of the $Ho^{3+}(^5I_7 \rightarrow ^5I_8)$ transitions in YSGG, YSAG, and YAG are:

Cr,Tm,Ho:YAG	2091 nm
Cr,Tm,Ho:YSAG	2095 nm 1944 nm
Cr,Tm,Ho:YSGG	2086 nm 1924 nm

It seems to be possible to get the Cr,Tm,Ho:YAG system cw lasing at 300K around 1.9 μm, too. Also, the experiments point out that it should be possible to achieve room-temperature tunable cw laser action between 1.9 and 2.1 μm in Cr,Tm,Ho:YSGG, YSAG and YAG.

6. Conclusion

In conclusion, we gave a review of the basic aspects of tunable Cr-garnet lasers and Cr-sensitized Nd,Tm,Ho-garnet lasers. The Cr-doped garnets are useful tunable lasers in the spectral region between 740 nm to 840 nm. The Cr-sensitized Nd-garnets are superior to Nd:YAG with respect to efficiency. The Tm- and Ho-lasers sensitized by Cr operate cw at 300 K at wavelengths around 1.8 μm and 2 μm and have interesting medical applications.

Acknowledgements We greatly acknowledge the support from the Bundesforschungsministerium für Forschung und Technologie (BMFT) of the Fed. Rep. of Germany.

References

[1] B. Struve, G. Huber, V.V. Laptev, I.A. Shcherbakov, E.V. Zharikov, Appl. Phys. B30, 117 (1983)
[2] B. Struve, G. Huber, J. Appl. Phys. 57, (1), 45 (1985)
[3] D. Pruss, G. Huber, A. Beimowski, V.V. Laptev, I.A. Shcherbakov, E.V. Zharikov, Appl. Phys. B28, 355 (1982)
[4] J. Drube, B. Struve, G. Huber, Optics Communications 50, No. 1, 45 (1984)
[5] B. Struve, G. Huber, Appl. Phys. B36, 195 (1985)
[6] J. Drube, G. Huber, D. Mateika, CLEO'86, paper ThB4, San Francisco, USA (1986)
[7] G. Litfin, W. Luhs, B. Struve, P. Fuhrberg, private communication, Spindler u. Hoyer (1986)
[8] L.F. Johnson, J.E. Geusic, and L.G. Van Uitert, Appl. Phys. Lett. 7, 127 (1965)
[9] E.V. Zharikov, V.V. Osiko, A.M. Prokhorov, and I.A. Shcherbakov, Izv. Akad. Nauk Ser. Physics 48, No. 7, 1330 (1984)
[10] E.V. Zharikov, S.P. Kalitin, V.V. Laptev, V.G. Ostroumov, Z.S.Saidov, V.A. Smirnov, and I.A. Shcherbakov, Sov. J. Quantum Electron. 16(1), 145 (1986)
[11] B.M. Antipenko, J. Tech. Phys. 54, Nr. 2, 385 (1984)
[12] D.P. Devor, and B.H. Soffer, IEEE J. Quantum Electron. QE-8,231(1972)
[13] E.P. Chicklis, C.S. Naiman, R.C. Folweiler, and J.C. Doherty, IEEE J. Quantum Electron. QE-8, 225 (1972)
[14] E.W. Duczynski, G. Huber, V.G. Ostroumov, and I.A. Shcherbakov, Appl. Phys. Lett. 48 (23), 1562 (1986)

Diode-Pumped 2 μm Holmium Laser

L. Esterowitz[1], *R. Allen*[1], *L. Goldberg*[1], *J.F. Weller*[1], *M. Storm*[2], *and I. Abella*[3]

[1]Naval Research Laboratory, Washington, DC 20375, USA
[2]Sachs Freeman Associates
[3]University of Chicago

A high-efficiency laser diode array pumped Nd:YAG solid-state laser has recently been demonstrated.[1] We report for the first time a diode array pumped rare-earth laser other than Nd. Laser action was observed on the trivalent Holium $^5I_7 \text{-->} ^5I_8$, 2.1 μm transition. Laser sources in this wavelength region are required in infrared optical fiber communication systems.

Laser action was achieved with a 10 mm length YAG rod sensitized with 60% Er and 3% Tm in addition to the 2% Ho activator concentration. The plano convex rod was polished and coated. A 2.1 μm high reflectance coating transmitting at 785 nm was placed on the flat surface and a 99.5% reflector at 2.0-2.1 μm was coated onto the curved surface. The YAG rod was cooled to 77^{0}K and end pumped with a 100 mW CW multi-transverse mode GaAlAs Spectra Diode Lab laser diode array (Fig. 1). The temperature of the diode was adjusted so that the diode wavelength was centered at 785.5 nm. The spectral width of its multi-longitudinal mode output was approximately 10A. The rare earth absorption band centered at 785.5 nm was sufficiently broad to assure efficient coupling of the multi-longitudinal mode output of the array with the gain medium.

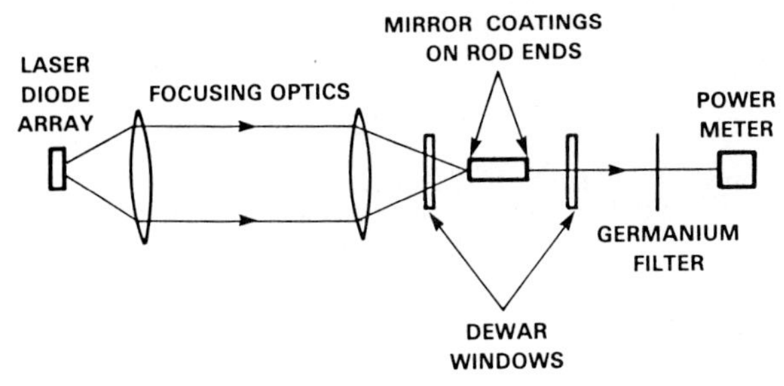

Fig. 1 Schematic diagram of laser-pumped Ho:YAG laser with end pumped geometry.

The YAG resonator which is formed by the coated laser rod ends had a fundamental spatial mode beam waist diameter of 115 μm. The lasing threshold occurred at 2.7 mW of incident pump power. With the diode operating at 100 mW we were able to collect and deliver to the YAG rod 34 mW which accounts for the overall collection efficiency and transmission of the various optical elements in the path. At this pump power, 5.6 mW of Ho emission was obtained at 2.1 μm. This yields an optical-to-optical power conversion slope efficiency of approximately 19% (see Fig. 2). Considering the multi-transverse mode nature of the diode pump and the less than optimum mode matching, the measured conversion efficiency was quite good.

The same diode with its spectral output wavelength tuned to 793 nm was used to pump a sample of Nd:YAG in a similar optical configuration as the Ho:YAG. The CW threshold obtained for the Nd 1.06 μm emission was somewhat higher than that obtained for the Ho emission. The lower threshold for Ho is expected since the longer lifetime (5 msec) as compared to Nd (260 μsec) more than compensates for the slightly lower gain.

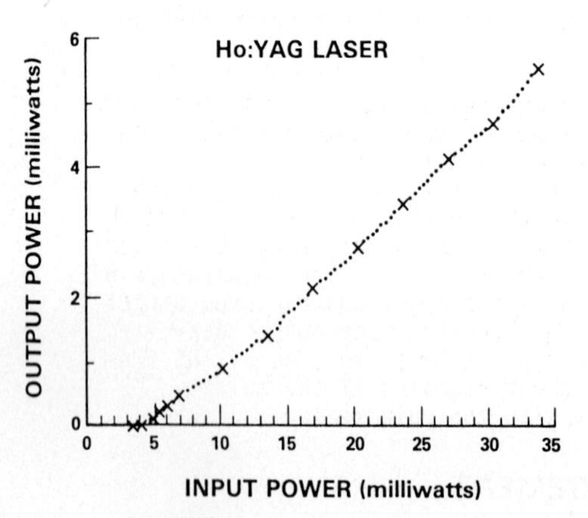

Fig. 2 Ho:YAG laser output power vs. semiconductor laser input power.

References

1. D. L. Sipes: In Applied Physics Letters 47 (2), p. 74, 15 July 1985 "Highly Efficient Neodymium: Yttrium Garnet Laser End-Pumped by a Semiconductor Laser Array."

Laser Operation on the Erbium and Holmium Transitions in the Yttrium-Scandium-Gallium Garnet Crystals

I.A. Shcherbakov

General Physics Institute of the USSR Academy of Sciences,
38 Vavilov Street, SU-117942 Moscow, USSR

The effective energy transfer from chromium ions not only to Nd^{3+}, but also to Ho^{3+}, Er^{3+}, Tm^{3+}, Yb^{3+} ions, is realized in gallium garnet crystals. It enables one to propose a number of new active media schemes and to obtain efficient laser operation. One of such media is yttrium-scandium-gallium garnet doped with chromium and erbium (YSGG:Cr^{3+}, Er^{3+}). Its composition was specially chosen with the aim of obtaining laser action on the erbium $^4I_{11/2} \longrightarrow {}^4I_{13/2}$ transition in the spectral region around 3 μm/1,2/. The spectral and luminescence investigations have shown that the highest gain coefficient at room-temperature is reached at the wavelength $\lambda = 2.794 \pm 0.005$ μm. The lifetime of $^4I_{11/2}$ state is 1.4 msec; and of the $^4I_{13/2}$ state, 6.5 msec. The energy transfer from chromium ions is the result of the static dipole-dipole interaction with its microparameter of $2 \times 10^{-39} cm^6 sec^{-1}$ and it finishes mainly on $^4I_{9/2}$ and $^4I_{11/2}$ levels of erbium ions. It turns out, however, that backtransfer also takes place, that is the energy transfer from erbium to chromium ions starting from erbium $^4S_{3/2}$ and $^4F_{9/2}$ states. The backtransfer is also static and dipole-dipole. Its constants are $2 \times 10^{-38} cm^6 sec^{-1}$ and $10^{-38} cm^6 sec^{-1}$ for $^4F_{9/2}$ and $^4S_{3/2}$, respectively. The energy transfer from Er^{3+} to Cr^{3+} ions is also possible from the higher erbium states. We must emphasize that the $Er^{3+} \longrightarrow Cr^{3+}$ energy transfer does not lead to any noticeable diminishing of the Er^{3+} ions $^4I_{11/2}$ upper laser level population, because of the energy received by chromium from erbium practically all (with the proper erbium concentrations) is returned to Er^{3+} ions. The presence of the direct and the back energy transfer between Cr^{3+} and Er^{3+} ions makes the picture of population and depletion of different erbium levels exceedingly complex. At high pumping densities it includes also the interaction of Er^{3+} ions in excited states. (See Fig. 1.)

Our estimates show that erbium and chromium ion concentrations can be varied over a wide range without noticeable deterioration of the energy transfer conditions to the upper 3 μm laser level. Moreover, the chromium ion concentrations should be chosen only from the point of view of good pumping of the active element; the erbium ion concentration should not be reduced below $5 \times 10^{20} cm^{-3}$.

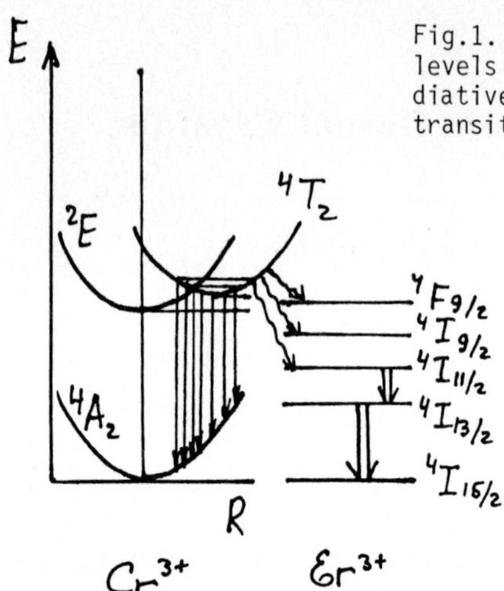

Fig.1. The energy level diagram for the low levels of Er^{3+} and Cr^{3+}. ↘ indicate the nonradiative Cr^{3+} Er^{3+} transitions; and ⇊ the laser transitions (3 μm and 1.5 μm)

A Ø5.9 x 70 mm active element was used in the laser experiment. The chromium and erbium concentrations were $2 \times 10^{20} cm^{-3}$ and $4.7 \times 10^{21} cm^{-3}$, respectively. The ends of the crystal rod were not coated. A quartz block with liquid cooling was used as a flashlamp housing. A 10 cm length resonator with one flat and one spherical (r=1 m) 100% mirror was used in the free-running regime. The results are given in Fig. 2. The slope efficiency was 1.3%, and the total efficiency was 1% with the pump energy 230 J.

The same crystal and the same flashlamp housing were used in the Q-switching regime. The Q-switching is obtained by utilizing a $LiNbO_3$ elec-

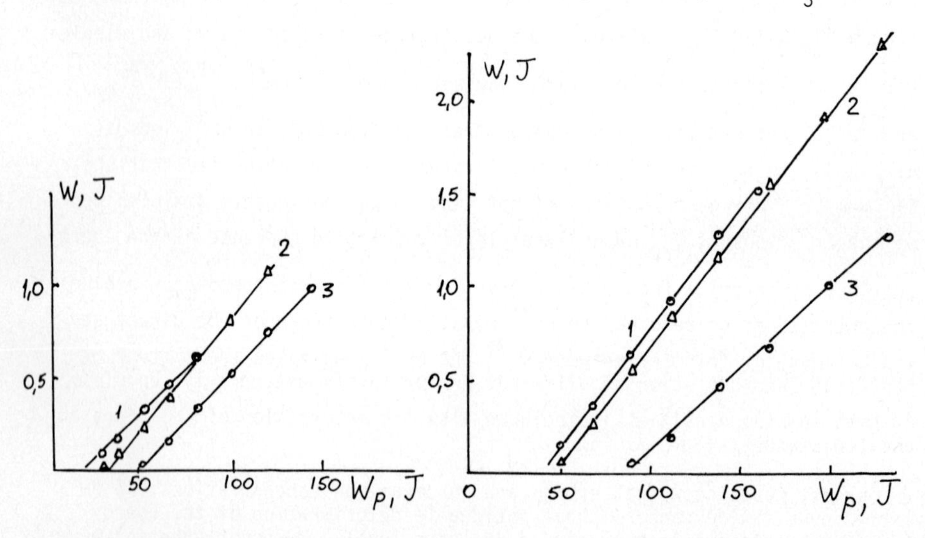

Fig.2. Laser pulse energy behaviour in the free running regime versus pumping pulse energy. The pumping pulse duration: a) 300 μs, b) 1600 μs; R = 80%(1), 60%(2) and 30%(3)

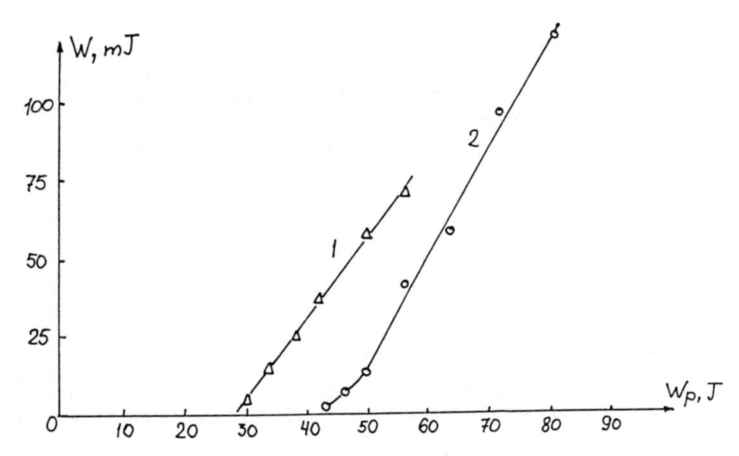

Fig.3. Laser oscillation energy in the Q-switching regime as a function of pumping energy

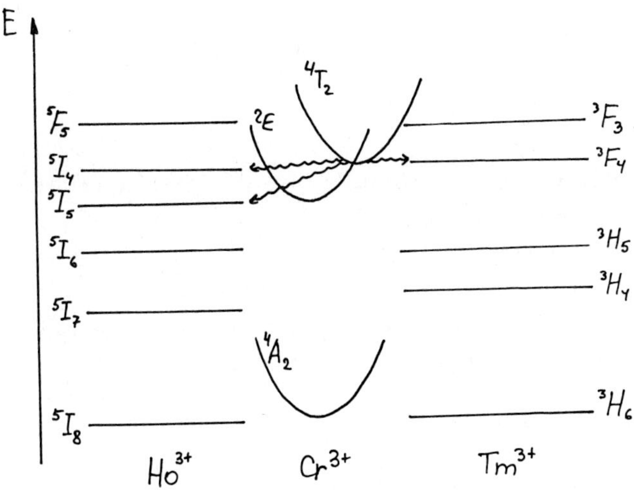

Fig.4. The energy level diagram of the low levels of Cr^{3+}, Ho^{3+} and Tm^{3+} ions

tro-optical shutter at Brewster angle. The results are given in Fig. 3. The pulse duration was 50 nsec; the oscillation wavelength was λ = 2.79 μm.

The energy transfer from chromium to Ho^{3+} and Tm^{3+} ions is on the whole also static and dipole-dipole with the transfer constants $3 \times 10^{-40} cm^6 sec^{-1}$ and $4 \times 10^{-39} cm^6 sec^{-1}$, respectively (Fig. 4)/3/. Only sufficiently high Ho and Tm concentrations ($>8 \times 10^{20} cm^{-3}$) can provide effective energy transfer with these microparameters. It is important to note that to obtain laser action on the transitions to the ground state of $Ho^{3+}(^5I_7 \rightarrow {}^5I_8)$ and $Tm^{3+}(^3H_4 \rightarrow {}^3H_6)$, one should work with low active species concentrations so as to provide low resonant losses. But then the effectiveness of the energy transfer from chromium ions is also very low. In this case it is necessary to use a third dopant. Its concentration should be high to make the energy transfer from chromium and the subsequent energy migration to the active ions effective. For example we may consider the following schemes:

1) $Cr^{3+} \rightarrow Er^{3+} \rightarrow Tm^{3+}$
2) $Cr^{3+} \rightarrow Er^{3+} \rightarrow Ho^{3+}$
3) $Cr^{3+} \rightarrow Tm^{3+} \rightarrow Ho^{3+}$
4) $Cr^{3+} \rightarrow Yb^{3+} \rightarrow Er^{3+}$
5) $Cr^{3+} \rightarrow Tm^{3+} \rightarrow Dy^{3+}$

and some variants thereof.

The lifetime of Ho^{3+} $^5I_7$ state and Tm^{3+} $^3H_4$ state at concentrations $<8 \times 10^{20}$ and at T = 300 K is $\simeq$ 13 msec in our crystals. The Ho^{3+} and Tm^{3+} ions in the YSGG crystals also have other metastable states that can play the role of the upper lasing levels. These, for example, are $^5I_6$ Ho^{3+} and $^3F_4$ Tm^{3+} levels which produce transitions in the IR region:

$^5I_6 \rightarrow {}^5I_7$ Ho^{3+} ($\lambda \sim 2.9\mu m$), $^3F_4 \rightarrow {}^3H_5$ Tm^{3+} ($\lambda \sim 2.3\mu m$),

$^3F_4 \rightarrow {}^3H_4$ Tm^{3+} ($\lambda \sim 1.5\mu m$).

The measurements carried out in our laboratory give the lifetime values of Ho^{3+} ($8 \times 10^{20} cm^{-3}$) $^5I_6$ level $\simeq$ 600 μsec and Tm^{3+} ($2 \times 10^{20} cm^{-3}$) $^3F_4$ level $\simeq$ 900 μsec in the YSGG crystal at T = 300 K. There exists a high probability of cross-relaxation processes from $^3F_4$ level of Tm^{3+}. With increased Tm^{3+} concentration, the lifetime of $^3F_4$ level shortens, giving 450 msec at the thulium concentration of $3.2 \times 10^{20} cm^{-3}$ /3/.

The energy spectrum of Ho^{3+} ions and in particular Tm^{3+} ions consists of relatively small number of lines. This is, in effect, an obstacle to wideband optical pumping into their own absorption bands. The possibility of exciting these ions through pumping into the Cr^{3+} absorption bands mostly removes this problem.

In previous work /4/ the free-running regime was realized. The laser action was demonstrated on the $^5I_7 \rightarrow {}^5I_8$ Ho^{3+} ion transition in the yttrium-scandium-gallium garnet (YSGG) at room-temperature with flashlamp excitation. A cylindrical Ø5.5 x 74 mm rod was used. The concentrations of the active ions of Cr^{3+}, Tm^{3+} and Ho^{3+} were 2.5×10^{20}, 8×10^{20} and $0.5 \times 10^{20} cm^{-3}$, respectively.

The concentration of chromium ions which served as the main absorber of the pump lamp energy was chosen so as to provide good pumping of the active element. The chosen Tm^{3+} ion concentration provided the following: firstly, a Cr^{3+} - Tm^{3+} energy transfer quantum yield close to unity; secondly, an

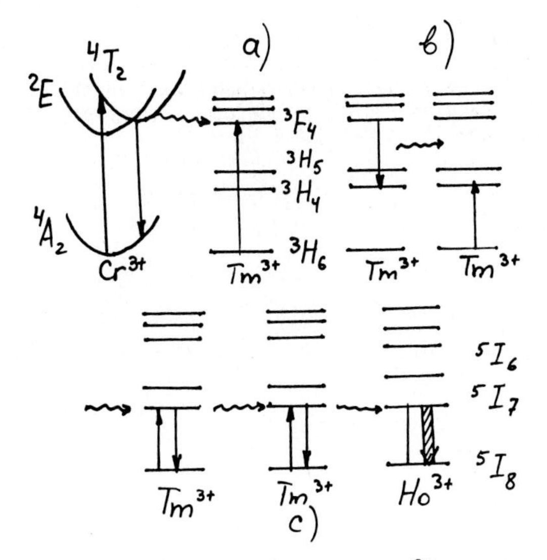

Fig.5. Three stages of energy transformation in the YSGG:Cr^{3+}, Tm^{3+}, Ho^{3+} crystal. a) $Cr^{3+} \rightarrow Tm^{3+}$ energy transfer. b) The cross-relaxation "exchange" of excitation in the Tm^{3+} ions. c) The energy migration between Tm^{3+} ions with subsequent transfer to the upper lasing level $^5I_7$ of Ho^{3+} ions. The energy level diagram was drawn according to [6]

effective cross-relaxation of Tm^{3+} ions consisting of the exchange of one high-energy quantum for two of lesser energy in the excited $^3H_4$ state; and thirdly, an effective energy migration among Tm^{3+} ions on the upper laser level $^5I_7$ with chromium ions. With Ho^{3+} ion concentrations of $5 \times 10^{19} cm^{-3}$ the resonant losses on the laser wavelength have the value of $0.1\ cm^{-1}$. It should be emphasized that in the present scheme, the heating is considerably decreased because the process of cross-relaxation between the Tm^{3+} electronic states provides the excitation transfer from the $^3F_4$ state to the $^3H_4$ state with the quantum yield close to 2 /5/, practically without the Stokes shift heating (Fig. 5).

The ends of the active element in our experiments were not coated. The pumping of the crystal was accomplished with a xenon flashlamp in a quartz silver-plated housing Ø 30 x 90 mm. The illuminated part of the active element was 60 mm. The pumping pulse duration was 1.4 msec in our experiment. Moreover, the pumping pulse shape was close to square.

The 260 mm laser resonator was formed by a 100% spherical (R = 1.5m) gold mirror and a 0.38 mm thick silica plane-parallel plate reflecting as a Fresnel reflector.

Fig. 6 represents the laser oscillation energy dependence on the pumping pulse energy at the repetition rate of 1 Hz. The oscillation threshold was 50 J. The slope efficiency was 3.12%. The oscillation energy of 5.3 J was obtained at the pumping energy of 250 J, corresponding to an efficiency of 2.1%. The oscillation wavelength was λ_L = 2.088 μm.

It is important that the concentrations of all types of ions may be varied over a very wide range. On the other hand, the preliminary results of the energy-transfer investigation show that in the crystal of the proposed composition a strong incoherent interaction between ions in $^5I_7$ and

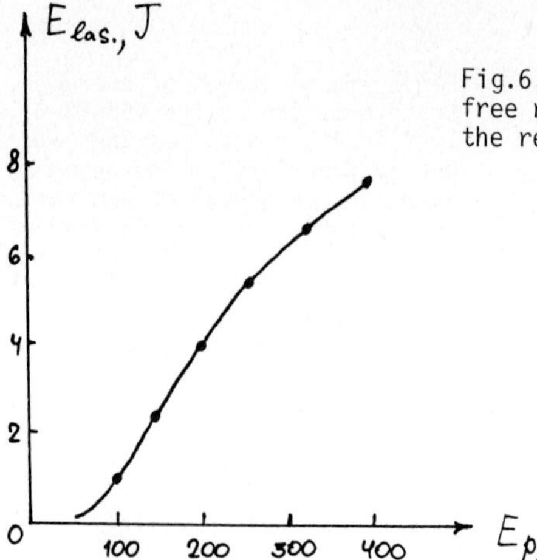

Fig.6. Laser pulse energy behaviour in the free running regime versus pump energy at the repetition rate of 1 Hz

$^3H_4$ states of holmium and thulium takes place. Thus, some unified level consisting of $^5I_7$ level of Ho^{3+} and $^3H_4$ level of Tm^{3+} plays the role of the upper lasing level. The thulium ions serve as energy accumulators at the $^3H_4$ state. At room-temperature, by strong incoherent interaction, this state instantly interchanges excitation with $^5I_7$ state of holmium ions on the scale of $^5I_7$ and $^3H_4$ decay rates. Estimates show that the holmium ion concentrations may be substantially lowered without adverse effect on the processes of energy accumulation at the higher lasing level $^5I_7$ in the quasistationary regime. The resonant losses are also decreased (down to $\lesssim 10^{-2}cm^{-1}$).

We should point out that for YSGG:Cr^{3+}, Tm^{3+}, Ho^{3+} crystals the functions of all three types of ions are independent. This independence and the capacity of YSGG to accept dopant ions to high concentration without change of structure as well as the high energy transfer efficiency of the gallium garnets allows us to choose the right composition of the host and the optimal dopant concentrations separately for each channel of energy transfer. This flexibility allows us to reach the high energy characteristics of lasers based on the chromium-doped rare-earth gallium garnet crystals.

Note, that operationally, for a lowering of the oscillation threshold or, for successive Q-switching the back energy transfer $Ho^{3+} \rightarrow Tm^{3+}$ on the final stage of the pumping energy transformation is a harmful effect. It can be suppressed by changing Tm^{3+} ions to Er^{3+} ions. Then, the third stage of the energy transfer will become essentially nonresonant and therefore, irreversible. The considerable lifetime of the $^4I_{9/2}$ erbium ion level in the gallium garnet crystals (~3 μsec) gives us a hope that the multiphoton relaxation channel at sufficiently high erbium concentrations will be destroyed by the ${}^4I_{9/2} \rightarrow {}^4I_{13/2}$; ${}^4I_{15/2} \rightarrow {}^4I_{13/2}$ cross-relaxation process, as it takes place in the fluoride crystals.

Thus, in the gallium garnet crystals, doped with Cr^{3+}, Er^{3+}, and Ho^{3+} as in the realized scheme, a strict division of the functions of the active ions may be obtained and the Stokes-shift heating considerably suppressed. This is important for lasers operating with high average powers.

REFERENCES

1. E. V. Zharikov, S. P. Kalitin, V. V. Laptev, V. V. Osiko, A. M. Prokhorov, V. A. Smirnov, I. A. Scherbakov: Preprint FIAN, M., No. 196 (1983).
2. E. V. Zharikov, N. N. Il'ichev, S. P. Kalitin, V. V. Laptev, A. A. Mulyutin, V. V. Osiko, P. O. Pashinin, A. M. Prokhorov, Z. S. Saidov, V. A. Smirnov, A. F. Umyskov, I. A. Scherbakov: Kvantovaya Electron., 13, No. 5, 973-979 (1986).
3. E. V. Zharikov, S. P. Kalitin, V. V. Laptev, V. G. Ostroumov, Z. S. Saidov, V. A. Smirnov, I. A. Scherbakov: Kvantovaya Electron., 13, No. 1, 216 (1986).
4. A. N. Alpat'ev, E. V. Zharikov, S. P. Kalitin, V. V. Laptev, V. V. Osiko, V. G. Ostroumov, A. M. Prokhorov, Z. S. Saidov, V. A. Smirnov, I. T. Sorokina, A. F. Umyskov, I. A. Scherbakov: Preprint FIAN, M., No. 26 (1986) (Kvant. Electron. to be publ.).
5. B. M. Antipenko: JTP 54, 11, 385-388 (1984).
6. G. H. Diere, H. M. Crosswhite: Appl. Optics 2, 675-686 (1963).
7. B. M. Antipenko, A. A. Mak, L. K. Sukhareva: Pis'ma v ZETP 10, 9, 513-517 (1984).

Operation of the High Dopant Density Er:YAG at 2.94 μm

M. Bass[1], *W.Q. Shi*[1], *R. Kurtz*[1], *M. Kokta*[2], *and H. Diegl*[3]

[1]Center for Laser Studies, University of Southern California, Los Angeles, CA 90089, USA
[2]Union Carbide Corporation, Washougal, WA 98671, USA
[3]Allied Corporation, Westlake Village, CA, USA

1. Introduction

Free-running, pulsed, flashlamp-excited operation of 50 and 33% Er doped YAG lasers is reported at 2.94 μm. This laser was described by researchers in the Soviet Union as early as 1975.[1] Since then there have been a number of further reports concerning this material all published by Soviet scientists. This paper represents, to our knowledge, the first publication outside of the Soviet Union about high dopant density Er:YAG laser operation. In addition to confirming some of the performance properties described earlier, this paper presents the unusual temporal waveforms of the Er:YAG, 2.94 μm laser. An outline is given of possible pumping and relaxation processes which may contribute to the laser's operation.

Er:YAG does not lase well at 2.94 μm when the concentration of Er is the usual 1%. However, when larger concentrations are used (generally over 15%) operation at this wavelength can be quite efficient. The relevant spectroscopic data for lasing at this wavelength are given below for 50% Er:YAG:

Upper laser level	$^4I_{11/2}$
Lower laser level	$^4I_{13/2}$
Upper level lifetime	100 μsec
Lower level lifetime	2 msec
Emission cross-section	2.6×10^{-20} cm^2
Pump absorption bands	0.80 μm ($^4I_{15/2}$ - $^4I_{9/2}$)
	0.65 μm ($^4I_{15/2}$ - $^4F_{9/2}$)
	0.54 μm ($^4I_{15/2}$ - $^4S_{3/2}$)
	0.52 μm ($^4I_{15/2}$ - $^2H_{11/2}$)
	0.49 μm ($^4I_{15/2}$ - $^4F_{7/2}$)
	0.45 μm ($^4I_{15/2}$ - $^4F_{5/2}$)
	0.44 μm ($^4I_{15/2}$ - $^4F_{3/2}$)
	0.41 μm ($^4I_{15/2}$ - $^2H_{9/2}$)
	0.38 μm ($^4I_{15/2}$ - $^4G_{11/2}$)

2. Laser Tests

The laser tests were conducted in a water-cooled, double elliptical pump cavity. This cavity had been designed for pumping Alexandrite laser rods and so had the desired silver-backed pyrex reflector. While the laser may work in a gold-plated cavity, the many visible, blue and near uv pump bands suggest better efficiency is possible with a silver pump reflector. The rods reported in this paper were 6.25 mm in diameter and 120 mm long. They were obtained from Union Carbide Corp. and Crystal Optics Research Inc. The flashlamps were 6.5 mm bore diameter, xenon-filled lamps from ILC and, in the pump cavity used, were able to pump 96.5 mm of the rod. Two different pump pulse durations were used; one, called the short pump pulse, was 120 μsec long at full width at half maximum (FWHM) and the other, called the long pump pulse, was 170 μsec FWHM. The resonators mirrors were spaced only 25 cm apart to provide a resonator that was relatively insensitive to thermal lensing in the laser rod. The 100% reflector used was an uncoated, polished copper mirror with a 5 m radius of curvature. Several flat output mirrors were tested but best performance was observed for all rods with a 75% reflector. All tests were conducted with no intracavity apertures and so represent long pulse, multimode lasing in which the whole rod aperture was filled with many oscillating modes.

The performance of the Er:YAG lasers tested is summarized in Figs. 1 and 2. Fig 1 a compares the performance of the 33% doped rod when pumped with the two different pump pulses mentioned above. The

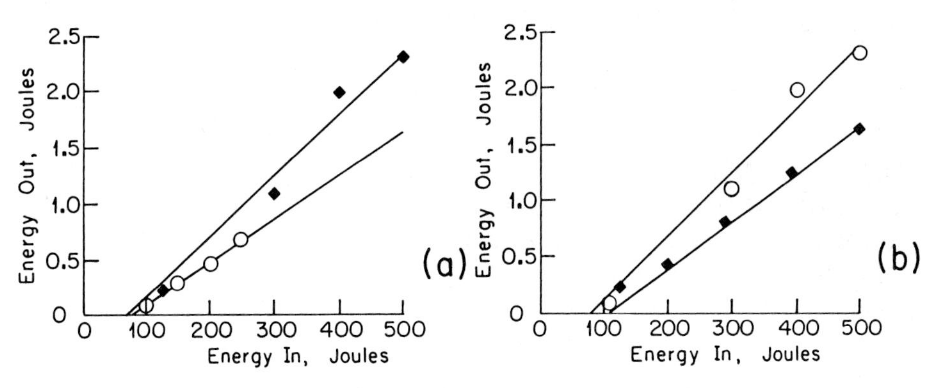

Fig. 1 A. Long pulse laser output energy of 33% doped Er:YAG at 2.94 μm versus input energy. Pump pulse duration =120 μsec for o and 170 μsec for ♦.

B. Long pulse laser output energy versus input energy for two different Er dopant densities in YAG. o = 33% and ♦ = 50%.

improvement in efficiency as the pump pulse duration is increased agrees with observations reported previously[2] and suggests that the reported 3-5% efficiencies are realizable. Fig. 1b shows the relative performance of the 33 and 50% doped rods. when pumped with the long pump pulse. The difference observed may be specific to the pump pulse waveform and pump cavity used in this work. The performance differences may depend on these parameters and so this data should not be taken as a firm preference for the 33% rod over the 50% rod. More research is necessary with excitation conditions properly tailored to the rod to be used. Both rods showed excellent optical quality when observed through crossed polarizers and no significant scattering of a HeNe beam could be detected. It should be noted that the 0.63 μm HeNe beam is attenuated in Er:YAG but the orange HeNe line is transmitted and should be used.

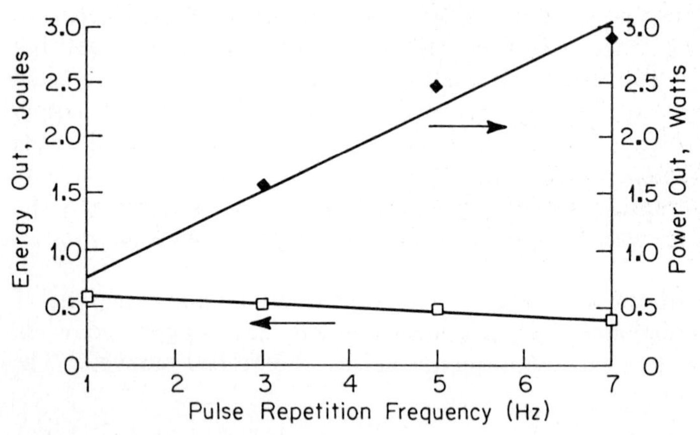

Fig. 2 Energy and power output of the 33% Er:YAG versus pump repetition rate using the 120 μsec duration pump pulse.

Fig. 2 shows the average power obtained using the 33% rod and the short pump pulse. These tests were limited by power supply considerations and it is clear that much higher average powers are possible. When the long pump pulse was used the maximum power supply repetition frequency was 3 Hz. At this prf over 5 W of average power was observed. The drop-off in energy per pulse indicates the onset of thermal lensing in the medium. Since no corrective measures were taken other than using a short resonator this problem can benefit from further laser engineering.

A major design consideration in developing this laser is preparing reliable, damage-resistant optical coatings. The major coating consideration is that the material not contain any water when it is set down and that it not adsorb water from the atmosphere afterwards. The absorption of water at 2.94 μm is maximum and so any water in a coating causes unacceptably high losses and low damage threshold. High reflection coatings were reliably produced by coaters who had experience in making coatings for HF and DF lasers. On the other hand, only one manufacturer made reliable AR coatings for the laser rods. These coatings performed reliably at pulse energies of over 1.7 J. AR coatings supplied by another manufacturer damaged at outputs of less than 250 mJ. In fact, the data given in Figs. 1 and 2 were taken with uncoated rods to avoid the issue of coating properties.

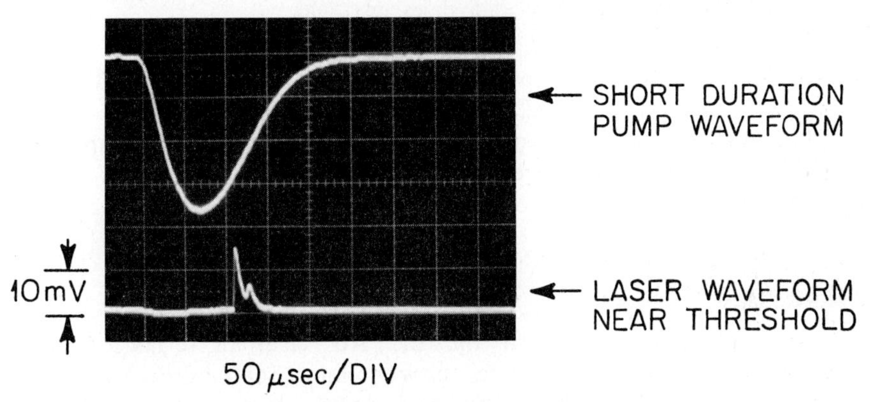

Fig. 3 Er:YAG laser waveform near threshold.

The waveform of the laser emission of Er:YAG at 2.94 μm is somewhat like that of any conventional optically pumped solid-state laser near threshold. Fig 3 shows the observed waveform at threshold using the short pump pulse. The important point is that there are relaxation oscillation (RO) spikes and that lasing begins after the peak of the pump pulse. When this laser is pumped well above threshold with either the long or the short pump pulse the output waveform is as shown in Fig. 4. The Er:YAG laser waveforms shown in Figs. 3 and 4 were obtained with a Judson Infrared J-10 InSb detector, cooled to 77 °K, with a specified response time of 500 nsec. These waveforms were subsequently verified with a 50 nsec response time Judson Infrared Model J-12-18C-R250U InAs room temperature detector.

In the case of Fig. 4 lasing begins before the peak of the pump and ends when the pump is nearly over. In between it operates nearly

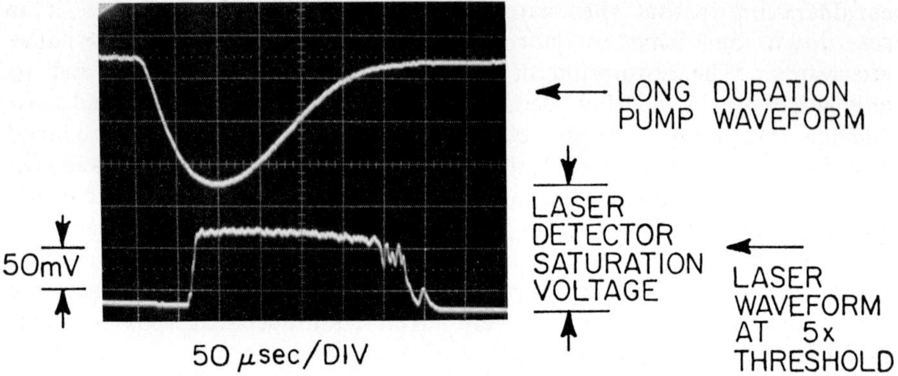

Fig. 4 Er:YAG laser waveform at 5 times threshold.

continuously without RO spikes typical of optically pumped solid state lasers. This type of RO spike does not appear until the pump has fallen well below threshold. It is expected that the improved performance reported for longer pump pulses is related to this mode of lasing in which some process operates to alter and extend the population inversion.

3. Discussion

The long pulse performance of the Er:YAG laser in terms of both energy per pulse and average power has been shown to be comparable to other optically pumped solid-state lasers. Higher outputs and efficiencies can be expected using longer duration pump pulses.[3]

The unusual waveforms observed at high output energies are the result of a combination of processes. These include contributions to the inversion due to:

1, the usual "four-level" pumping process involving excitation from the $^4I_{15/2}$ ground state,

2, "three-level" pumping from the long-lived lower laser level, the $^4I_{13/2}$ state, and

3, cross-relaxation between nearby Er^{3+} ions. The cross-relaxation process is one in which an ion in the $^4I_{13/2}$ state relaxes to the $^4I_{15/2}$ state while, simultaneously, a neighboring ion in the $^4I_{13/2}$ jumps to the $^4I_{9/2}$ state.

When in the $^4I_{9/2}$ state the ion rapidly relaxes into the upper laser level. The process of cross-relaxation is particularly important in high concentration material and is discussed in detail by Bagdasarov et al.[4] An analysis is in progress of the relative importance of the several processes of excitation and relaxation as they contribute to the observed quasi-continuous lasing. This work will be reported at a later date.

The 2.94 μm operation of the Er:YAG laser is sufficiently interesting to warrant further study. It is a material which can be grown easily and lases well. It has potential application in surgery and as a source to drive a variety of infrared sources at important wavelengths. Since it is compatible with existing Nd:YAG laser systems exploration of its potential is straightforward. As has been pointed out in the present work and in the work of the Soviet scientists, optimum efficiency requires longer duration pump sources and so conversion of Nd:Glass or ruby lasers may lead to better performance. Reliable optical coatings for 2.94 μm requires that they be specified to contain and adsorb no water. As a result of this work it is clear that the Er:YAG laser is available to anyone who wishes to use it.

Acknowledgment

The work at U.S.C. was supported by AFOSR Contract No. AFOSR-84-0378.

References

[1]E. V. Zharikov, V. I. Zhekov, L. A. Kulevskii, T. M. Murina, V. V. Osiko, A. M. Prokhorov, A. D. Savel'ev, V. V. Smirnov, B. P. Starikov, and M. I. Timoshenko, Sov. J. of Quantum Electronics 4, 1039 (1975)
[2] V. I. Zhekov, V. A. Lobachev, T. M. Murina and A. M. Prokhorov, Sov. J. Quantum Electron. 13, 1235 (1984)
[3]I. A. Scherbakov, General Physics Institute, Moscow, USSR, seminar presented at U.S.C., June 1986
[4] Kh. S. Bagdasarov, V. I. Zhekov, V. A. Lobachev, A. A. Manenkov, T. M. Murina and A. M. Prokhorov, Izvestiya Akademii Nauk SSSR, 48, 1765, (1984)

Part IX

Neodymium Lasers

The $YAlO_3$:Er Laser

H.P. Weber and W. Lüthy

Institute of Applied Physics, University of Bern, Sidlerstrasse 5,
CH-3012 Bern, Switzerland

Absorption and fluorescence properties of 1.25% and 50% $YAlO_3:Er^{3+}$ are summarized. Laser properties, particularly the polarization dependence, at room temperature are described. New laser lines around 1.7 μm are reported, and the laser transitions in the 3 μm range are discussed.

1. Introduction

Investigations in erbium solid-state lasers are motivated mainly by the possibilities that these lasers offer for medical applications. Precise microscopic cutting of tissue can best be performed in a regime of very high absorption of laser light. In this case, the deposited energy is concentrated near the surface; and the desired interaction, e.g., evaporation or cutting, can be performed at lower power levels than in the case of large penetration depth. Furthermore, steep temperature gradients can be obtained leading to a high localization of the heated zone and a minimum thermal influence in neighboring tissue. Another goal is to be able to control the penetration depth to achieve hemostatic coagulation.

Since biological material mainly consists of water, it is helpful to look for a laser source that can emit at various wavelengths which are absorbed to different degrees by water. The absorption spectrum of water is shown in Fig. 1.

Water highly absorbs in the wavelength regions shorter than 200 nm and around 3 μm. Whereas, irradiation with UV or even VUV radiation can lead to

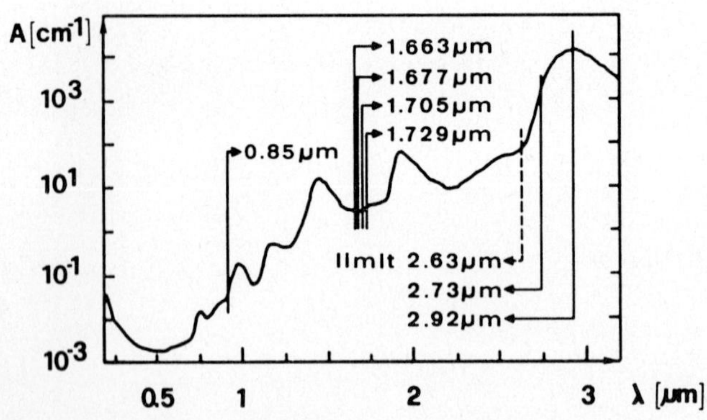

Fig. 1: Optical density $A = \log_{10} T$ (T = transmission of a 1 cm layer) of distilled water in the wavelength range from 0.19 μm to 3.2 μm

photolytic reactions and dissociate molecules in an undesired way, three (3) μm irradiation seems to be a more suitable method for medical applications because only thermal effects can be expected. In addition, 3.0 μm light is suitable for transmission through special optical fibers.

Numerous lasers are capable of emitting near 3 μm. Favorable candidates in view of high power, simple construction, and high reliability are the erbium solid-state lasers. Trivalent Er^{3+} can be doped into different hosts such as YAG, $YAlO_3$, and many others [1,2]. As shown in Fig. 1, erbium shows laser action at several shorter wavelengths, covering a wide range of penetration in water. The dashed line indicates the shortest wavelength of all the possible transitions around 3 μm; which, in contrast to the solid lines has as yet not shown laser emission. In our work, we used $YAlO_3$:Er crystals with erbium concentrations of 1.25% and 50%, respectively.

2. Material Characteristics

$YAlO_3$ has an orthorhombic crystal structure with the space group D_{2h}^{16}-Pbnm [3,4]. Doping with Er leads to a partial replacement of Y^{3+} ions with Er^{3+} ions. The Y^{3+} and Er^{3+} ions are surrounded by 12 oxygen ions and are located at a site of C_s symmetry. Unlike the cubic YAG:Er crystals with D_2 site symmetry, the absorption and emission of light by rare-earth ions in $YAlO_3$ is strongly polarization dependent. We used crystals that were cut with their rod axis along the crystallographic b-axis. The dimensions of the 1.25% and 50% Er^{3+} rods were 6.35 mm diameter by 75 mm length and 5 mm diameter by 60 mm length, respectively. In contrast to the cubic symmetry of YAG, which leads to no optical birefringence, $YAlO_3$ is a birefringent crystal.

3. Absorption Spectroscopy

A lot of work has already been performed in the absorption spectroscopy of $YAlO_3$:Er [5,6,7,8]. Up to now, however, no information has been presented on the effects of high doping levels (50%) and the dependence of the absorption on the polarization of the light with respect to the a- and c-axes. Therefore, a new measurement of the absorption spectrum was performed in the wavelength range between 350 nm and 2 μm. Different samples of 50% Er:$YAlO_3$ were prepared. The samples were cut and polished perpendicular to the crystallographic b-axis. The thickness varied between 0.067 cm and 0.38 cm. The measurement was performed with a spectrophotometer (Perkin Elmer Lambda 9) equipped with a Glan-Thompson polarizer. More than 250 absorption peaks were absorbed and evaluated for polarizations with the E-vector oscillating parallel to the a-axis and parallel to the c-axis. The influence of the high dopant concentration was clearly visible. We observed slight changes in the width of the spectral lines, negligible shifts in the wavelengths, and significant changes in the relative absorption intensities. Additionally, a very strong polarization dependence was found. Depending on the polarization direction with respect to the crystallographic axes of the laser rod, the relative strengths of the absorption lines vary considerably. An example is given in Fig. 2.

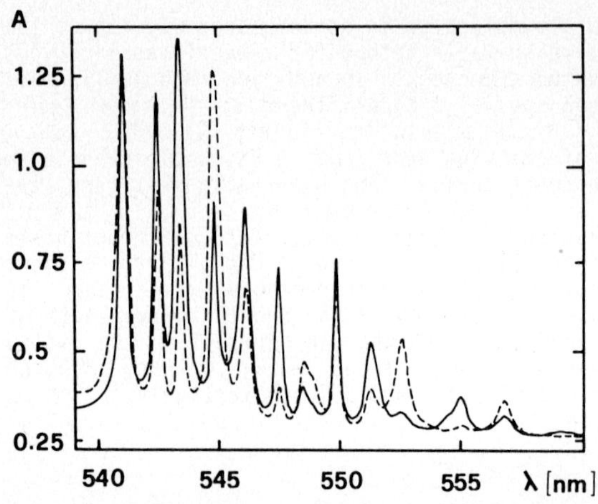

Fig. 2: Absorption spectrum (A=optical density for a 0.067 cm thick sample) in the wavelength range from 540 nm to 560 nm for two directions of the polarization. The solid line gives the absorption of light with the E-vector parallel to the a-axis and the dashed line with the E-vector parallel to the c-axis.

A report that describes the properties of the absorption spectra in more detail is in preparation [9]. Based on the absorption spectra, it is possible to give the precise energy levels of the Er^{3+} ions in Fig. 3.

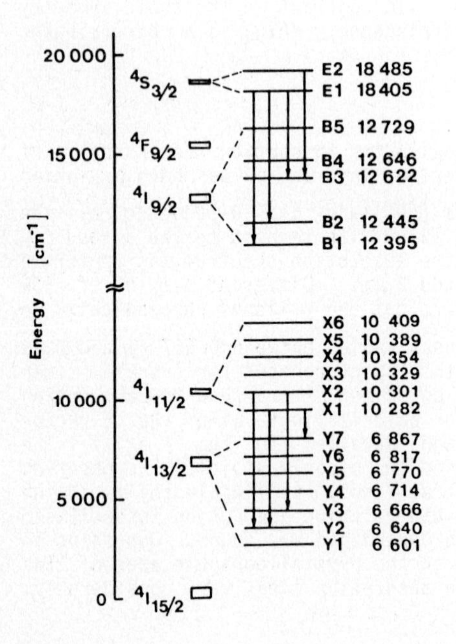

Fig. 3: Energy-level scheme of $YAlO_3$:Er. On the righthand side of the figure some of the multiplets and some laser transitions are shown in an expanded energy scale

4. Fluorescence Spectroscopy

Fluorescence spectra were obtained using excitation from different laser lines from an Ar^+-ion laser. Even though the wavelengths of these laser lines do not correspond to the strong absorptions of the $Er:YAlO_3$ rods, strong fluorescence is still seen.

To obtain more information on the lifetime of the Er^{3+} levels, fluorescence was also excited with a 1 ns pulse from a thallium iodide laser at 377.6 nm. This wavelength corresponds to the strong $^4I_{15/2} \rightarrow {}^4G_{11/2}$ absorption of Er^{3+} and leads to a cascade of fluorescence transitions. The experimental arrangement is shown in Fig. 4.

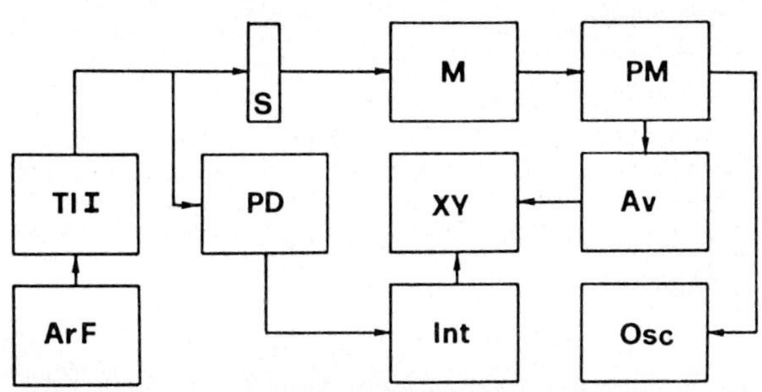

Fig. 4: Experimental setup for the measurement of fluorescence lifetimes and transitions. ArF: Argon-Fluoride excimer laser; TlI: thallium iodide laser; S: Sample; M: Monochromator; PM: Photomultiplier; Av: Signal averager; XY: XY-recorder; Osc: Oscilloscope; PD: Photodiode; Int: Pulse-integrator

An ArF excimer laser (Lambda Physik EMG 500) was used to excite a photolytically-pumped thallium iodide laser. This laser outputs two lines at 535 nm and 377.6 nm. The two laser lines were separated with the aid of a prism, and the sample illuminated using the 377.6 nm line. Fluorescence lines were detected with a double monochromator (Jobin Yvon HRD 1) and fast photomultiplier (Hamamatsu R 955). Spectra were recorded with a signal averager and a XY-plotter; the fluorescence lifetimes were measured with an oscilloscope. The results are shown in Fig. 5.

The fluorescence transitions shown in Fig. 5 consist of a large number of transitions between Stark-split levels. For example, the transition between $^4S_{3/2}$ (with two Stark-levels) and the ground state $^4I_{15/2}$ (with eight Stark-levels) leads to a total of 16 fluorescence lines, which are shown in Fig. 6.

Excitation of the $^2H_{9/2}$ emitting level occurs via nonradiative decay of the excited $^4G_{11/2}$ level resulting in emission (τ = 180 ns) to the $^4I_{13/2}$ and $^4I_{15/2}$ levels. Population of the $^4S_{3/2}$ state is very important for laser action. The lifetimes of the two Stark levels of this multiplet at

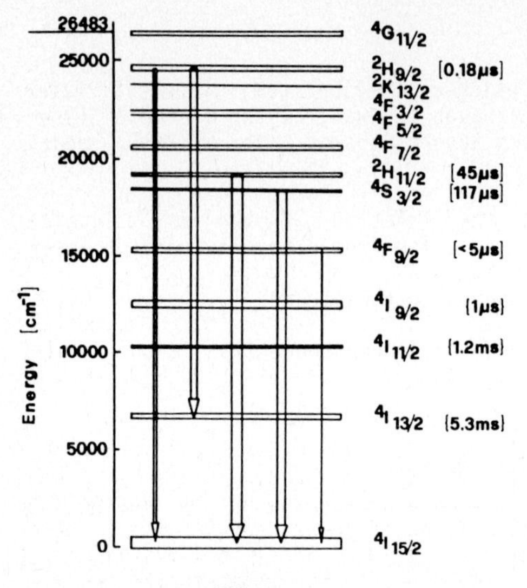

Fig. 5: Fluorescence lines measured in the sensitive wavelength range of a Hamamatsu R 955 photomultiplier. The width of the arrows indicate the relative intensities of the respective transitions. Measured lifetimes are in [ ] and literature values in { }

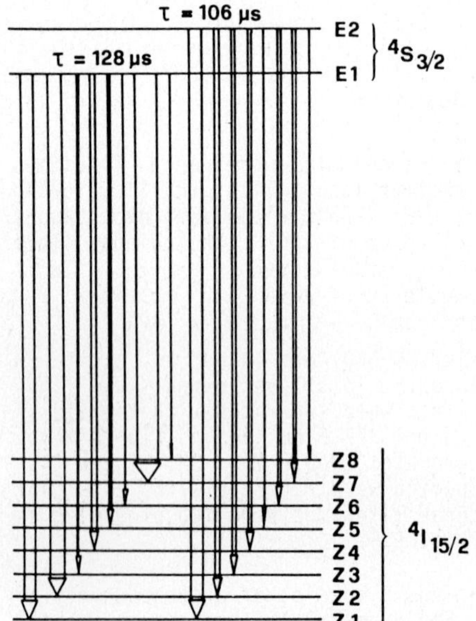

Fig. 6: Fluorescence transitions between $^4S_{3/2}$ and $^4I_{15/2}$ states. As in Fig. 5, the widths of the arrows are a measure of the intensity of the fluorescence lines

ambient temperature are 128 μs for E_1 and 106 μs for E_2. Pumping the $^4S_{3/2}$ level leads to lasing at 1.7 μm ($^4S_{3/2} \rightarrow {}^4I_{9/2}$). With a subsequent radiationless $^4I_{9/2} \rightarrow {}^4I_{11/2}$ transition, the $^4I_{11/2}$ level is populated. From this level, 3 μm laser transitions $^4I_{11/2} \rightarrow {}^4I_{13/2}$ can occur.

5. Lasing of $Er:YAlO_3$

Previous laser experiments with Er^{3+} in $YAlO_3$ have been performed [5,7,10,11,12,13,14].

As mentioned above, lasing has been reported at ambient temperature between the levels of $^4S_{3/2} \rightarrow {}^4I_{9/2}$ at 1.663 μm in $YAlO_3$:Er crystals with doping concentrations in the order of 1%. Furthermore, lasing $^4I_{11/2} \rightarrow {}^4I_{13/2}$ at 2.7 μm has also been achieved at room temperature. Since the $^4S_{3/2}$ state consists of two levels and the $^4I_{9/2}$ state of five levels, a total number of ten laser lines might be expected in 1.7 μm region. The transition probabilities of these 10 lines have been estimated theoretically [15,16,17]. Our most recent experiments [18] show that, besides the known line at 1.663 μm, three new laser lines can be obtained by tuning the resonator with an etalon. The 1.663 μm line is polarized with the E-vector parallel to the crystallographic a-axis and the new lines at 1.677 μm, 1.706 μm, and 1.729 μm are polarized with the E-vectors parallel to the c-axis. According to theoretical predications [15], the 4 laser lines reach threshold in the sequence of their branching ratio. However, one exception is found. The 1.642 μm line which is expected to appear with the second highest intensity, is not found in the laser emission. The reason for this behavior is found in the reabsorption of this line by a resonance transition from $^4I_{15/2}$ (Z8) $\rightarrow {}^4I_{13/2}$ (Y1) at 1.642 μm.

Besides 1.7 μm lasing can also be obtained simultaneously in the wavelength region of 2.7 μm. As it can be seen from Fig. 1, the wavelength of 2.7 μm is not in the maximum of the absorption curve of aqueous media. For several medical applications it would be of advantage to work more closely to the maximum at a wavelength of about 2.9 μm. This wavelength is obtained if very highly doped $YAlO_3$:Er laser rods were used.

In our experiments we worked with rods of 50% Er doping concentration. In this case the emission wavelength is 2.92 μm and corresponds with the absorption maximum in water. The laser threshold was 45 J with a rod of 60 mm length and 5 mm diameter operated in a short resonator consisting of a mirror with reflectivity R = 1 and 2 m radius of curvature, and a flat mirror of R = 0.95. With 100 Joules pump energy it produces 100 μs spiking pulses of up to 30 mJ energy (see Fig. 7).

By optimizing the laser system these ratios are expected to be considerably improved. In YAG it has been observed [19] that the irregular spiking of the short resonator configuration tends to become more regular with the use of long resonator configurations. The explanation is found [19] in the higher laser threshold with larger resonator length and the more prominent role of cross-relaxational interaction (see below).

6. Resonant Phenomena

With high doping concentrations, resonant nearfield interaction processes between closest neighbor erbium ions become very effective. Thereby the oscillator strengths, transition frequencies and linewidths as well as the fluorescent lifetimes of the absorption as well as of the emission lines are modified. As a consequence, the properties of the laser transitions are also significantly influenced.

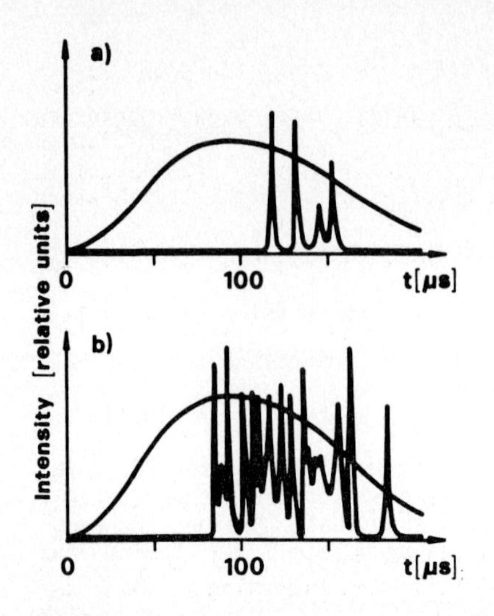

Fig. 7: Pump intensity and 2.92 μm emission at a pump level of
a) E/Eth = 1.7,
b) E/Eth = 2.29

Due to the large number of energy levels and the additional Stark splitting, coincidences are found in the wavelengths of different transitions between these levels that can give rise to resonant phenomena. Such coincidences are found between $^2H_{11/2} \rightarrow {}^4I_{9/2}$ and $^4I_{15/2} \rightarrow {}^4I_{13/2}$, between the levels $^2H_{11/2} \rightarrow {}^4I_{13/2}$ and $^4I_{15/2} \rightarrow {}^4I_{9/2}$ or between the levels $^4I_{13/2} \rightarrow {}^4I_{15/2}$ and $^4I_{13/2} \rightarrow {}^4I_{9/2}$ [19,20].

The combination of these resonances with the ion-ion interactions at high concentration give rise to cross-relaxation processes as shown in Fig. 8.

The balance of the resonant exchange of energy between all the participating levels is made irreversible due to the effect of fast depletion (1 μs [13]) of the $^4I_{9/2}$ level, as has been pointed out previously

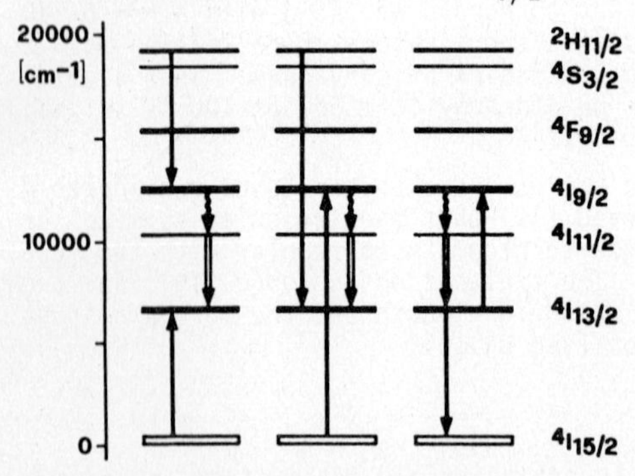

Fig. 8: a)-c) resonant transitions in the energy-level scheme of $YAlO_3$:Er together with the multiphonon transition $^4I_{9/2} \rightarrow {}^4I_{11/2}$ and the 3 μm laser transition $^4I_{11/2} \rightarrow {}^4I_{13/2}$

[19]. This level is deexcited rapidly by multiphonon transitions to $^4I_{11/2}$.

The irreversible mechanism seems to be responsible for the fact, that the 1.7 μm transition has not been observed for the 50% Er concentration materials. On the other hand the described mechanism plays an important role in the generation of population inversion for the 2.6 μm - 2.9 μm transitions $^4I_{11/2} \rightarrow {}^4I_{13/2}$ in highly doped crystals [12,21], since it can help to populate the upper laser level as well as to give rise to an effective depopulation (c.f. Fig. 8c) of the lower laser level.

In low concentration material (1%) the lifetime of the upper level of this transition $^4I_{11/2}$ is shorter (1.2 ms) than that of the lower level $^4I_{13/2}$ (5.3 ms) [13]. The long lifetime of the lower laser level leads to a saturation of the transition in materials of low doping level and it is therefore impossible to achieve 3 μm pulse durations exceeding about 1 ms. In highly doped material, however, with the aid of the cross-relaxation of Fig. 8 c), this limit in pulse duration could be eliminated [20] and even steady-state 3 μm emission in YAG:Er could be demonstrated [22, 23, 24].

7. Conclusion

In conclusion, we have performed new measurements of polarization-dependent absorption spectroscopy of $YAlO_3$:Er at room temperature and with high doping concentrations. We have further investigated the fluorescence spectroscopy and given a value of the individual lifetimes of the two $^4S_{3/2}$ levels. Three new laser lines have been found in the 1.7 μm band and for the first time lasing in a $YAlO_3$:Er crystal with 50% doping has been achieved.

It has further been shown, that the erbium laser operating at 2.92 μm has 90% absorption within 1 μm in water, corresponding to a penetration depth at 1/e of 0.43 μm. At 2.72 μm the penetration depth is 1.2 μm and at 1.7 μm it is 1.7 mm. For medical applications, the erbium laser may therefore be applicable not only for very precise cutting in the high absorption region but also for coagulating relatively large volume, where bleeding has to be stopped. This could all be performed with the same laser rod by just modifying the resonator properties.

Acknowledgements

We would like to thank M. Datwyler, P.-D. Henchoz, M. Stalder, S. Schnell, and J. Frauchiger for their contributions to the presented data. We would like to thank Hj. Weder and B. Wicki for their help with the figures. This work was supported in part by the Swiss Commission for the Encouragement of Scientific Research.

1. A. A. Kaminskii, T. I. Butaeva, A. O. Ivanov, I. V. Mochalov, A. G. Petrosian, A. I. Rogov, and V. A. Fedorov: Sov. Tech. Phys. Lett. 2, (9) 308-310 (1976).

2. A. A. Kaminskii "Laser Crystals": Springer-Verlag, Berlin, Heidelberg, New York, p. 172 ff (1981).

3. R. Diehl and G. Brandt: Mat. Res. Bull. 10, 85-90 (1975).

4. B. Dischler and W. Wettling: J. Phys. D: Appl. Phys. 17, 1115-1124 (1983).

5. M. J. Weber, M. Bass, and G. A. deMars: J. Appl. Phys. 42 (1), 301-305 (1971).

6. V. L. Donlan, and A. A. Santiago: Jr. J of Chem Phys. 47 (11), 4717-4723 (1972).

7. M. J. Weber, M. Bass, T. E. Varitimos, and D. P. Bua: IEEE J. of Quantum Electron. QE-9 (11), 1079-1086 (1973).

8. I. V. Mochalov: Phys. Stat. Sol. (a) 55, 79-87 (1979).

9. S. Schnell, P.-D. Henchoz, and W. Luthy: (in preparation).

10. W. Falkenstein: Verhandlungen DPG 17, 454 (1982).

11. M. J. Weber, M. Bass, G. A. deMars, K. Andringa, and R. R. Monchamp: IEEE J. of Quantum Electron, QE-6, 654 (1970).

12. A. A. Kaminskii, V. A. Fedorov, A. O. Ivanov, I. V. Mochalov, L. I. Krutova: Sov. Phys. Dokl. 27, 725-727 (1982).

13. A. A. Kaminskii: Sov. Phys. Dokl. 27, 1039-1041 (1982).

14. A. A. Kaminskii, V. A. Fedorov, and I. V. Mochalov: Sov. Phys. Dokl. 25, 744-746 (1980).

15. K. K. Deb: SPIE 335 Advanced Laser Technology and Applications 31-35 (1982).

16. M. J. Weber, T. E. Varitimos, and B. H. Matsinger: Phys. Rev. B 8 (1), 47-53 (1973).

17. J. A. Detrio: Phys. Rev. B 11 (3), 1257-1259 (1975).

18. M. Datwyler, W. Luthy, and H. P. Weber: (in preparation).

19. A. M. Prokhorov: Laser Spectroscopy VI, H. P. Weber and W. Luthy Editors, 427-429, Springer-Verlag, Berlin, Heidelberg, New York, Tokyo (1983).

20. V. I. Zhekov, B. V. Zubov, V. A. Lobachev, T. M. Murina, A. M. Prokhorov, and A. F. Shevel: Sov. J. Quantum Electron. 10 (4), 428-430 (1980).

21. V. I. Zhekov, V. A. Lobachev, T. M. Murina, and A. M. Prokhorov: Sov. J. Quantum Electron. 11 (2), 279-281 (1981).

22. Kh.S. Bagdasarov, V. I. Zhekov, V. A. Lobachev, T. M. Murina, and A. M. Prokhorov: Sov. J. Quantum Electron. 13 (2), 262-263 (1983).

23. V. I. Zhekov, V. A. Lobachev, T. M. Murina, and A. M. Prokhorov: Sov. J. Quantum Electron. 13 (9), 1235-1237 (1983).

24. V. I. Zhekov, V. A. Lobachev, T. M. Murina, and A. M. Prokhorov: Sov. J. Quantum Electron. 14 (1), 128-130 (1984).

Comparative Performance of Nd-Doped Solid-State Laser Materials

T. Driscoll, G. Hansen, R. Stone, M. Peressini, and H. Hoffman

Research & Development Division, Lockheed Missiles & Space Company., Inc., 3251 Hanover Street, Palo Alto, CA 94304, USA

1. INTRODUCTION

One particular application under consideration for some of the Nd-doped materials investigated is an efficient laser pump source for Ti-doped solid-state laser hosts. For currently available Ti-doped materials, low doping results in relatively high threshold power leading to efficient operation only near the damage threshold. This is especially true for operation at the lower gain wavelengths of the Ti:sapphire tuning range. Longer pump pulse duration may alleviate the damage problem. A longer-pulse-duration (>20 ns) Q-switched source constructed by using a material with gain lower than YAG (glass or YLF), operating near threshold, is therefore of particular interest. We have recently constructed a Nd:YLF acousto-optically Q-switched oscillator/amplifier which produces 150 mJ in a 100-ns-duration pulse with TEM_{00}-mode beam quality. This beam at 1.047 µm is frequency doubled in KTP, yielding 75 mJ for pumping Ti-doped laser materials. Other candidate materials for this application are Nd:BEL and Nd:glass, both of which are currently under consideration as alternatives to Nd:YLF. The increasing interest in tunable solid-state laser materials has underscored the need to identify optimal pump sources.

In this paper we present preliminary data from performance measurements of Nd-doped solid-state materials. The objective of this effort was to measure, in a comparative fashion, the characteristics of several relatively new Nd:host materials, which affect their usefulness in a practical laser system. Consequently, the scope of the work was limited to those crystals that can be grown in high enough optical quality to merit consideration as near-term alternatives to glass and YAG systems. In addition, we only consider the rod-geometry performance of the materials. While slab geometry can compensate for the thermal effects, stringent requirements on surface figure, flatness, and parallelism of the TIR surfaces as well as difficulties in cavity design make slabs less attractive as near-term alternatives to YAG rods. Throughout the experimental investigations, those trades and issues associated with performance scaling to higher average powers were emphasized. For example, Nd:Cr:GSGG has the highest slope efficiency measured among the materials we have tested, but strong depolarization at moderate pump levels (15% single-pass loss at 200 W pump power) requires cumbersome resonator design or slab configuration to avoid losses in Q-switched operation or in a subsequent frequency conversion.

2. RESULTS

The materials tested are listed in Table 1.

Results for normal-mode slope efficiency at the fundamental laser transition of some of these laser materials, that are pumped at 2 Hz in a single-

Table 1 List of materials tested

Materials Tested*	Supplier	Dimensions [in.]	Notes
Nd:YAG	Airtron	1/4 x 3	Baseline
Nd:Cr:GSGG	Airtron	1/4 x 3	Nd = Cr = 2×10^{20} cm^{-3}
	Allied	1/4 x 3	
Nd:YLF	Airtron	1/4 x 3	0.5% Nd
	Sanders	4 mm x 3.5	
Nd:BEL	Allied	1/4 x 4	X-axis; 1079 nm
	Allied	1/4 x 3	Athermal; 1070 nm, 1079 nm
Nd:Glass	Schott	1/4 x 3	5% Nd LG760
	Kigre	1/4 x 3	Q-98

*All tested in CVI cavity with Ce-doped Xe lamp

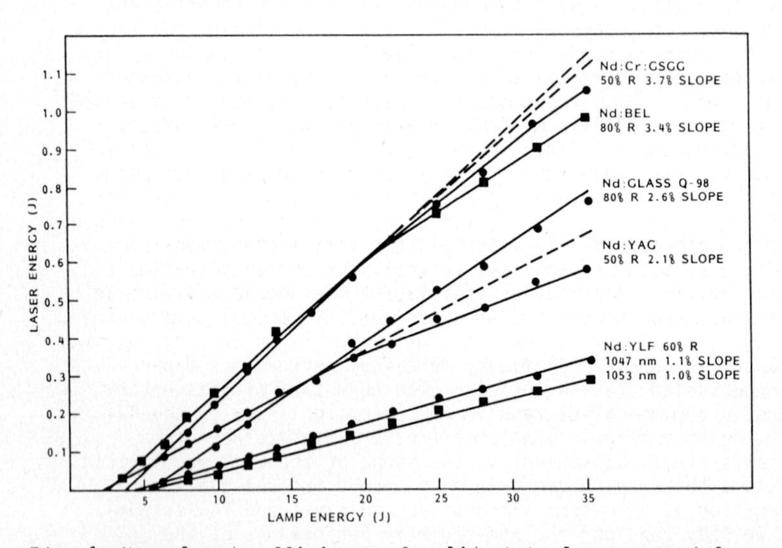

Fig. 1 Normal-mode efficiency of solid-state laser materials

lamp CVI cavity with dielectrically coated focusing reflectors, are shown in Fig. 1. The dashed lines are extrapolations of the linear least squares fit to the low-energy points. Deviation of laser energy from a straight line is caused by parasitics or blue spectral shift of the 80-μs duration flashlamp pulse at high energy. The outcoupler used gave the highest laser energy over the range of lamp energy covered, not necessarily the highest slope efficiency. Results for the two lines of an athermal Nd:BEL rod are shown in Fig. 2. The lower gain 1079-nm line was selected using an intracavity reflecting thin-film polarizer.

In a practical laser system, Q-switched pulses with Gaussian mode structure are often desired (perhaps for frequency conversion) at some moderate average power. Normal-mode slope efficiency then does not reflect the actual efficiency of a laser system, but the maximum effi-

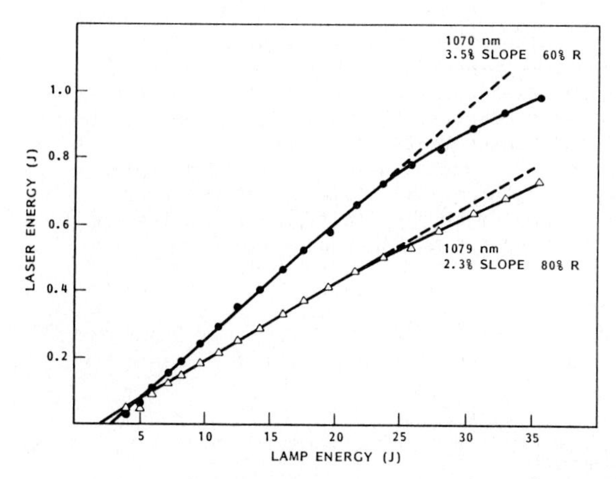

Fig. 2 Normal-mode efficiency of athermal Nd:BEL

ciency of the laser material. Loss of efficiency results from parasitic lasing, depolarization from thermally induced birefringence, diffraction, incomplete mode matching in the gain volume, and thermal lensing. As a measure of achievable lasing efficiency in a Q-switched, single-transverse mode, we have measured thermally induced depolarization and lensing properties of the candidate media listed in Table 1.

The depolarization shown in Fig. 3 was determined by measuring the transmission of a CW 1.06-μm beam through the pumped laser material situated between crossed polarizers. The probe beam illuminated the full aperture of the rod and the pump power was varied by changing the repetition rate at constant lamp energy. Note that depolarization is not constant across the radius of the rod; the effect of thermal birefringence in reducing laser efficiency depends on the beam size in the rod. The birefringent materials YLF and BEL are not expected to exhibit thermally induced depolarization.

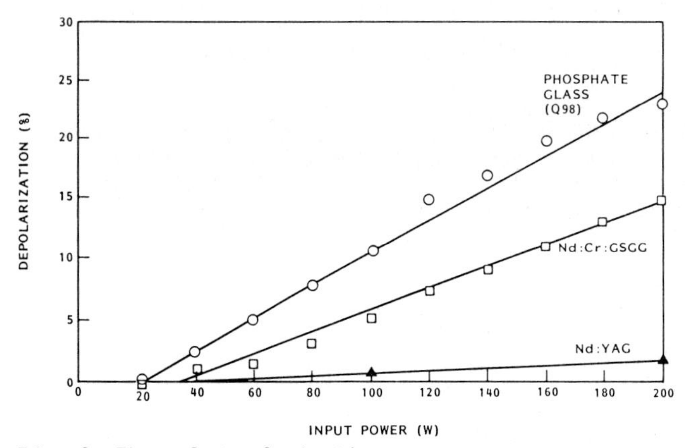

Fig. 3 Thermal depolarization

Shearing plates were used to measure the phase front of a CW probe beam at 1.06-μm passing through the pumped laser rod. Spherical phase changes cause a tilt and parallel displacement of fringes in the interfrogram.[1] The (biaxial) focusing power of the laser rod could then be determined. The results of these measurements are shown in Fig. 4. The flashlamp and cavity geometry are the same as used for the slope efficiency measurements. The focal length of GSGG could not be measured above 250 W pump power due to large nonspherical distortions in the rod. Other thermal lensing powers reported for Nd:YAG range from 0.5-1 diopter/kW.[2]

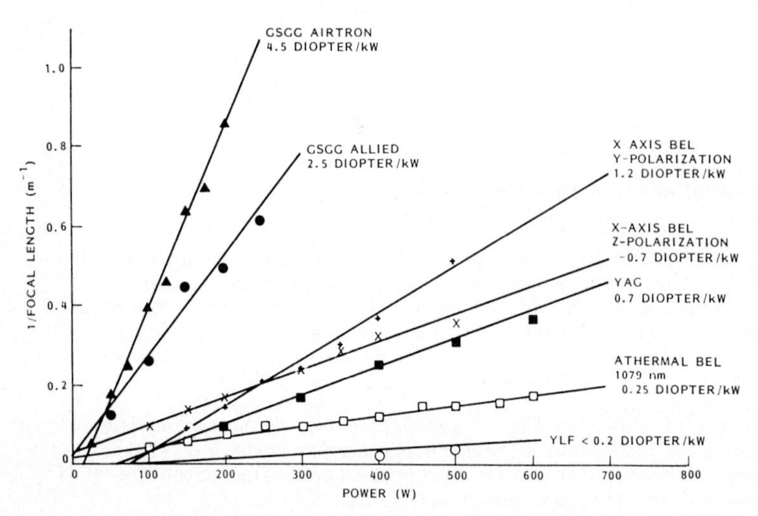

Fig. 4 Thermal focusing of solid-state laser materials

Additional measurements, such as Q-switched performance as a function of the repetition rate, are in progress. Although the materials we have may not achieve the highest performance reported, they are representative of typical samples,and the comparative procedure used has the advantage of eliminating variations due to pumping geometry and resonator optics.

3. CONCLUSION

A summary of the comparative measurements is shown in Table 2. GSGG lases most efficiently of all materials tested, but poor thermal properties limit its usefulness, in rod geometry, to the single-shot high-energy-per-pulse mode. Even a factor of two higher slope efficiency (compared to YAG) does not compensate for the much greater loss from thermal birefringence in a polarized resonator.

BEL is quite efficient in the x-axis cut with thermal lensing power comparable to YAG. The lack of depolarization offers a significant advantage for high average power applications. The thermal lensing powers shown in Table 2 for athermal BEL are for focusing in the x-direction with polarization of the probe beam either along the x-axis (1070 nm) or perpendicular to the x-axis (1079 nm). The values in parentheses are focusing power perpendicular to the x-axis for both polarizations. Some slight astigmatism is apparent from these results.

Table 2 Summary of Nd:doped materials performance

Material	Slope Efficiency (%)	Thermal Depolarization (%)	Thermal Lensing
YAG	2.1	1 (200 W)	0.7 Diopter/kW
GSGG	3.7	15 (200 W)	2.5 - 4.5
X-Axis BEL 1079 nm	3.4	-	1.2
Athermal BEL 1079 nm	2.3	-	-0.25 (0.28)
1070 nm	3.5	-	1.2 (1.6)
YLF 1047 nm	1.1	-	0.2
1053 nm	1.0	-	0.2
Glass	2.6	23 (200 W)	-

These BEL measurements coupled with lower gain than YAG makes it suitable for higher average power pumping of Ti:sapphire.

An extension of this comparison should include Q-switched performance, beam quality measurements, and evaluation of the potential of slab geometry for higher average powers.

REFERENCES

1. M. E. Riley and M. A. Gusinow, Applied Optics 16, 2753 (1977).
2. W. Koechner, "Solid-State Laser Engineering", Springer-Verlag, (1976) p. 355.

CW Tunable Laser Emission of $Nd^{3+}:Na_{0.4}Y_{0.6}F_{2.2}$

H. Chou, P. Albers, A. Cassanho, and H.P. Jenssen

Massachusetts Institute of Technology, Crystal Physics and Optical Electronics Laboratory, Cambridge, MA 02139, USA

1. Introduction

The combination of efficient laser diodes and optically pumped laser materials with high energy storage is currently of great interest. The requirements of the laser material for this application are somewhat different than with normal lamp excitation. First, the fluorescence lifetime of the upper laser level should be long in order to integrate the limited peak power from the laser diode. Second, since pumping by a laser diode can be near-resonant with the upper laser level, the thermal and mechanical requirements of the laser material are not quite as strict.

We here report on a Nd^{3+} laser host crystal ($Na_{0.4}Y_{0.6}F_{2.2}$) that meets the first requirement, τ = 800 μs. In addition, the emission is relatively broad and the laser emission may therefore be tuned both in the 1.05 μm and the 1.30 μm regions. Laser action has previously been reported in this system by Bagadasarov et al [1].

2. Sample Preparation

A $Na_{0.4}Y_{0.6}F_{2.2}$ single crystal was grown by the Czochralski technique from material synthesized under HF atmosphere according to the reaction $4NaF + 6YF_3 \rightarrow 10Na_{0.4}Y_{0.6}F_{2.2}$ with 1 mol % NdF_3 in the melt. The crystal retained the high-temperature cubic structure with a lattice constant a = 5.52Å when cooled to RT in agreement with observations in reference [1].

3. Spectroscopic Properties

The crystal structure of this material is a modification of the fluorite (CaF_2) structure. As shown in Fig. 1, the basic fluorite structure is cubic, with cation (Ca^{2+}) sites at the corners and the face centers of a cube and the anion (F^-) sites along the diagonals. Each cation is at the center of eight anions situated at the corners of a surrounding cube; and each anion has about it a tetrahedron of cations. In $Na_{0.4}Y_{0.6}F_{2.2}$, 40% of the Ca^{2+} ions are replaced by Na^+ and the remaining 60% by Y^{3+}. Due to the requirement of charge neutrality, there is an excess of 10% F^-. These excess F^- ions occupy interstitial sites. Also due to the requirement of local charge compensation, the remaining F^- ions are expected to be displaced from their normal positions. At present the precise nature of these displacements is not yet understood, and work is underway to study the site symmetry of the dopant ions. In any event, it is expected that there will be more than one type of dopant site. Some of the spectroscopic data to be presented below clearly show this.

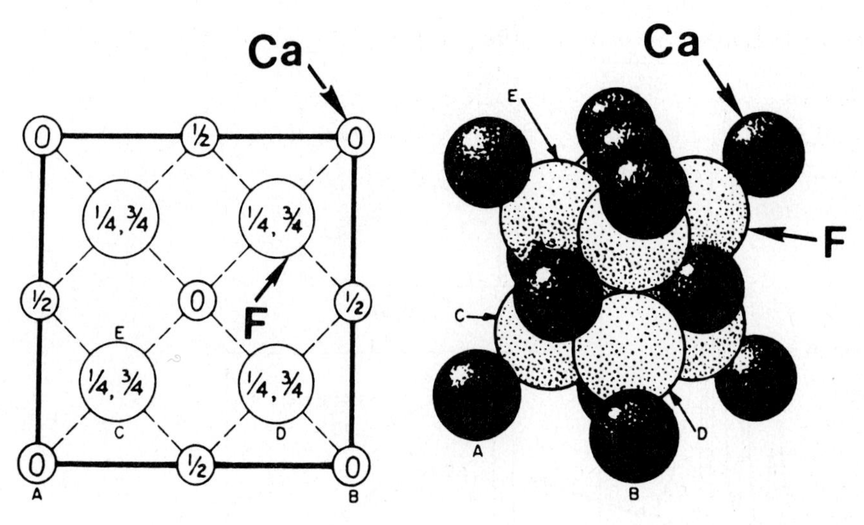

Fig. 1. The basic fluorite (CaF_2) structure

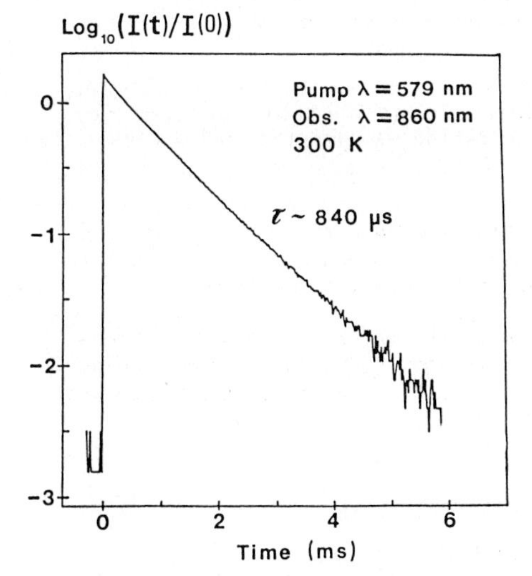

Fig. 2 Fluorescence decay of Nd $^4F_{3/2}$ after pulsed excitation

Figure 2 shows the fluorescence decay of the Nd $^4F_{3/2}$ level after pulsed excitation. The decay is not strictly exponential, but an approximate lifetime of 840 μs may be assigned. This is considerably longer than the fluorescence lifetime of Nd:YAG (230 μs) and Nd:YLF (500 μs). The main contribution to the non-exponential nature of the fluorescence decay is the simultaneous excitation of dopant ions at different sites at a given pump wavelength. Evidence of multiple dopant sites is shown in Fig. 3. Fig. 3(a) shows the room-temperature and 10K absorption spectra of the material between 540 and 600 nm, corresponding to the transition Nd $^4I_{9/2} \rightarrow {}^4G_{5/2}, {}^2G_{7/2}$. Two pump wavelengths, corresponding to points A and B in the figure, are chosen for pulsed excitation. At 10K, the fluorescence decay rate depends strongly on

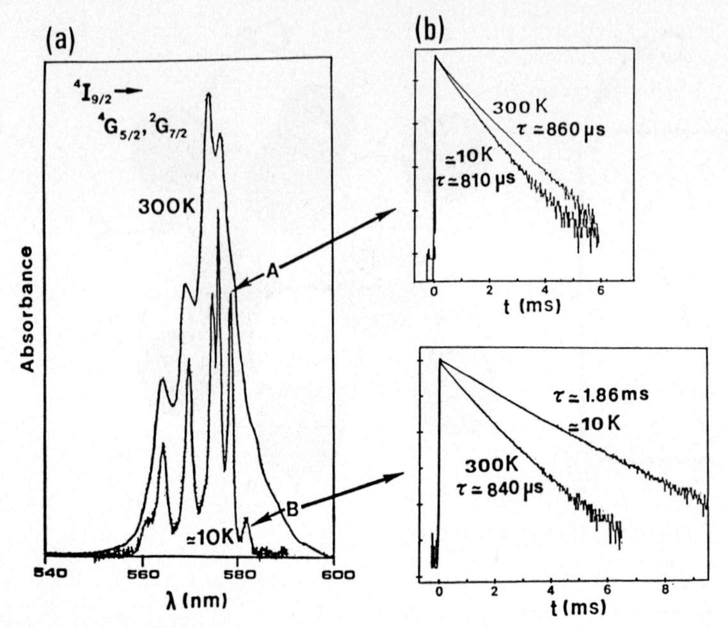

Fig. 3. Dependence of fluorescence decay rate on excitation wavelength: (a) room-temperature and low-temperature absorption spectra; (b) fluorescence decays at the indicated excitation wavelengths

the pump wavelength (see Fig. 3) indicating the selective excitation of different dopant sites. At room temperature, this dependence is no longer as pronounced due to the "smearing out" of the absorption spectrum.

The stimulated emission cross-section of the material between 1.0 and 1.45 μm is obtained by the "integral β-τ method" [2] and is shown in Fig. 4. The fluorescent quantum efficiency of the Nd $^4F_{3/2}$ level is assumed to be

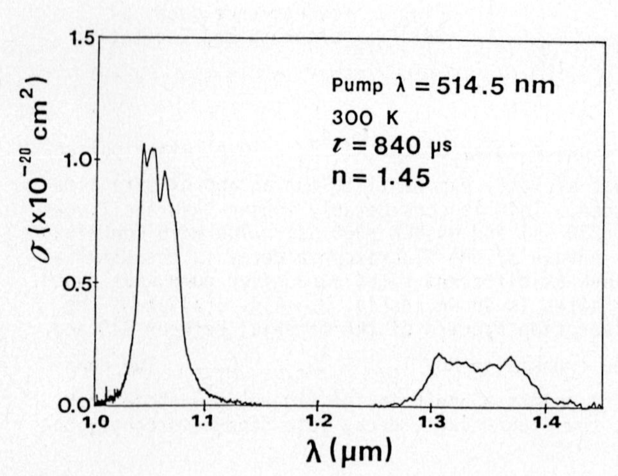

Fig. 4. Stimulated emission cross-section of $Nd^{3+}:Na_{0.4}Y_{0.6}F_{2.2}$

unity. The refractive index has not been accurately measured but is assumed to be 1.45. A single fluorescence lifetime of 840 μs was used to obtain the cross-section. We are now refining the measurements to separate out the cross-sections for the different dopant sites. The peak cross-section in Fig. 4 is at 1.04 μm, and is about 10^{-20} cm^2. The emission bands are broad due to different dopant sites and phonon broadening, indicating the potential for tunable laser emission.

4. Laser Results

The experimental arrangement for laser measurements is shown in Fig. 5. In the free-running setup, Fig. 5(a), a 514 nm argon laser beam is focused into a nearly concentric resonator of 20 cm in length. The pumped volume of the crystal is approximately 160 μm in diameter and 8 mm in length (the absorption coefficient at the pump wavelength is 0.15 cm^{-1}). In the tunable resonator, Fig. 5(b), an anti-reflection coated lens and a fused-silica prism are inserted between the crystal and the output mirror. Tuning is achieved by rotating the flat output mirror. The distance between the prism and the output mirror is about 50 cm.

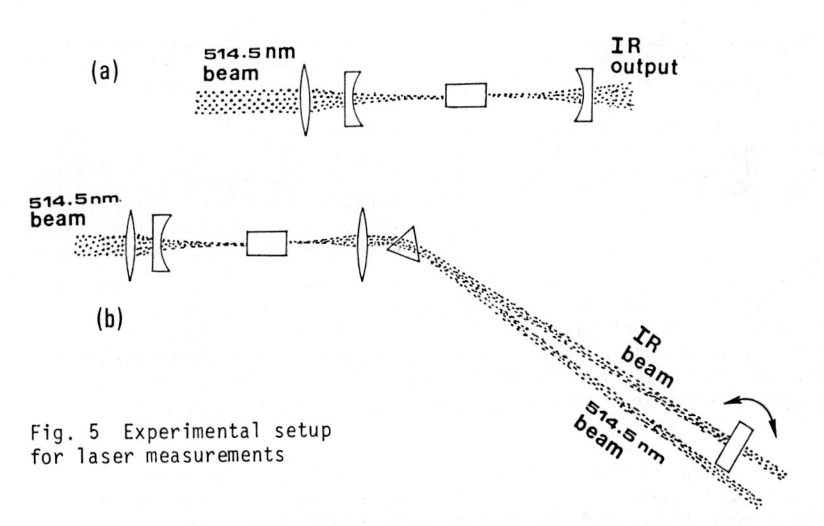

Fig. 5 Experimental setup for laser measurements

The output-vs-input characteristics for free-running operation are shown in Fig. 6. From the slope efficiencies of the curves, the round-trip loss in the resonator can be estimated from the equation

$$\frac{T}{T + L} = \frac{\lambda_L}{\lambda_p}\eta ,$$

where T is the output mirror transmission, λ_L the lasing wavelength, λ_p the pump wavelength, η the slope efficiency and L the resonator loss to be estimated. Our results show that L = 1.6 (± 0.1)%.Since this L includes scattering losses at the interfaces as well as the mode mismatch between the pumped volume and the lasing volume, it is not a parameter characteristic of the material alone. A parameter more representative of the material property is the distributed loss of the material.

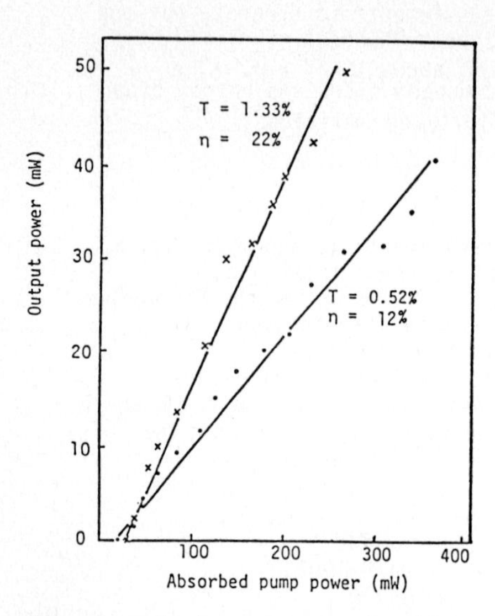

Fig. 6 Output power as a function of absorbed pump power. T is the output mirror transmissivity and η the slope efficiency

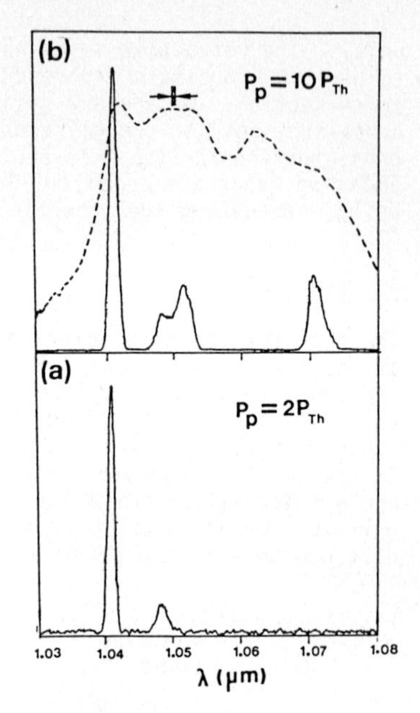

Fig. 7 Spectral characteristics of $Nd^{3+}:Na_{0.4}Y_{0.6}F_{2.2}$ in free-running operation. The dashed curve is the stimulated emission cross-section spectrum

Therefore it is useful to compare the above value of L with that obtained by using a different laser medium in the same setup. This was done with a Nd:BEL crystal [3], and we obtained L = 1.74%. These low loss values indicate that the roundtrip resonator loss is probably limited by the experimental arrangement and not by the loss in the material.

The spectral characteristics of the free-running laser are shown in Fig. 7. At low pump powers, Fig. 7(a), the laser emission appears primarily at two wavelengths. With increasing pump power, Fig. 7(b), a third peak appears around 1.07 μm. A comparison of these spectra with the spectrum of the stimulated emission cross-section (dashed curve in Fig. 7(b)) shows that the first two lasing peaks correspond well to the peaks in the cross-section, but at 1.06 μm, there is a peak in the cross-section which seems to produce no lasing. This latter peak may be associated with ions of the longer fluorescence lifetime in Fig. 3 (b). Since our cross-section measurements assumed the shorter lifetime of 840 μs, the actual cross-section for this peak may be considerable lower than indicated in Fig. 7.

The tuning curve is shown in Fig. 8. This is obtained by fixing the pump power at twice the threshold pump power and measuring the relative intensity of the laser output peak as the resonator is tuned. The positions of the three peaks agree with those in free-running operation. The regions between the peaks were not accessible: as the resonator is tuned

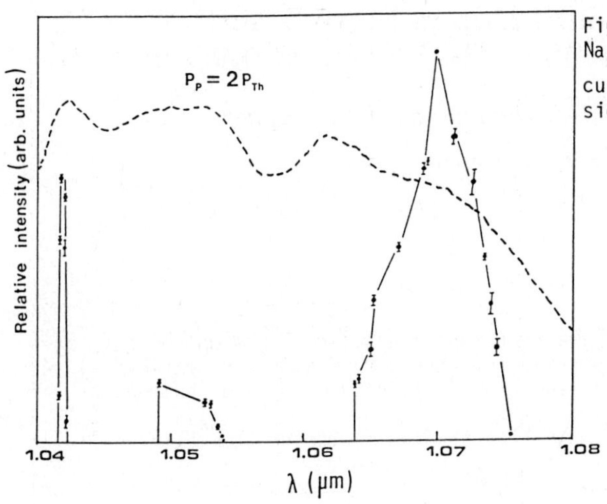

Fig. 8 Tuning curve of Nd^{3+}: $Na_{0.4}Y_{0.6}F_{2.2}$. The dashed curve is the stimulated emission cross-section spectrum

across these regions, at some point the lasing wavelength changes abruptly.

5. Summary

Nd^{3+}:$Na_{0.4}Y_{0.6}F_{2.2}$ has a long fluorescence lifetime of about 800 μs and is ideally suited to applications where a long storage time is required. It also has broadband emission and therefore tunable laser output. This material has good optical quality and low loss. A large boule (1 cm in diameter and 8 cm in length) has been grown and shows very good doping and optical uniformity.

Spectroscopic data revealed the existence of multiple dopant sites. The precise nature of these sites is now being investigated.

Acknowledgements

This work was supported by the Lawrence Livermore National Laboratory (Subcontract No. 5960505). A. Cassanho acknowledges partial support from FAPESP (Brazil).

References

1. Kh. S. Bagadasarov, A. A. Kaminskii, and B. P. Sobolev: Sov. Phys-Crystallography, **13**, 779 (1969).
2. B. Aull and H. P. Jenssen: IEEE J. Quant. Electronics, **QE-18**, 925 (1982).
3. H. P. Jenssen, R. F. Begley, R. Webb, and R. C. Morris: J. Appl. Phys. **47**, 1496 (1976).

Monomode Neodymium-Doped Fibre Laser: Tunable Continuous-Wave Oscillation at 0.9 μm

I.P. Alcock, A.I. Ferguson, D.C. Hanna, and A.C. Tropper

Physics Department, University of Southampton,
Southampton, SO9 5MH, UK

Recent experiments have shown that monomode silica fibre which has rare earth impurity ions incorporated into the core region can exhibit laser action with low threshold and high efficiency when end pumped by an external laser source [1,2]. Initial studies featuring the $^4F_{3/2} \rightarrow {}^4I_{11/2}$ transition around 1.08 μm of Nd^{3+} ions in monomode silica fibre investigated this sytem under conditions of Q-switching [3] and active modelocking [4] and demonstrated tuning over the range 1.07 μm - 1.14 μm [2]. A number of different sources have been used to pump fibre lasers including a cw dye laser [1,3,4], and GaAs diode laser [1], and an Ar^+ ion laser [2].

In this paper, we describe the first demonstration of cw operation of the $^4F_{3/2} \rightarrow {}^4I_{9/2}$ transition of Nd^{3+} ions in glass around 0.9 μm using an end pumped monomode fibre laser of this type. The configuration of the fibre laser cavity is shown in Fig. 1. Five hundred and eighty (580) nm wavelength light from a R6G dye laser is launched into the fibre through a high reflector optically contacted to the cleaved end of a Nd^{3+}-doped fibre (core diameter ~6 μm). An intracavity microscope objective (x10) produced a collimated beam at the output coupler. Tuning of the laser output is accomplished by a 2-plate birefringent filter of the type used in cw dye lasers. The insertion loss of such a filter is negligible when used at Brewster's angle with appropriately polarized light. However since the

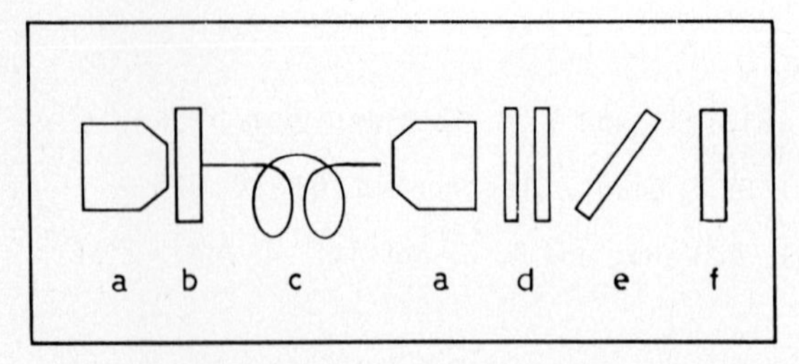

Fig.1. Fibre laser cavity: a) x 10 microscope objective; b) high reflector; c) doped fibre; d) $\lambda/4$ plates; e) birefringent filter; f) output coupler

fibre does not preserve polarization, it is necessary to insert two λ/4 plates into the cavity between the fibre and filter to convert elliptically polarized output of the fibre into the correct linear polarization.

For the $^4F_{3/2} \rightarrow {}^4I_{11/2}$ Nd^{3+} 4-level laser transition, the fibre length is chosen simply to ensure that the available pump power is absorbed. For the $^4F_{3/2} \rightarrow {}^4I_{9/2}$ Nd^{3+} transition, the situation is complicated by ground state reabsorption of the laser radiation, and, in general, shorter lengths of fibre must be used. The data presented here were obtained using a 1.2 m length of fibre which absorbed ~53 mW out of the available pump power of 130 mW.

Without any tuning elements in the cavity, the laser could be operated either at 906 nm or at 935 nm--wavelengths corresponding to the two peaks in the fluorescence curve (Fig. 2). Either peak could be selected by a small focusing adjustment (~5 μm) of the intracavity microscope objective. The linewidth at 935 nm was 1 nm in this configuration. With the birefringent filter and λ/4 plates in place, the laser output could be tuned smoothly from 900 nm to 945 nm. The variation of power output with wavelength is shown in Fig. 2.

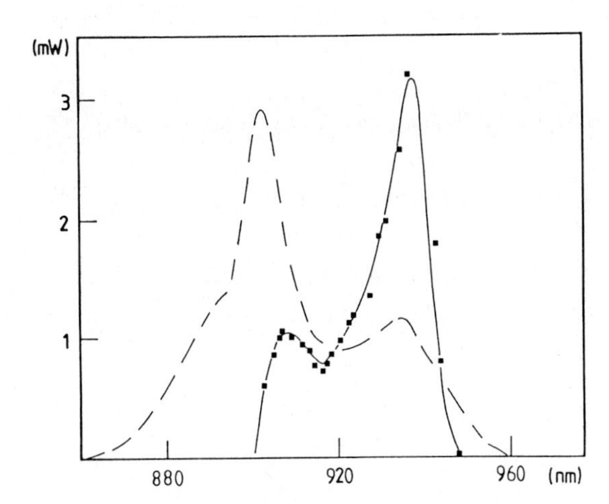

Fig.2. $^4F_{3/2}$ $^4I_{9/2}$ tuning curve (solid line) and fluorescence spectrum (broken line, arbitrary units)

Small rotations of the λ/4 plates were needed to maintain the correct polarization over the entire tuning range. With the tuning elements in the cavity, the laser linewidth is 0.06 nm. A cw output power of 3.2 mw was measured at the peak of the gain curve.

In conclusion, we have observed cw laser operation on the $^4F_{3/2} \rightarrow {}^4I_{9/2}$ Nd^{3+} transition which has hitherto only been shown to lase in pulsed mode in glass. This illustrates an essential property of glass lasers in fibre form: that confinement of the pump radiation field within a core of small transverse dimensions produces intense pumping without associated thermal problems [5]. Another instance has been provided by the first demonstration of cw laser operation on the 1.54 μm line in an Er^{3+}-doped fibre [2]. We anticipate that it should be possible to achieve lasing on other transitions which have not hitherto exhibited laser action in a glass host. The special technique used to fabricate these doped fibres can be adapted to a very wide range of impurity types [6]. In addition we have applied a simple technique for tuning fibre lasers which has the merit of producing polarized laser output from a polarization non-preserving fibre [7].

This work has been supported by a grant from the UK SERC and also under a JOERS program. The authors are grateful to the Optical Fibre Group in the Department of Electronics and Information Engineering for supplying us with the fibre.

References

1. R. J. Mears, L. Reekie, S. B. Poole, and D. N. Payne, Electron. Lett. 21, 738 (1985).

2. L. Reekie, R. J. Mears, S. B. Poole, and D. N. Payne: Journal of Lightwave Technology 4 956 (1986).

3. I. P. Alcock, A. C. Tropper, A. I. Ferguson, and D. C. Hanna, Electron. Lett. 22, 85 (1986).

4. I. P. Alcock, A. I. Ferguson, D. C. Hanna, and A. C. Tropper, Electron. Lett. 22, 268 (1986).

5. I. P. Alcock, A. I. Ferguson, D. C. Hanna, and A. C. Tropper, Opt. Comm., in press.

6. S. B. Poole, D. N. Payne, and M. E. Fermann, Electron. Lett. 21, 737 (1985).

7. I. P. Alcock, A. I. Ferguson, D. C. Hanna, and A. C. Tropper, Opt. Lett. in press.

Nonradiative Processes and Blue Emission in Nd:YLF

T.Y. Fan and R.L. Byer

Edward L. Ginzton Laboratory, Stanford University,
Stanford, CA 94305, USA

1 Introduction

In addition to one-photon absorption and fluorescence in rare earth and transition metal ion-doped crystals and glasses, a number of other processes can occur such as two-step excitation and energy transfer which may affect laser operation. For example, concentration quenching in Nd^{3+} doped materials is an energy-transfer phenomenon, and energy transfer or two-step excitation can convert infrared photons to visible fluorescence [1] or visible lasers[2]. It has also been shown that energy-transfer upconversion (ETU), also known as Auger recombination, can increase laser threshold and change laser dynamics [3]. Both sequential two-photon excitation (STEP) and energy-transfer upconversion (ETU) have been identified in Nd^{3+} doped materials. These two processes are shown in Fig. 1. STEP involves absorption of one photon to an excited-state, followed by excited-state absorption (ESA) of a second photon to an even higher-lying level. In ETU, two nearby ions which are in excited-states interact causing one ion to relax while the other is simultaneously excited to a higher state. This is identical to concentration quenching except that one of the ions is in the ground state in concentration quenching. The transition rate for ESA increases with excited-state population density as does the rate for ETU. These processes may be especially important in resonant pumping because the excited-state densities can be quite high.

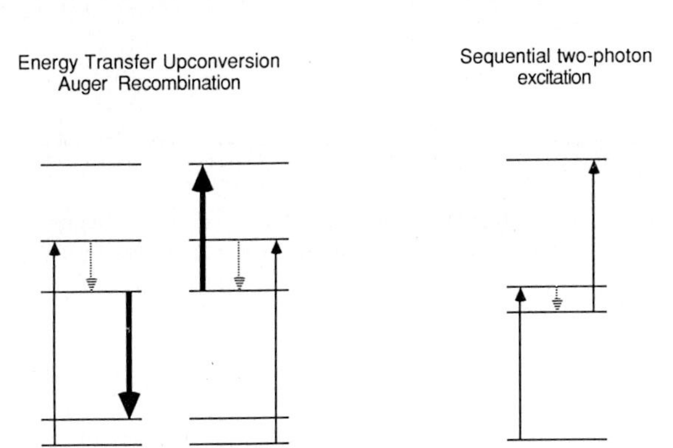

Figure 1 ETU and STEP illustrated. In ETU on left, two ions are excited and relax to metastable states as shown by the light solid arrows and dashed arrows respectively. The ions then interact causing one to relax and the other to be excited simultaneously. In STEP on the right, an ion is excited by one photon, relaxes to a metastable level then is excited to a higher lying state by a second photon

Both STEP and ETU were intially identified in Nd:YLF during work on cw resonantly pumped lasers[4]. Strong blue and near uv fluorescence occurred when the pump source was a Rhodamine 590 dye laser operating at 589 nm, and weak yellow fluorescence occurred when the pump source was a diode laser operating at 791 nm. Both the blue and yellow fluorescence decreased sharply when the laser cavity was aligned, indicating that the $^4F_{3/2}$ upper laser level was important in the process. Blue and uv fluorescence has been observed in other Nd^{3+}-doped crystals by STEP in Nd:YAG[5], both STEP and ETU in $Nd:LaF_3$ [6], and by two-photon excitation in Nd:YLF [7] (all under pulsed excitation). ETU has been identified in $NdCl_3$ [8], NdP_5O_{14} [9], and $LiNdP_4O_{12}$ [10] (with yellow fluorescence noted in $NdCl_3$).

2 Fluorescence

The polarized fluorescence spectrum in the blue has been measured by pumping with a cw Rhodamine 590 laser via STEP. The upper metastable level was assigned to the $^4D_{3/2}$ and $^4D_{5/2}$ manifolds. Fluorescence was observed around 358 nm, 383 nm, 413 nm, and 450 nm which correspond to terminal manifolds $^4I_{9/2}$, $^4I_{11/2}$, $^4I_{13/2}$, and $^4I_{15/2}$. This fluorescence spectrum is shown in Fig. 2. The fluorescent lifetime was measured to be 23±2 μs and, therefore, based on the multiphonon relaxation rate for YLF [11], the radiative quantum efficiency is ~40%.

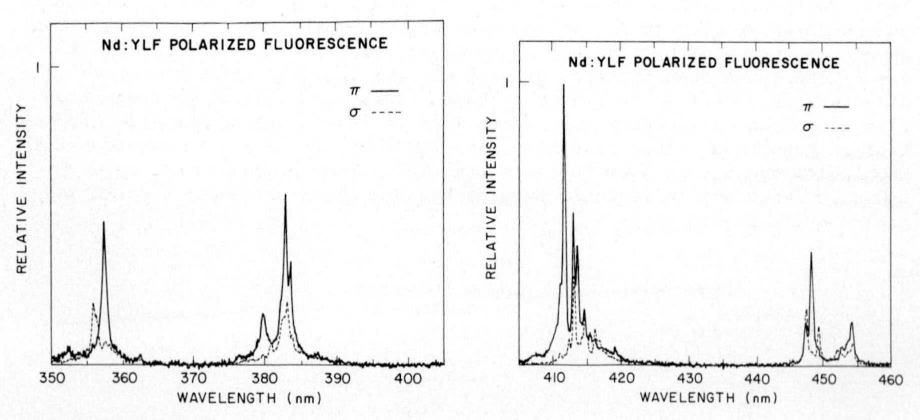

Figure 2 Fluorescence spectra for both π (E||c) and σ (E⊥c) polarizations. The spectra were normalized for the spectral response of the measurement system

An Ar^+ laser at 514.5 nm was used to excite fluorescence near 585 and 650 nm, which were assigned to the $^4G_{7/2} \rightarrow {}^4I_{11/2}$ and $^4G_{7/2} \rightarrow {}^4I_{13/2}$ transitions, respectively. We have tentatively assigned this to be the same emission as that excited by the diode laser, but in Ar^+ laser pumping the excitation is by typical one-photon absorption as opposed to an upconversion process. The fluorescence lifetime has not been measured for these transitions, but most of the relaxation (~90%) should be nonradiative based on radiative transition rates calculated from Judd-Ofelt theory for $^4G_{7/2}$ and multiphonon relaxation rates. This implies that the fluorescent lifetime is essentially equal to the inverse of the multiphonon relaxation rate for YLF [11] which is ~10^5 sec^{-1}. Fluorescence from the upper states ($^4G_{7/2}$, $^4D_{3/2}$, and $^4D_{5/2}$), in Nd^{3+}-doped materials) is typically seen only under pulsed excitation because the relaxation rates are dominated by multiphonon relaxation, which increases rapidly with increasing host phonon energies. YLF has a lower multiphonon relaxation rate than many other Nd doped crystals [12] (about one order of magnitude less than YAG) because of its relatively low phonon energies, so more of the relaxation is radiative, allowing fluorescence to be easily measured in a cw regime.

3 Analysis and Discussion

STEP and ETU have different wavelength, temporal, and dopant concentration dependences which allow the two processes to be distinguished. In the present case, we have used excitation spectra where the pump wavelength is scanned and the STEP or ETU induced fluorescence is monitored. If ETU occurs, the relative excitation should go as the square of the one-photon absorption (for dipole-dipole interaction) since absorption of photons on only one transition is required, and the transition rate is proportional to the excited state population. On the other hand, if STEP occurs, the excitation spectrum may appear very different from the one-photon absorption spectrum because two absorption transitions are involved, which are at the same wavelength. The excitation spectra of the blue fluorescence was measured by monitoring the $^4D_{3/2}$, $^4D_{5/2} \rightarrow {}^4I_{13/2}$ transition while the Rhodamine 590 laser wavelength was tuned in the 575 - 608 nm range. The excitation spectra for both π and σ polarizations indicate that the mechanism for the blue fluorescence is STEP with the second absorption being the $^4F_{3/2} \rightarrow {}^4D_{5/2}$ transition. Figure 3a summarizes the energy levels involved in STEP.

In the case of diode laser excitation, a slightly different technique was used. Yellow emission was noted for pumping both at 791 nm and 747 nm. For these pump wavelengths STEP can be ruled out because there is no higher energy level to which ESA can occur. The closest energy level is $^2K_{15/2}$ for both wavelengths which is off-resonance by ~120 cm^{-1} for 791 nm pump and ~920 cm^{-1} for 747 nm pump [13]. However for ETU, there are energy levels

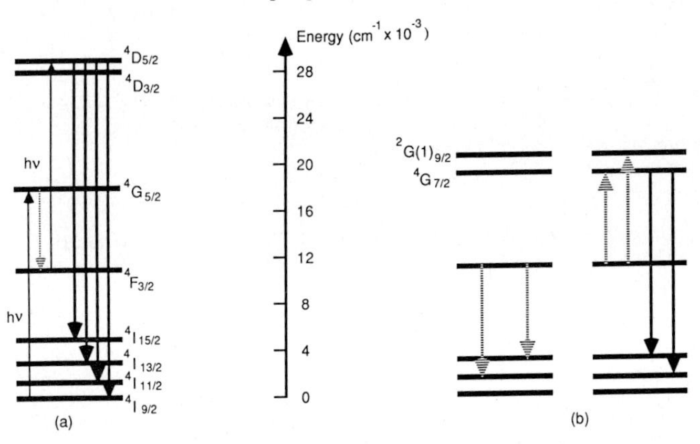

Figure 3 (a) STEP energy level assignments showing excitation in light solid arrows, relaxation in dashed arrows, and subsequent fluorescence in heavy arrow. (b) Tentative ETU energy level assignments showing interaction between ions in dashed arrows and fluorescence in solid arrows

which are in resonance for the interaction between ions, so ETU is likely. Figure 3b is shows the tentative assignment of energy levels involved in ETU.

It may be possible to achieve laser oscillation on the $^4D_{3/2}$, $^4D_{5/2}$ - ^4I_J transitions by the traditional pumping approach. From the fluorescence spectra and estimated radiative lifetime, the peak effective stimulated emission cross-section was calculated to be ~5 x 10^{-20} cm^2 at 411.7 nm in the π polarization. It even may be possible to achieve laser oscillation by STEP pumping. The pump rates can be calculated from the one-photon absorption coefficient and the effective ESA cross-section from $^4F_{3/2}$, which was measured to be 2 (± 1) x 10^{-20} cm^2 at 587.4 nm in the π polarization. These quantities are sufficient to calculate a laser threshold under the assumptions of no ESA from $^4D_{3/2}$, $^4D_{5/2}$ and some level of cavity loss. Assuming the round-trip cavity loss to be 0.5%, the pump and cavity spot sizes to be 20 microns, and the pump double-passed, an estimate of 70 mW threshold was calculated for a 587.4 nm pump.

Both STEP and ETU are likely to occur to varying degrees in other Nd^{3+} doped materials just as with concentration quenching. In fact, we have also observed faint blue emission under cw Rhodamine 590 pumping of Nd:YAG and Nd-doped phosphate glass. Since we have not analyzed the processes involved in these two cases, it appears likely that STEP is the promising mechanism. As previously mentioned, all the above effects can increase laser threshold and affect excited-state dynamics. STEP should cause no difficulties as long as the pump is tuned away from the ESA. On the other hand, it may be important to model ETU in high-gain amplifiers since small differences in upper-state population can lead to large differences in small signal gain. Accordingly, cross-sections for ETU need to be measured before the magnitude of the effect can be determined.

4 Summary

Two excited-state processes, STEP and ETU, are identified in Nd:YLF. These processes may occur in other Nd^{3+} doped crystals and glasses as well. Further measurements are needed to determine the importance of these effects on laser action. However, it may be that laser operation from the $^4F_{3/2}$ manifold should be only slightly affected in lightly-doped crystals (i.e., not stoichiometric). Finally, efficient blue emissions are observed in Nd:YLF, which may serve as new laser transitions.

This research was supported by the U.S. Office of Naval Research and NASA. The authors would also like to thank H.P. Jenssen for providing the Nd:YLF used in this experiment.

References

1. F. Auzel, Proc. IEEE 61, 758 (1973).
2. L.F. Johnson and H.J. Guggenheim, Appl. Phys. Lett. 19, 44 (1971).
3. K. Otsuka and K. Kubodera, IEEE J. Quantum Electron. QE-16, 538(1980).
4. T.Y. Fan, G.J. Dixon, and R.L. Byer, Opt. Lett. 11, 204 (1986).
5. G.J. Quarles, G.E. Venikouas, and R.C. Powell, Phys. Rev. B 31, 6935 (1985).

6. B.R. Reddy and P. Venkateswarlu, J. Chem. Phys. 79, 5845 (1983).
7. Z. Song, S. Lia, Y. Gui, J. Jiang, D. Hua, S. Wang and Y. Huo, Acta Phys. Sin. 33, 1023(1984).
8. W.D. Partlow, Phys. Rev. Lett. 21, 90 (1968).
9. M. Blatte, H.G. Danielmeyer, and R. Ulrich, Appl. Phys. 1, 275(1973).
10. K. Otsuka and K. Kubodera, IEEE J. Quantum Electron. QE-16, 419 (1980).
11. H.P. Jenssen, Technical Report 16 (MIT Crystal Physics Laboratory, 1971), p. 83.
12. W.F. Krupke, in *IEEE Region Six Conference Record* (IEEE, New York, 1974), p. 21.
13. A.A.S. da Gama, G.F. de Sa, P. Porcher, and P. Caro, J. Chem. Phys. 75, 2583 (1981).

Part X

Applications and Nonlinear Optics

Future Lidar Platforms in Space

L.V. Taylor and R.R. Nelms

NASA Langley Research Center, Mail Stop 476, Hampton, VA 23665, USA

1. INTRODUCTION

Current Agency plans for the NASA Space Station Program include development of a laser facility to conduct scientific experiments from a polar-orbiting platform. A recommended set of experiments for the facility is to improve scientific understanding of the Earth's atmosphere utilizing techniques of laser atmospheric backscatter (optical radar), Differential Absorption Lidar (DIAL), altimetry and retro-ranging. Laser atmospheric backscatter experiments have a long heritage as remote sensing experiments to measure the vertical profiles of atmospheric aerosols, the altitude and optical thickness of clouds, and the height of the planetary boundary layer (PBL). These experiments have been conducted since the early seventies from ground-based facilities, and more recently from low-altitude aircraft. For optical backscatter experiments, wavelength tunable lasers have not been essential, and these experiments have been conducted using conventional ruby lasers, and Nd:YAG lasers at the fundamental and harmonic frequencies (i.e. 1064 nm; 532 nm; 355 nm; 694 nm; 347 nm). The basic principle of optical radar when compared to conventional radar depends upon the fact that at optical frequencies, the wavelength is approximately five orders of magnitude smaller and approaches the size of atmospheric particulates and molecules. This enhances the backscatter cross section considerably, and through range gating techniques provides a capability to measure the vertical profiles of atmospheric aerosols with a resolution of better than 1 km. The narrow field of view provided by the coherence of the laser beam, in principle, allows a horizontal resolution on the order of 50 meters from space.

Two lidar experiments currently being developed by NASA to apply lidar for measuring atmospheric constituents will be discussed, highlighting subsystem techniques that are being developed to provide the high accuracy and precision required to meet the scientific objectives. One project is the Lidar Atmospheric Sensing Experiment (LASE) to measure H_2O profiles from a NASA U-2 aircraft (ER-2). The other project is the Lidar In-Space Technology Experiment (LITE) to measure aerosol backscatter from the Space Transportation System (STS).

This paper will also discuss the Laser Atmospheric Sounder and Altimetry (LASA) lidar facility for the EOS mission in the early 1990's. The science for this facility has been developed by the LASA Working Group. The technology for the LASA Lidar facility will be verified on the LASE and LITE experiments.

2. LASE EXPERIMENT[1]

The LASE program is the first step in the overall NASA effort to develop and demonstrate an autonomous tunable DIAL laser instrument for airborne and spaceborne flight experiments. Performance criteria of a DIAL instrument to measure water vapor and aerosol vertical profiles in the atmosphere have been defined through extensive development of mathematical simulations. One of the objectives of the LASE program is to verify and validate these mathematical simulations, and conduct scientific investigations of lower tropospheric water vapor and aerosols on a broad spatial scale.

In this paper, a brief description of the lidar platform, the instrument and its subsystems will be presented. The scope of the instrument development is to design a modular instrument to measure water vapor and aerosol in the atmosphere with a fine vertical resolution ranging from 0 to 16 km altitude during daytime and nighttime. The modular design for the platform is intended to permit future experiments to be performed with new laser technology as it evolves from the laboratory stage.

A required design feature of the LASE instrument is minimum pilot interaction in the instrument operation, i.e., nearly autonomous operation. Prior to each flight, the experimenter selects two H_2O absorption lines whose wavelengths are programed into a real-only computer memory for in-flight selection by the pilot in a prespecified manner. The pilot control of the instrument consists of power on/off, standby, operate, line 1, and line 2. A monitor light will indicate to the pilot when the mission should be terminated for predetermined instrument conditions that preclude further operations.

A block diagram of the LASE instrument is shown in Fig. 1. The interfaces of all subsystems are with the command and data subsystem (CDS). The LASE instrument consists of the following subsystems: laser,

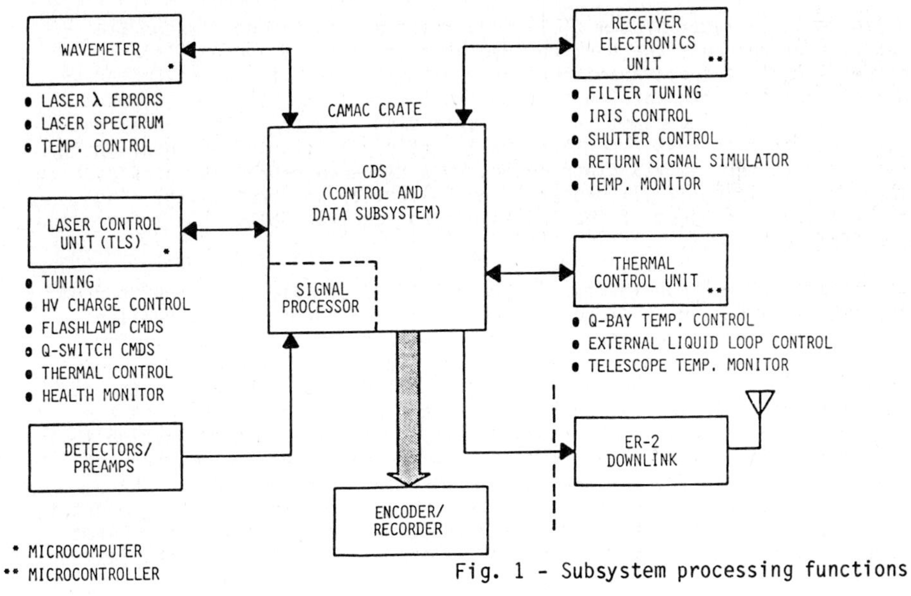

Fig. 1 - Subsystem processing functions

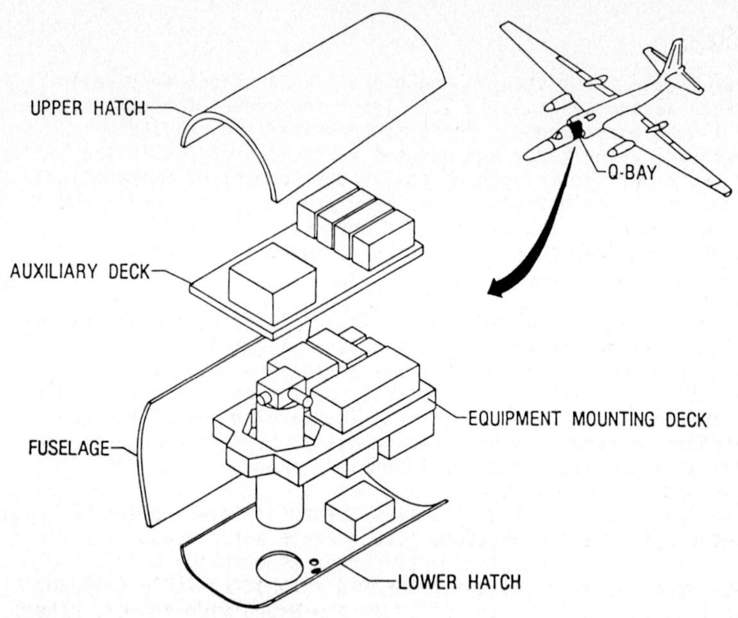

Fig. 2 - Lidar atmospheric sensing experiment - ER-2 installation

wavemeter, control and data system, receiver, and thermal control. All data generated in the instrument are formated by the CDS and recorded onboard on a 14-track recorder. All data can also be down-linked to the ground for real-time monitoring when the aircraft is within range of the ground receiver. Figure 2 is representative of the mounting of the instrument into the Q-bay of the ER-2. The data recorder is mounted in the nose section of the ER-2.

The tunable laser system (TLS) contains dual, multimode Alexandrite lasers that generate the on-line (H_2O absorption line) and off-line pulses. Part of the transmitted laser beam is routed to the wavemeter by mirrors for centroid wavelength and relative mode amplitude measurement, the latter being important in error correcting schemes in the postflight data reduction process. Another small part of the transmitted beam energy is routed to the CDS via fiber optics to a time-coherent phase generator in the CDS for timing of all onboard data collection and processing. The overall tuning process of the TLS is controlled internally with information provided by the CDS.

The wavemeter shown in Fig. 3 receives 10 mJ of laser beam inputs from the TLS via a system of mirrors. The laser beam enters an integrating sphere with outputs to each of three Fabry-Perot interferometers. The integrating sphere is necessary to assure that the multimode laser beam uniformly illuminates the interferometers. The output of each of the interferometers is focused onto a photodiode array for retrieval of the fringe patterns. Using algorithms developed, the transmitted wavelength is computed in the wavemeter electronics. Also the difference between the desired wavelength and the transmitted wavelength is computed in the wavemeter electronics and stored in the CDS and used in the tuning process of the TLS. The third stage of the wavemeter is a spherical Fabry-Perot interferometer that provides a spectral energy distribution of the laser

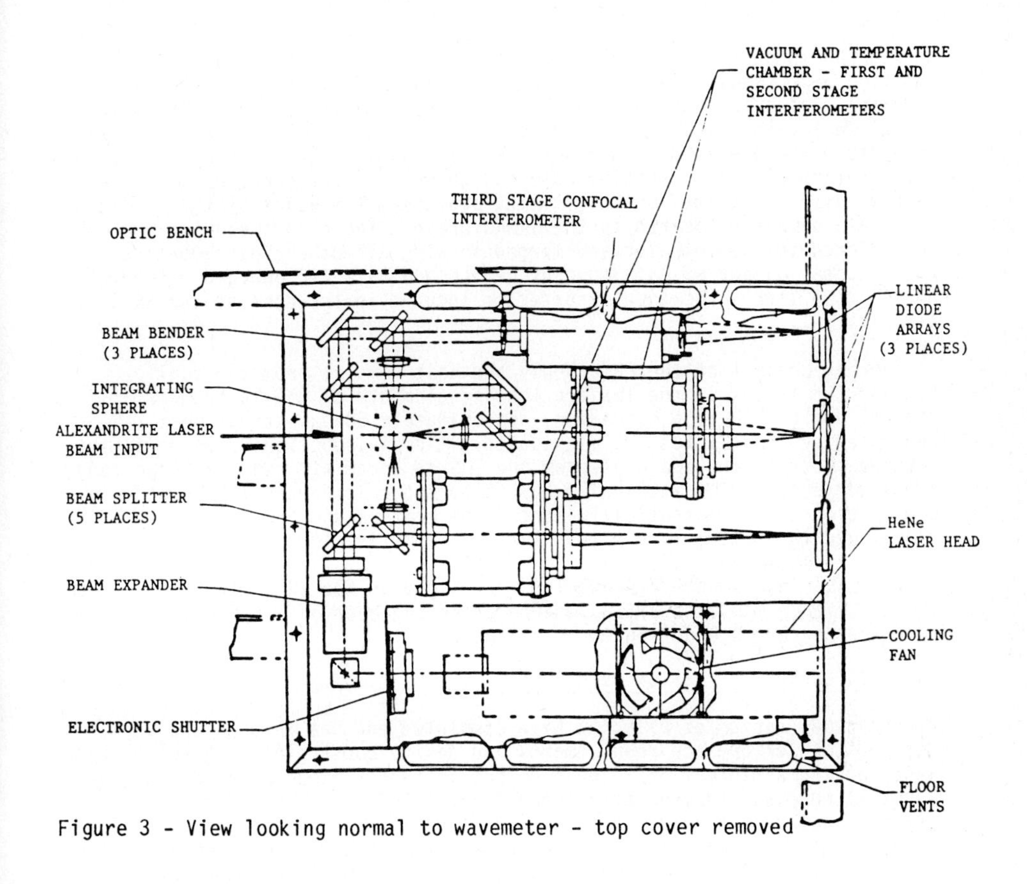

Figure 3 - View looking normal to wavemeter - top cover removed

outputs to a relative accuracy of 10 percent for overall error reduction schemes in the processing of science data. Included in the wavemeter housing is a stable He-Ne laser to be used to track the stability of the Fabry-Perot interferometers during mission operations. The reference laser output from the interferometers will be used to provide corrections to the in-flight software that computes the wavelength centroid of the Alexandrite laser pulse.

The telescope/receiver system is a Dall-Kirkham design. The field of view is adjustable remotely by an electronic field stop at the focal point of the telescope. A bandpass filter is used in the aft optics to minimize background noise during daytime missions. To accommodate the in-flight selection of two different preselected H_2O absorption lines, the narrowband filter can be tilted up to 10° for centerband shifts of approximately 2.5 nm. The science detectors are avalanche photodiodes (APD's) for a high-gain and a low-gain channel. The high-gain channel detects the on-line and off-line atmospheric backscatter signal for use in the range-resolved H_2O vapor DIAL computation. The low-gain channel detects the cloud return and ground return signal amplitudes. The ground return signal is used to derive the total H_2O column content. Also included in the aft optics is a quad detector that will be used to align the telescope and monitor alignments during flights.

The CDS is on the ER-2 auxiliary deck located above the equipment mounting deck (EMD). The CDS is contained in CAMAC crates using plug-in modules. The output of the two science detector APD's are each input to a 12-bit analog-to-digital (A/D) converter, Transiac model TR2012. The A/D converters are operated at a 5-MHz word rate during the short bursts of the lidar return signal. The time coherent generator, triggered with the laser output pulse, initiates the digitizing process. The heart of the CDS is a LSI-11/75 computer located in the CAMAC crate. The high-speed science data is buffered in the computer and merged in with all other instrument data with a total output of 93.33K words per second rate for onboard recording and RF down link. The tape recorder is located in the nose of the ER-2 aircraft.

Thermal control of the ER-2 Q-bay and heat removal from the dual Alexandrite lasers in the TLS are the key elements of the instrument thermal design. During operation, approximately 3000 watts of heat generated by the TLS must be dissipated. A dual-loop liquid system in the instrument transfers the heat from the TLS to a ram air heat exchanger on the lower hatch. The internal Q-bay temperature is controlled by a combination of thermostatically-controlled heaters and fans to evenly distribute the temperature within a range from 15°C to 40°C.

The LASE instrument is scheduled for flight on the ER-2 aircraft in late 1988. Science missions will commence after engineering verification of hardware performance during early flights.

3. LITE EXPERIMENT DESIGN OVERVIEW[2]

A conceptual design of LITE has been completed and has through analyses defined the performance requirements and design constraints. A solid-state laser having a minimum of 1 joule output at the fundamental wavelength of 1.064 microns and incorporating second and third harmonic generating crystals would be necessary to meet the lidar system performance requirements. In addition, a 1-meter telescope would be necessary to provide the sensitivity required to make a high signal-to-noise ratio measurements of the return signals.

The solid-state laser requires a large amount of electrical power (approximately 2 kilowatts) the majority of which must be dissipated in the form of heat, therefore, power and coolant services must be approved by the Space Transportation System (STS) Orbiter. The STS Orbiter will also provide a number of command and data management services necessary to conduct the experiment. A principal design task was to utilize these services in an efficient and cost-effective approach through the selection of the most suitable Spacelab pallet (experiment carrier) system.

The conceptual design process resulted in the selection of the Spacelab Enhanced MDM Pallet (EMP) as the experiment carrier. The EMP will provide all mechanical, thermal, electrical, and command and data management services necessary to support the experiment in an efficient and cost-effective manner. An autonomous environmental data subsystem was included to provide measurements of the environment to which the LITE is subjected during the STS launch, orbit insertion, on-orbit, reentry, and landing phases.

In addition to the above requirements and constraints, it was also essential that the overall program costs be minimized by utilizing existing space-qualified hardware (or existing space hardware designs). Another

Figure 4 - LITE configuration

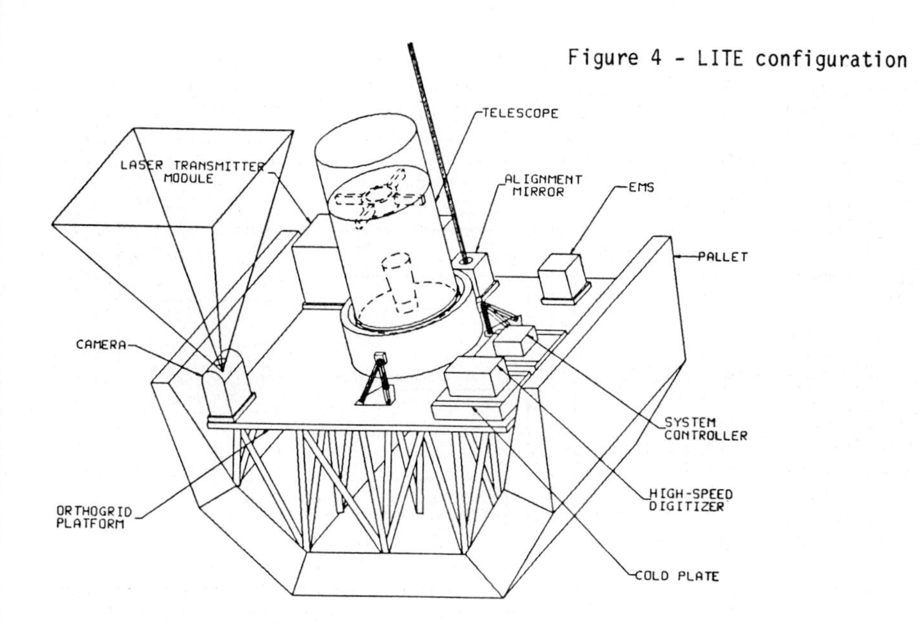

requirement was to configure the system design to accommodate possible modification on later flights (e.g., different lasers, detector packages, or processing electronics).

3.1 Experiment Configuration

The LITE lidar platform is of a modular configuration (Fig. 4) that facilitates component or subsystem replacement during field servicing or upgrades on future missions. The LITE design is modular and consists of a Laser Transmitter Module (LTM), telescope receiver, aft-optics assembly, electronics subsystem, alignment system, environmental measurement system and a platform.

The laser transmitter module (Fig. 5) consists of a flashlamp-pumped, Q-switched Nd:YAG laser with a fundamental output of 1.064 microns. The fundamental output energy is in excess of 1 Joule per pulse, with a selectable pulse rate of either one or ten pulses per second. In addition the module includes second and third harmonic generation which convert part of the fundamental energy into harmonic energy at 0.532 and 0.355 microns. The net energy at 1.064 microns is 200 mJ, 400 mJ at 0.532 microns, and 150 mJ at 0.355 microns. The laser is housed in a sealed container and cooled via a heat transfer unit which is connected to the EMP coolant system.

The telescope receiver was originally designed for the NASA Orbiting Astronomical Observatory (OAO) spacecraft, and was the prototype for the Goddard Experiment Package (GEP) of the OAO. The telescope section of the GEP was obtained from the Goddard Space flight Center and the necessary refurbishment for the LITE application is presently in process. The telescope has a 38-inch diameter primary mirror and is a Ritchey-Chretien form of Cassegrain configuration. The primary mirror features a beryllium substrate with a Kanigen overcoat and has an aluminum reflecting surface. The secondary mirror is fused quartz with an aluminum reflecting surface.

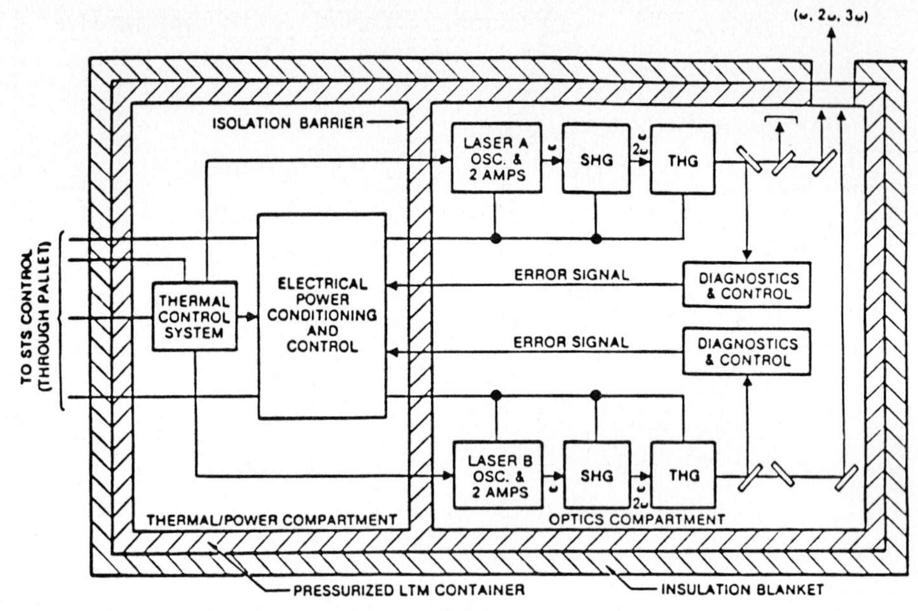

Figure 5 - LITE laser transmitter module

The aft-optics assembly contains the optical detector components which receive and separate the backscattered laser signals. The optical assembly is mounted on an optical bench and suspended under the primary mirror. The assembly consists of lenses, mirrors, optical filters, beamsplitters and detectors in a configuration which separates and filters the three spectral components of the return signal. There is a photomultiplier tube for the 0.532 and 0.355 micron wavelength channels and a silicon avalanche photodiode detector for the 1.064 micron wavelength channel. The optics assembly design contains a variable field stop controlled either by command or auto sequence operation. A movable mirror just beyond the field stop will redirect the return signals into a separate optical train for automatic realignment of the laser output signal to the telescope optical axis.

The alignment system consists of a two-axis mirror that maintains co-alignment between the laser transmitter and the telescope receiver. This collinearity is accomplished through automatically directing the outgoing laser beam by a closed-loop feedback system that uses a four quadrant detector to determine the location of the backscattered laser signal. The alignment between the laser transmitter beam and the telescope receiver optical axis is required to be within 50 microradians in order to achieve the required co-alignment. The system is also required to passively hold the preflight alignment throughout the mission and fail to the initial mirror position as a primary failure mode. Prior to each data pass, the system will check the alignment and make the required corrections when needed to reestablish a laser transmitter and telescope receiver co-alignment.

The system electronics (Fig. 6) consists of a system computer unit and a high-speed digitizer unit. The computer controls and monitors the

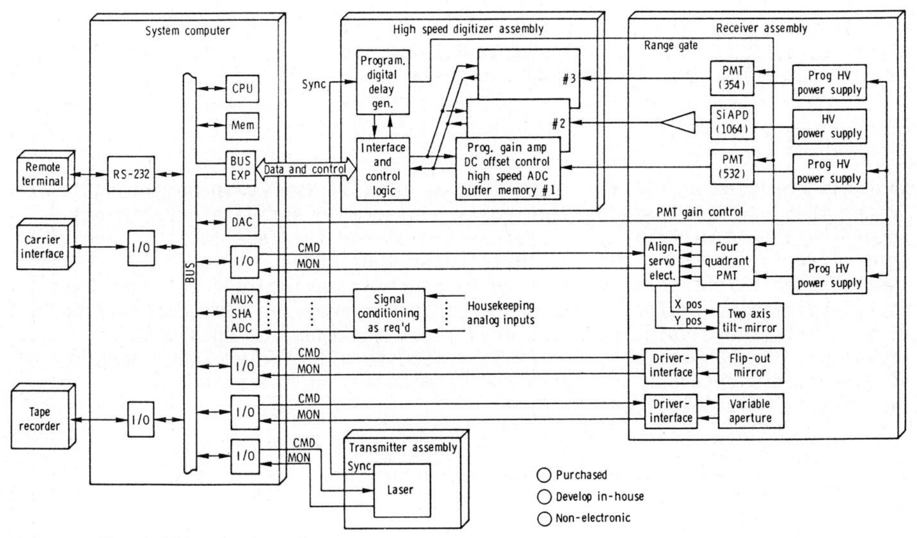

Figure 6 - LITE electronics

operation of the transmitter and receiver assemblies and high-speed digitizer. The computer also provides the interface to the STS Orbiter for uplink commands and downlink data services. The high-speed digitizer unit contains an analog-to-digital converter and a buffer memory which allows data from the detectors to be acquired and stored at a very high rate during the data-taking mode and reproduced at a lower rate for downlinking. The programmable digital delay generator provides the range gate signal to gate on the signal detectors just prior to the signal return.

The environmental measurement system is a completely autonomous system which is activated by a motion switch 6 seconds prior to STS launch. The system will measure and record environmental data on its internal tape recorder during the ascent, on-orbit, and landing phases of the STS mission. These measurements will be used to assess the in-flight environmental effects on the LITE instrument, to compare the flight data to the engineering mathematical design models, and to provide design guidelines for future STS and free-flying lidar experiments. The measurements to be made are vibration, acoustical, strain, contamination, and temperature. The system is completely autonomous requiring no external electrical power, commands, coolant, or data recording.

3.2 The Experiment Platform

The experiment platform is a support structure that serves as the LITE optical and equipment bench. The platform was designed by NASA for use with the Spacelab 3-meter pallets and sections of the structure have flown on past STS missions. The design is a 2-1/4-inch thick aluminum plate with 2-inch by 2-inch square cutouts in an orthogrid pattern. The plates are constructed in quarter sections to allow ease of handling. The LITE will use four quarter sections to extend the full pallet length. The orthogrid structure is attached to the pallet with a series of adjustable struts that keep the platform structurally rigid. The structure is required to maintain rigidity to assure alignment between the laser transmitter, the alignment mirror, and the telescope receiver.

4. LASA LIDAR FACILITY

NASA will provide a LASA lidar facility as part of the Earth Observing System (EOS) on the Space Station polar platform. Candidate experiments scoped by the atmospheric and lidar scientist for the LASA facility are summarized in Table I. Most of these experiments have demonstrated flight performance in a terrestrial environment but none have any spaceflight certification. The experiments proposed can generally be divided into those using backscattered lidar and those using a DIAL technique. Quite possibly, the final selection will include some merging of experiments once the synergism and commonality of the selected experiments have been identified.

	Airborne demonstrated	
Altimetry	X	
Cloud top height	X	
PBL Height	X	
Stratospheric aerosols	X	
Cloud properties	X	
Tropospheric aerosols	X	LIDAR
H_2O Column	X	DIAL
Surface pressure		
O_3 Column	X	
H_2O Profile	X	
Pressure profile		
Temperature profile		
O_3 Profile	X	

Table I - LASA science panel experiments

Parallel Phase B studies are underway to define the concept for the Space Station polar platform. These studies are expected to be completed by the end of the year and should provide a single concept for the polar platform. The concept shown in Fig. 7 is one of the RCA concepts for the polar platform and represents no particular compliment of experiments. The concept illustrates an integral core carrier and payload section for the first launch which would provide about 2500 kilograms of payload capability. A second launch would add an additional payload carrier section, replace the propulsion module, and offer an opportunity to repair or replace degraded hardware. The second payload section would provide about 2100 kilograms of payload capacity. Provisions are expected for a third payload carrier but each additional payload carrier will provide proportionally less net payload due to the increased propulsion requirements.

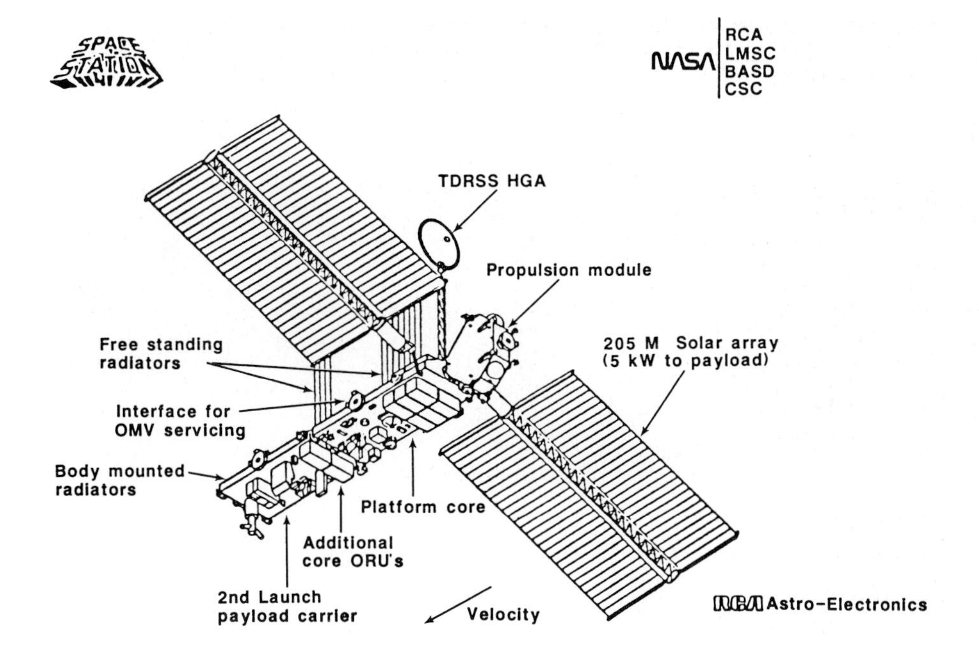

Figure 7 - Polar platform (2nd launch configuration)

The LASA facility will provide a telescope, optical bench and data handling system and will provide all interfaces to the polar platform. The selected experiments are expected to provide the laser system, detectors, back optics and specialized electronics since these subsystems are peculiar to the experiment. In order to scope the LASA facility, the experiments in Table I have been used as a representative selection. The experiments span some six separate laser technologies and list telescope apertures from 0.1 to 2.0 meters. Experiment electrical power requirements range from a few hundred watts to 2000 watts. A variety of studies have been made and some are still ongoing to establish a set of LASA facility requirements until an experiment selection is made.

Current weight estimates are summarized in Table II for the LASA facility including a baseline estimate for experiment components. This weight of 4214 pounds makes the LASA facility one of the heavier EOS facilities. A goal of 3300 pounds has been set for the facility and appears to be achievable. The three areas being closely studied now are the laser system, telescope and thermal radiators. It is prudent at this point in the program to have a weight contingency of about 20 percent and that is included in the weight budget. Further studies are underway to refine the engineering aspects of candidate laser systems and 300-400 pounds weight reductions are within reason. However, this weight savings cannot be fully assessed until selection of experiments.
The studies with the lightweighting techniques for telescopes may offer another 300 pounds. The engineering assessment of thermal radiators is complicated by the fact that parallel Phase B definition studies are still in progress on the platform and details are still evolving on the thermal design.

Table II - Current EOS LASA facility baseline

• EXPERIMENT	WEIGHT (LBS)		POWER (Watts)
➤ Laser	1100		
Backoptics	100		
Detectors/sig. cond	50	1250	
• FACILITY			
➤ Telescope	750		
Backoptics	100		
Data systems	110		
Power conversion	50		
Star tracker (2)	42		
Inertial ref. unit	35		
Thermal control	75		
➤ Thermal radiators	325		
Plumbing	200		
Optical bench	250		
Optical bench mounting	175		
Cabling	150	2262	
• CONTINGENCY (20%)		702	
• TOTAL		**4214**	2500
• GOAL		**3300**	

Table III - Laser systems options

TECHNOLOGY	WEIGHT (LBS)	POWER
Nd:YAG 3λ Flashlamp	1050	2025
Diode	710	435
Alexandrite Flashlamp	1210	2000
Ti: Sapphire Diode	1005	1175
Excimer 2λ	1670	1335
CO_2 Pulse	1675	980
CO_2 CW	903	1300
Baseline	1100	2000

Table III lists the laser technology proposed by the experiments of Table I. Studies have been made to establish envelope parameters to assess the impact on the LASA facility. The studies considered the lasers to have outputs of 0.5 to 1.0 joule per transmitted wavelength. Conservative engineering assessments on how to deal with the thermal issues, launch environments and requirements to package hardware in "standard orbital replacement units" have resulted in the weights shown. Studies are continuing to refine these numbers and weights are expected to come down slightly. The electrical power requirements for most of the lasers are substantial and will require some type of duty cycle operation in order to operate within LASA facility power budgets. The inefficiency of these lasers hurts doubly however, in that it requires significant systems to dissipate the heat. The availability of the more efficient diode array pumping techniques not only saves electrical power but reduces thermal systems requirements which ripples through the structural design and yields significant system and facility weight savings.

The availability of a suitable solid-state tunable laser offers the possibility of having one laser system that meets most of the experiment requirements of the experiments of Table I. The DIAL technique would be satisfied by having one transmitted wavelength on a molecular absorption wavelength and a second wavelength transmitted at a separated but close wavelength. The second wavelength could also be used through harmonic generation to provide double and/or triple wavelengths that would satisfy the backscattered lidar techniques. The incorporation of a short transmitted pulse capability would also satisfy the altimetry portion of the experiment. Such a laser would have the characteristics in Table IV. Eye safety considerations might limit transmitted energies in the visible spectrum.

The receiver telescope requirements for the experiments of Table I are satisfied by a primary aperture of 1.25 meters and a field-of-view of 0.1 milliradians. The telescope does not require "imaging" quality and is only required to collect light onto a detector. Studies have also been conducted on telescopes as large as 1.8 meters, that being the largest that can practically be accommodated in the LASA facility. Table V summarizes some of the weight tradeoffs for various primary apertures and technologies. The ultralow expansion (ULE) quartz is a conventional technology and represents a baseline point for the LASA facility. The beryllium hot isostatic pressing (Be HIP) technology has not been demonstrated for these apertures and the largest currently available autoclave may limit single component primary mirrors to 1.3-1.5 meters. An

Table IV - Summary solid-state laser parameters

• **Tuning range**	**λ On – H_2O, O_2 Absorption lines; ≤0.5 picometer** **λ Off – λ On + 30–100 picometers; ≤ 2 picometer**
• **Spectral purity**	**99% of energy ≤1pm (λ On and λ Off)**
• **Efficiency**	**> 3%**
• **Energy per pulse**	**0.5–1.0 Joule**
• **Repitition rate**	**10 Hertz**
• **Pulse length**	**< 100 Nanoseconds**
• **Lifetime**	**> 10^8 Shots**
• **Harmonic generation and space hardened**	

Table V - Telescope options

TECHNOLOGY	APERTURE	WEIGHT (LBS)
Ultra low expansion quartz	1.0 M	490
	1.25 M	750 --- BASELINE
	1.8 M	1400
Beryllium hot isostatic pressing	1.0 M	355
	1.25 M	480
	1.8 M	766
LIGHTWEIGHTED*	1.25 M	375
	1.8 M	497

*Preliminary weight estimates

evolving lightweighted technology for large mirrors involves the fitting of the components together. The core would be composed of pyrex tubing and the faceplate would be sagged pyrex sheet material. Studies are starting on the concepts for such a telescope and the weight savings are significant. The use of the 1.8-meter telescope although very attractive from an experiment viewpoint, especially if the light weighting techniques are successful, must await more definition of the polar platform.

The altimetry portion of the experiment places some stringent requirements on attitude and position measurements for the facility. The principle function of the experiment is to measure the growth or change of the polar ice region. This is accomplished by measuring the range from the polar orbiting platform to the polar ice to a "few" centimeters accuracy. Pointing knowledge must be measured to a few arcseconds for each of the transmitted laser pulses. Additionally, the position or ephemeris for the platform must be known for each laser shot to a few tens of meters. Orbital altitude of the polar paltform to a few 10's of centimeters can be achieved by ground-based lidar tracking ranges. The position data can be provided by the Global Positioning System and is expected to be provided by the polar platform core system. The "few" arcsecond pointing knowledge will require a startracking and inertial reference unit on the LASA facility optical bench. The thermal warpage of the polar platform and payload structure would make it very difficult to transfer an arcsecond quality measurement from the platform core to the LASA facility optical bench. Studies are being made to determine if it is practical to optically tie the LASA facility optical bench to some point on the platform core and place the startracking and inertial reference unit function in the platform core and making the measurements available to other EOS experiments.

Concepts for the LASA facility are shown in Figs. 8 and 9. As noted above, the baseline telescope is 1.25 meters and is shown as a fixed, nadir staring telescope. The telescope is tied to the bottom of the optical bench and the experiment and remaining facility modules are attached to the top side of the optical bench. Transmission of the laser pulse is through

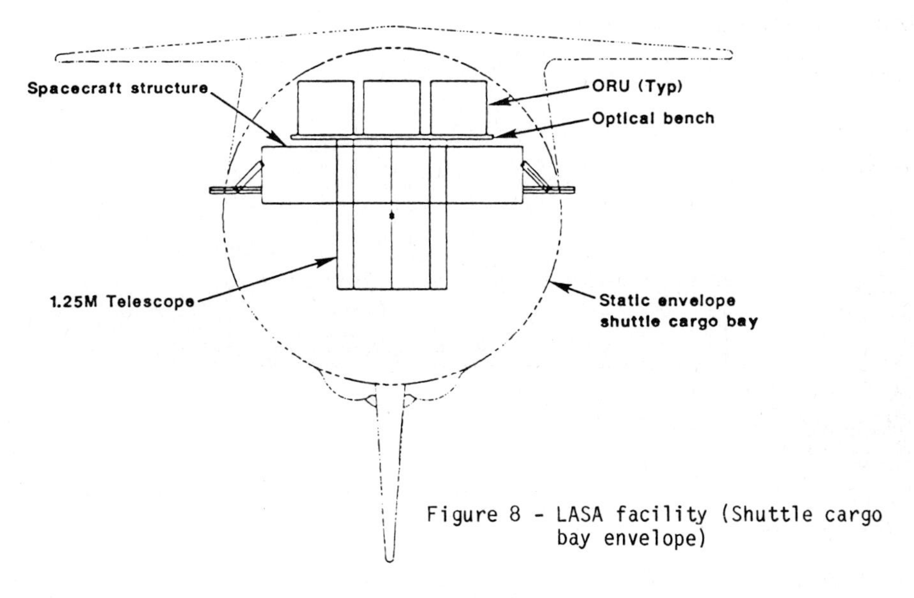

Figure 8 - LASA facility (Shuttle cargo bay envelope)

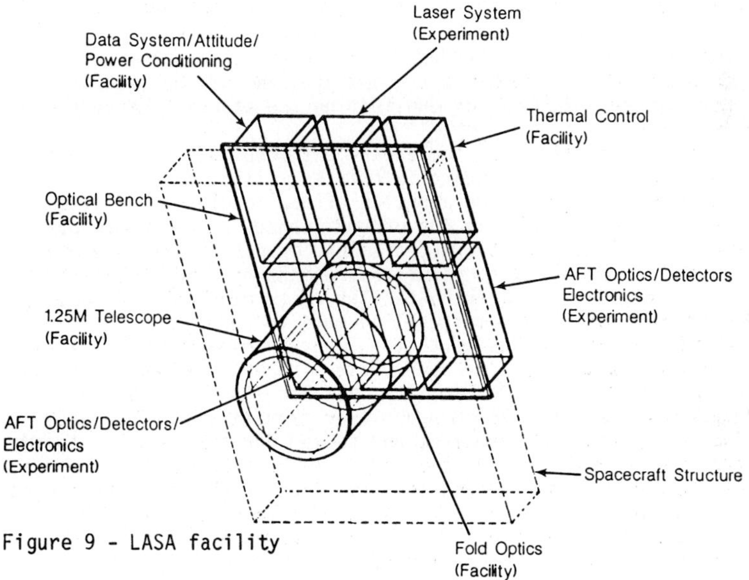

Figure 9 - LASA facility

the bottom of the module and optical bench. The concept accommodates up to three experiment modules. The star tracking and inertial reference units are mounted in the facility data handling and control module and require a field of view to the side of the velocity vector. Each of the modules is assumed to be a standard "orbital replacement unit" (ORU). The concepts being developed for serviceability and repair on-orbit center around the replacement of the ORU's using the shuttle remote manipulator system (RMS) with astronaut extravehicular activity (EVA) as a backup function. The

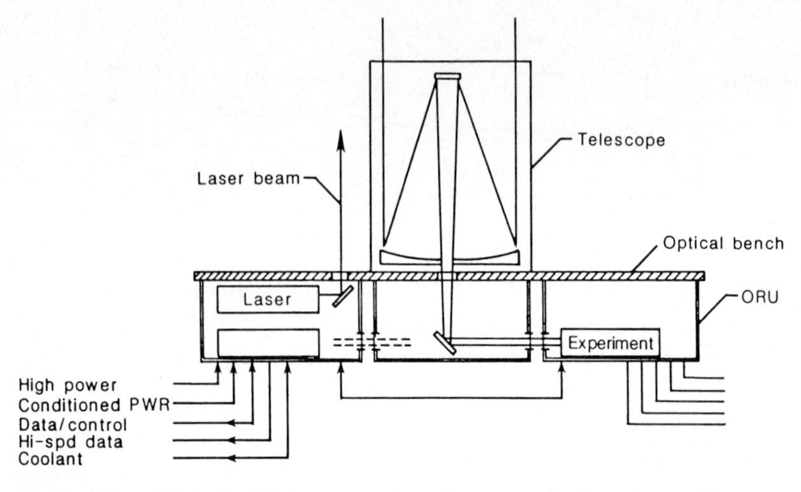

Figure 10 - LASA facility orbital replacement unit interfaces

exact dimensions for an ORU are still under study, however, they are not small units and sizes of 0.75 by 0.75 by 1.5 meters are being discussed. The on-orbit repair for smaller assemblies, although not ruled out entirely, is not very practical. Standard facilities, procedures and storage will be provided for the ORU's and any program with nonstandard modules will have to bear the cost of duplicating the standard services.

A concept for the interfacing of the experiment modules into the LASA facility is shown in Fig. 10. Standard interfaces will be developed to permit future on-orbit change-out or addition of new modules. The primary power source for the modules will be the unconditioned, high-voltage power provided by the polar platform. A limited amount of conditioned power will be provided by the LASA facility to permit operations of inter-module control and communications. LASA intra-facility data handling and control will be via a serial bus. Provisions will be made to permit synchronization of experiment modules to the laser-transmitted pulse to the nanosecond range. A dedicated, high-speed data line will be provided from the experiment modules to the LASA facility data handling and control module to accommodate burst data from each laser transmission. The data from the LASA facility will be formatted and queued for transmission to the platform. Data rates for the LASA facility during operational periods are expected to be 1-2 megabits per second and less than 50 kilobits per second during nontransmission periods. Optical ports will be used to permit steering of the received light signal to appropriate experiment modules. A fluid coolant loop will provide thermal control for the modules.

Evolving laser technology and experiments will be accommodated by on-orbit addition or exchange of ORU's. As shown in Figs. 8 and 9 provisions are made for three experiment modules and all three positions may not be filled at the initial orbit capability. A scanning capability has been discussed for the telescope to provide for additional coverage and could be accommodated by retrieving the facility and mounting the telescope horizontally and using a 45 degree scanning fold mirror. This option was investigated for the initial orbit capability and deferred due to weight considerations.

REFERENCES

1. W.R. Vaughan, E.V. Browell, W.M. Hall, J.J. Degnan, R.D. Averill, J.G. Wells, D.E. Hinton, and J.H. Goad: "A lidar instrument to measure H_2O and aerosol profiles from the NASA ER-2 aircraft." Presented at the SPIE's 1986 Quebec Symposium on Optical and Optoelectronic Applied Sciences and Engineering, Quebec City, Quebec, Canada, June 2-6, 1986.

2. H.E. Poole, J.W. Cox, R.H. Couch, and W.H. Fuller, Jr: "A lidar technology experiment from Space Shuttle: Lidar In-Space Technology Experiment (LITE)." Presented at the SPIE's 1986 Quebec Symposium on Optical and Optoelectronic Applied Sciences and Engineering, Quebec City, Quebec, Canada, June 2-6, 1986.

Preliminary Study of a Tunable Narrow Line Double Pulse Alexandrite Laser for Meteorological DIAL Applications

C. Loth[1], *J. Pelon*[2], *P.H. Flamant*[1], *and G. Megie*[2]

[1]Laboratoire de Météorologie Dynamique du CNRS, Ecole Polytechnique, F-91128 Palaiseau, France
[2]Service d'Aéronomie du CNRS, BP 3, F-91370 Verrières le Buisson, France

We report on narrow line emission (0.5 pm) of an alexandrite oscillator with 20 mJ per pulse using a new cavity arrangement. A main non-dispersive resonator acting as a forced oscillator is continuously self-injected by a narrow line from an auxiliary cavity. In a separate experiment we also report on (broadband) double-pulse emission within 100 microseconds with an equal 50 mJ - energy in each pulse.

I - INTRODUCTION

Tunable narrow line laser emissions with a high spectral purity and a high output energy are required for meteorological applications of the Differential Absorption Laser (dial) method to water vapor, temperature and pressure measurements [1-3]. New vibronic solid-state materials such as alexandrite, Cr^{3+}: GSAG and titanium-sapphire [4-5] are available now, which make the Dial measurements from an air platform very attractive in the near future. The requirements on the laser source are two-fold : i) a restriction on laser linewidth stability, and tunability with a high spectral purity (99.9% of the emitted energy within the laser linewidth) ; ii) a requirement on the time separation between the two pulses. These constraints are summarized in table 1. In addition, a laser energy of 50 to 100 mJ, in 200 ns or less and a 5 to 10 Hz pulse repetition frequency are also required.
Since the dial applications require the emission of two wavelengths, dual wavelength double pulse oscillators look very promising with respect to weight, electrical power, and volume for airborne and spaceborne lidar missions [6-7]).

Table 1 -

	H_2O	p,T
emission wavelength	725-732 nm	760-770 nm
on-line width	1 pm	0.3 pm
on-and off-wavelength separation	70 pm	70 pm
on-and off-temporal pulse separation	100-400 μs	

Recently, a new research program on vibronic-state laser has been launched in France by CNES and CNRS which is a follow-on of an extended program on laser-pumped dye laser for atmospheric applications. These developments are to be considered in the frame of two major airborne lidar programs : LEANDRE on a Fokker F-27, in France, supported by CNRS and CNES, and LASE on the ER-2 in the US supported by NASA in cooperation with CNES. In addition to a narrow line alexandrite laser, a high-resolution wavemeter to be associated to the lidar transmitter is also under study in France for the two programs mentioned previously.
In this paper, we present the preliminary results on an alexandrite laser to assess its potential for : i) an ultra narrow line emission (see table 1) ; ii) a double pulse emission within a single flashlamp excitation.

II - EXPERIMENTAL

A home-built alexandrite laser has been realized starting from an Allexite kit (Appolo Corp.). The alexandrite rod (6.35 mm-diameter, 102 mm-length, chromium concentration of 0.13 %) is pumped in a dual-elliptical reflector by two linear flashlamps.
An acousto-optics device (AOD) is used to actively Q-switch the alexandrite cavity. An AOD for Q-switching as compared to a Pockels cell presents several advantages : 1) a reduction in cavity losses, 2) an easier insertion into the resonator for optical alignment purposes, 3) a very low input driving signal resulting in low-level EMI noise.
The input energy is provided by a 120 μF capacitor bank and a high-voltage power supply available from Quantel - France. The pulse repetition frequency is set equal to 5 Hz. The flashlamp pulse duration is approximately 150 microseconds (FWHM). A simmer discharge (a low-intensity CW electrical current) improves the overall laser efficiency. Both flashlamps and the laser rod are temperature controlled by a re-circulating water-loop.

III - SINGLE-PULSE BROADBAND EMISSION

In the first experiment the alexandrite laser is tested for a single pulse emission, and no tuning element is inserted in the resonator (Fig. 1). The optical cavity, 86 cm-long, is closed at both ends by two flat wedged mirrors. A 3 mm-diameter diaphragm is inserted inside the cavity to improve the spatial energy distribution. We selected a mirror reflectivity 0.80 for the broadband emission experiments.
For a rod temperature equal to 45°C and a total electrical energy of 120 J in the two flashlamps we observed the optimum delay to maximize the output

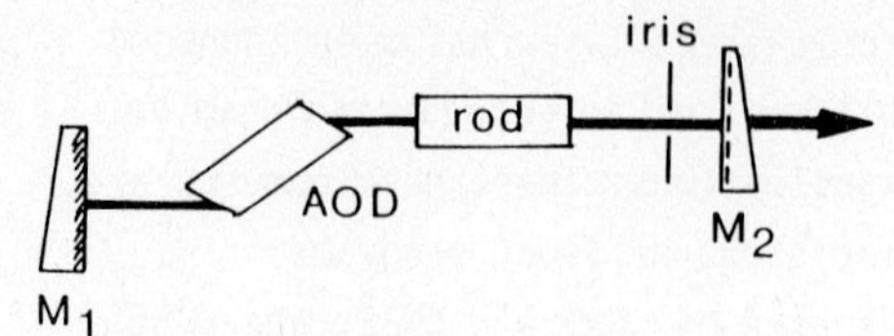

Figure 1 : Experimental set up of the Q-switch alexandrite oscillator used for broadband emission experiments

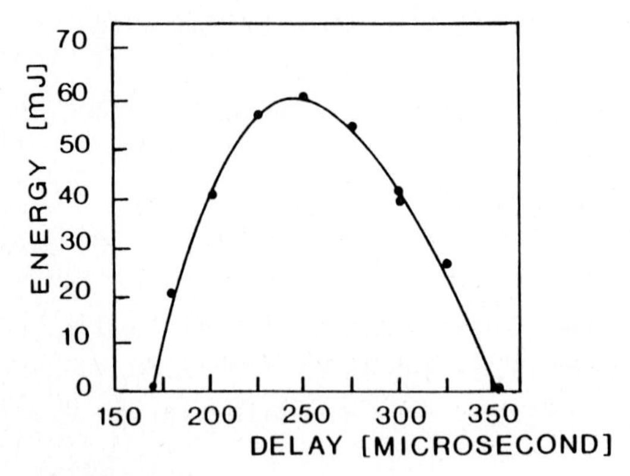

Figure 2 : Output laser energy as a function of the delay between the firing of the flashlamps and the RF signal on AOD (input energy 120 J rod temperature 45°C)

energy at about 250 μs after the beginning of the flashlamp excitation (Fig. 2). The delay corresponds to the maximum emitted fluorescence observed from the rod under the flashlamp excitation when the mirrors are removed.

With a non-selective resonator, a laser output energy larger than 100 mJ is easily achieved. As an example, the emitted energy is 120 mJ for an electrical input of 140 J and a rod temperature of 45°C. Then, the pulse duration (FWHM) is 200 ns.

The wavelength at the maximum of the laser emission spectrum shifts to the red from 750 nm to 751 nm as the temperature of the alexandrite rod is increased from 40 to 80°C. Also, the spectrum broadens from 1,5 nm to 3,5 nm. A comparison between Q-switch and long pulse (relaxation oscillations) emissions shows that for the same rod temperature the short pulse (broader emission spectrum) is slightly shifted to the blue. For a temperature of 40°C, the parameters are respectively 751.7 nm and 1.5 nm for a long pulse emission, and 751.1 nm and 5.2 nm for a short pulse emission.

IV - DOUBLE-PULSE BROADBAND EMISSION

A double-pulse emission can easily be obtained using the AOD twice, but it requires an adjustment of two delay times : i) for the first Q-switching and ii) between the two pulses.

A double pulse (broadband) emission with the same energy of about 50 mJ in the two successive pulses has been obtained in a single flashlamp excitation. The time delays are 180 μs for the first Q-switching and 90 μs between the two pulses. A total energy of 100 mJ compares well with the maximum single pulse energy of 110 mJ after a 250 μs delay.

V - NARROW-LINE SINGLE-PULSE EMISSION

An auxiliary cavity made of a diffraction grating (G) at a grazing incidence in conjunction with a mirror M3 (Fig. 3), is added to the main cavity (Fig. 1). In this arrangment the M1 M2 main cavity is continuously self-injected by the low-intensity retro-reflected narrow-line for the auxiliary cavity GM3. The auxiliary cavity length is varied from 9 cm to 80 cm. The gold coated grating (2400 lines per mm) is 65 mm long. It is used at incidence angles larger than 85°. The output energy is a few tens of milli-Joules.

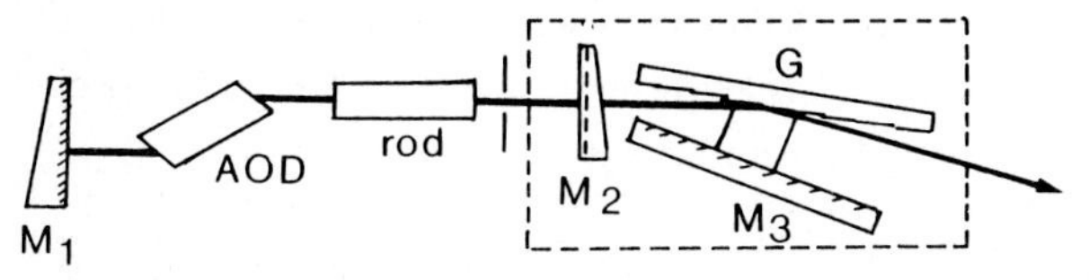

Figure 3 : Schematic of the narrow-band Alexandrite oscillator

The narrow line emissions are analyzed using either a high-resolution spectrometer and an optical multichannel analyser (resolution of 0.0017 nm at 750 nm) or a multi-beam Fizeau interferometer illuminated through a beam expander (X10). The etalon plates (55 mm-diameter, reflectivity of 0.92, flatness of λ/50) are 12 cm apart in a mechanically stable structure. The free spectral range (FSR) is 0.0023 nm at 750 nm. The fringe pattern is recorded using a 512 photodiode array.

The potential for a broad tunability at medium resolution is studied for an incidence angle of 85° on the grating. A 4 pm-laser line is tuned over more than 25 nm with a spectral purity better than 90 % (Fig. 4). The electrical input energy is 150 J and the rod temperature is 45°C.

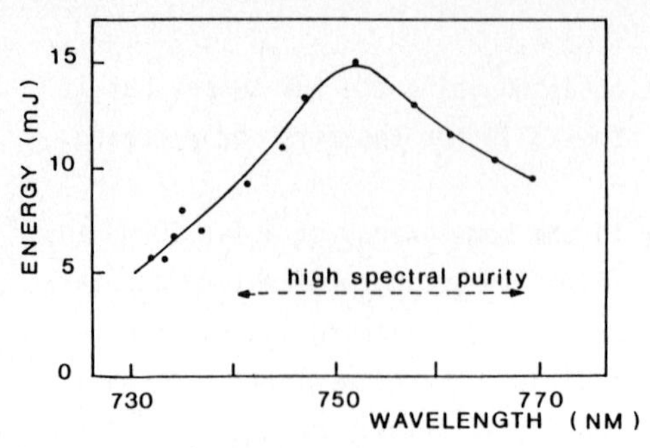

Figure 4 : Output laser energy as a function of wavelength for the self-injected oscillator (see Fig. 3). The laser linewidth is 4 pm. (input energy 150 J, rod temperature 45°C)

The auxiliary cavity is 80 cm long. The laser linewidth we measured is smaller than the theoretical value associated to a double-pass grating arrangement and a diffraction-limited beam. This linewidth is approximately given by the laser-operating wavelength divided by the total number of grooves on the diffraction grating covered by the laser beam.
A dual-mode laser emission (with an equivalent linewidth of 0.5 pm) is observed when the auxiliary cavity length is reduced to 9.3 cm total and when the grating is set at a grazing incidence angle of 88°. The laser mode separation recorded with the Fizeau interferometer is 0.27 pm (Fig. 5 a).

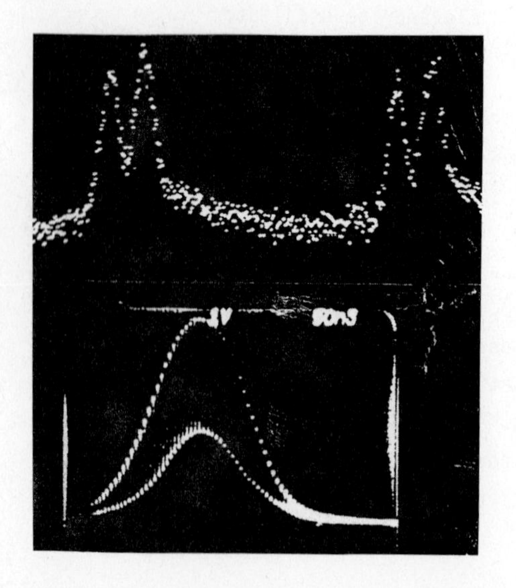

Figure 5 : Dual mode laser emission as recorded : a) by a Fizeau interferometer with a 12 cm spacing between plates ($\delta\lambda$ = 2.3 pm between FSR) b) by a fast photodiode (temporal beating of the pulse shape)

It is in agreement with a value of 0.29 pm calculated from the main resonator cavity length. Also, a strong modulation of the pulse envelope is also observed due to mode beating (Fig. 5 b). The output energy is 20 mJ (for an input of 180 J).

During the same experiment, we also observed quasi-single longitudinal mode emissions and smooth temporal envelopes, but the results suffer a great lack of stability. The recorded mode linewidth 0.14 pm (75 MHz), is limited by the apparatus function. It corresponds to an experimental finesse Fizeau interferometer equal to 17. This value is in agreement with an apparatus function determined by the reflective finesse of 38, the defect finesse of approximately 25 and the finite number of reflections which reduces the experimental finesse at a value smaller than the expected theoretical limit.

Future work will be devoted to the implementation and improvement of the grating resonator described above for high-energy dual wavelength double pulse alexandrite laser.

Acknowledgments :

This work is supported in France by the Centre National d'Etudes Spatiales and the Centre National de la Recherche Scientifique. We thank Dr Claude Cahen at Electricité de France for fruitful discussions, and also thank Quantel France for helpful discussions.

References :

[1] C. CAHEN, G. MEGIE, P. FLAMANT : J. Appl. Meteor. 21, 1506 (1982).

[2] G. MEGIE : Appl. Optics. 19, 34 (1980).

[3] C.L. KORB, C.Y. WENG : J. Appl. Meteor. 21,1346 (1982).

[4] J. WALLING, O. PETERSON : IEEE J. QE-16, 119 (1980).

[5] See, e.g. : Tunable solid-state lasers for remote sensing, ed. by R.L. Byer, E.K. Gustafson, and R. Trebino, Optical Sciences. 51 (Springer, Berlin Heidelberg, 1985).

[6] G. MEGIE, P. FLAMANT, M. BOURDET, E. BROWELL, W. HALL, J. TALBOT in Proc. ESA Workshop on Space laser Applications and Technology ESA SP 202, 189 (1984).

[7] P.H. FLAMANT, C. LOTH, G. MEGIE, J. PELON : ESA Journal 9, 449 (1985).

Nonlinear Infrared Frequency Conversion in $AgGaS_2$ and $AgGaSe_2$

Y.X. Fan[1], R.C. Eckardt[1], R.L. Byer[1], R.K. Route[2], and R.S. Feigelson[2]

[1]Ginzton Laboratory, Stanford University, Stanford, CA 94305, USA
[2]Center for Materials Research, Stanford University, Stanford, CA 94305, USA

The nonlinear infrared materials silver thiogallate ($AgGaS_2$) and silver selenogallate ($AgGaSe_2$) have recently been grown in large-size high-optical-quality crystals which have allowed several significant demonstrations of nonlinear infrared frequency conversion. Both materials have been operated as tunable infrared parametric oscillators.[1,2] High efficiency second harmonic conversion of 10.6-μm radiation has been demonstrated in room temperature $AgGaSe_2$ using a low repetition rate pulsed CO_2 laser pump.[3] Recent preliminary results have shown that bulk $AgGaSe_2$ has high average power capability.[4] A few growth related problems remain to be resolved, such as improvement of the surface damage threshold and dielectric coating performance, and further reduction of scattering. With increased availability, these materials should find numerous applications in the generation of widely tunable high-average-power coherent infrared radiation.

The two compounds $AgGaS_2$ and $AgGaSe_2$ crystallize in the chalcopyrite structure, space group $\bar{4}2m$. They are negative uniaxial crystals, and their birefringence allows angle tuned phase matching for second harmonic generation of fundamental wavelengths between 1.8 to 11 μm in the thiogallate and 3.1 - 13 μm in the selenogallate. The regions of high transparency are similarly shifted with $AgGaS_2$ transmitting from 0.45 to 13 μm with some multiphonon absorption at wavelengths longer than 9 μm. The transparency range of $AgGaSe_2$ extends from 0.73 to 17 μm with multiphonon absorption shifted to wavelengths longer than 13 μm. $AgGaS_2$ is useful for mixing processes that involve visible or near-infrared radiation whereas $AgGaSe_2$ is superior for longer wavelength applications such as harmonic generation of CO_2 laser radiation and parametric oscillators that operate at wavelengths longer than 3 μm. The nonlinear optical coefficients are d_{36} = 18 x 10^{-12} m/V and 43 x 10^{-12} m/V respectively. Most of their properties have been known since the early 1970's when these crystals were identified as potential nonlinear infrared materials.[5,6,7] Early growth-related problems made it difficult to obtain high-quality crystals of adequate size. Similar problems existed with other nonlinear infrared materials and limited early progress in nonlinear infrared frequency conversion.

The growth-related problems of these two materials are similar, and experience first gained with the $AgGaS_2$ was directly applicable to the $AgGaSe_2$. Both are reactive and moderately volatile at their melting temperatures MP = 966°C and MP = 856°C respectively, and must be grown in sealed ampoules using the Bridgman-Stockbarger technique. Both have anomalous expansion along the c axis. Therefore, it is necessary to seed the growth in the c direction and to use precision-tapered ampoules to avoid cracking when cooling after growth. Large crystals solidify only with non-stoichiometric composition resulting in finely dispersed scattering centers and reduced infrared transmission in the as-grown state. Post-growth heat treatment is required to remove the scattering and provide high optical quality.[8] Boules of 2.8-cm diameter by 10-cm length have been successfully grown and heat treated for each material. The boule size of $AgGaSe_2$ is now being increased to 4-cm diameter. Our preliminary demonstrations of nonlinear infrared frequency conversion were performed with crystals approximately 2-cm long with 1-cm cross section entrance and exit surfaces. This is the largest size crystal that could be harvested at the phase matching angles from the 28-mm diameter boules.

Infrared parametric oscillation was demonstrated in $AgGaS_2$ with output continuously adjustable from 1.4 - 4.0 μm by angle tuning of the crystal. The pump source for this experiment was a 1.064-μm, Q-switched Nd:YAG laser with a 20-ns duration output pulse. The transverse distribution was Gaussian-like with a beam waist of 0.7 mm. Threshold for parametric oscillation was 1.5 mJ. Energy conversion efficiency of 16% was reached near degeneracy where the output wavelength was 2.13 μm. The infrared parametric oscillator performance would improve with the use of a pump beam with larger and flatter transverse distribution. Also, any improvement in the nonlinear material such as decreased losses, higher damage threshold or longer interaction length will be reflected in improved parametric oscillator performance. With such improvements the tuning range of the $AgGaS_2$ parametric oscillators could be extended to the transparency limit of the material, 13 μm.

Harmonic generation of CO_2 laser radiation has not been widely used because of the limitations of available nonlinear infrared materials. The improved-quality $AgGaSe_2$ grown recently is well suited for this harmonic conversion application. We have demonstrated 14% internal energy conversion efficiency using a 2-cm crystal of $AgGaSe_2$. The pump pulse, generated by a pulsed TEA laser, had a 75-ns gain-switched spike followed by a 650-ns nitrogen tail. Power conversion efficiency at the peak of the spike was calculated to be 60%. This measurement was performed with an uncoated crystal which had 20% reflection losses at each surface.

We have achieved parametric oscillation in $AgGaSe_2$ using two different pump sources. Output tuning over the range of 1.6 - 1.7 μm and 6.7 - 6.9 μm was obtained using a 1.34-μm Nd:YAG laser, and continuous tuning was obtained between 2.65 and

9.0 μm using a 2.05-μm Ho:YLF laser in a collaborative experiment with the Naval Research Laboratory.[2] Angle-tuned, 2-cm crystals were used in these measurements. Both lasers provided linearly polarized Q-switched pulses with Gaussian-like transverse intensity distributions. The crystal used in the Nd:YAG-pumped oscillator was tuned over the 77° - 90° range. Both end surfaces were antireflection coated with ThF_4. Conversion efficiency of 18% was obtained with the Nd:YAG pump and with the parametric oscillator output tuned to 1.66 μm. Conversion efficiency of the Ho:YLF-pumped parametric oscillator was slightly lower, perhaps due to the greater bandwidth and spectral instability in the output of that laser. The crystal used in the Ho:YLF-pumped oscillator was tuned over a 45° - 50° range. Broad-band antireflection coatings were applied to both ends of the crystal. This particular crystal had 1% internal losses due to scattering.

Nonlinear frequency conversion in these materials for both harmonic generation and parametric oscillation has been limited by surface damage caused by the intense infrared beams. Surface damage typically occurs at peak intensities of 13 MW/cm^2 for multiple exposures to Q-switch Nd:YAG or short pulse CO_2 laser pulses. In some cases the application of dielectric antireflection coatings has increased surface damage thresholds. However, changes in damage threshold with the application of these coatings has been inconsistent. An increase in damage threshold for 20-μs 1.06-μm pulses to 37 MW/cm^2 was obtained with one antireflection coating applied on $AgGaSe_2$. In many other cases the damage threshold was unchanged, particularly with broadband antireflection coatings. Some coatings demonstrated poor average-power capability when pump laser repetition rate was increased.

Bulk damage threshold of $AgGaSe_2$ has been found to be substantially higher than the surface damage threshold. In a recent measurement performed in collaboration with United Technologies Research Center, second harmonic generation of focused 10.6-μm CO_2 laser radiation was characterized.[4] The repetitively Q-switched continuously pumped laser had 8-W average power and approximately 5-kW peak power. This beam was focused to a diameter of 70 μm at the center of the crystal with no damage in several minutes exposure. The calculated peak intensity at the focus in the bulk material was 130 MW/cm^2, and the average power at the center of the beam was 200 kW/cm^2. Internal harmonic conversion efficiency was close to the theoretically predicted value indicating small thermal distortion in the crystal.

These demonstrations of nonlinear frequency conversion are preliminary in nature. The investigation of parametric oscillation and the high-average-power performance of $AgGaSe_2$ are continuing. Extending the growth to larger crystal sizes, improving quality by development of the heat treatment process, and the development of processing techniques for reducing surface damage thresholds also remain active areas of research.

The continued development of the chalcopyrite materials $AgGaS_2$ and $AgGaSe_2$ has provided valuable and much-needed nonlinear infrared materials. The preliminary results indicating high average power capability are particularly exciting. The achievement of broadly tunable parametric oscillation provides a tool with great potential for use in infrared spectroscopy and photochemistry. Support for this work has been provided by the U. S. Army Research Office, the Office of Naval Research, and N.A.S.A.

References.

1. Y. X. Fan, R. C. Eckardt, R. L. Byer, R. K. Route and R. S. Feigelson, Appl. Phys. Lett. **45**, 313, (1984).
2. R. C. Eckardt, Y. X. Fan, R. L. Byer, M. E. Storm, C. L. Marquardt and L. Esterowitz, "Tunable Infrared Parametric Oscillator Using Silver Gallium Selenide," paper MH2, Conference on Lasers and Electro-Optics, San Francisco, June 1986. (to be published in Appl. Phys. Lett.)
3. R.C. Eckardt, Y. X. Fan, R. L. Byer, R. K. Route, R. S. Feigelson, and J. van der Laan, Appl. Phys. Lett. **47**, 786, (1985).
4. L. Newman, R. C. Eckardt and R. L. Byer, unpublished.
5. D. S. Chemla, P. J. Kupecek, D. S. Robertson, and R. C. Smith, Opt. Commun. **3**, 39 (1971).
6. G. D. Boyd, H. Kasper, and J. H. McFee, IEEE J. Quantum Electron. **QE-7**, 563 (1971).
7. G. D. Boyd, H. Kasper, J. H. McFee, and F. G. Storz, IEEE J. Quantum Electron. **QE-8**, 900 (1972).
8. R. K. Route, R. S. Feigelson, R. J. Raymakers, and M. M. Choy, J. Cryst. Growth **33**, 329 (1976).

Non-Linear Conversion of 1.3 μm Nd:YLF Emission

H.H. Zenzie[1], *M. Thomas*[1], *C. Carey*[1], *E.P. Chicklis*[1], *and M. Knights*[2]
[1]Sanders Associates, Inc., CS 2035-M/S MER15-2250,
Nashua, NH 03061, USA
[2]Schwarz Electro-Optics, 45 Winthrop St., Concord, MA 01742, USA

Current interest in efficient blue solid-state laser sources has led to research aimed at converting laser radiation in the 1.3μ region to the blue (4300-4600Å). In this paper we describe laser operation on the $^4F_{3/2}$-$^4I_{13/2}$ laser transition in Nd:YLF including the prominent π and σ polarized lines, 1.321μ and 1.313μ, and continuous tuning over the weaker band between 1.362 and 1.37 microns. Small signal amplifier gain measurements of the various transitions between these multiplets are presented in addition to a Q-switched oscillator, emitting 90mJ at 1.3665μ and an oscillator amplifier chain emitting 0.5J at 1.313 microns.

The 1.3 micron emission of Nd materials can be shifted to the blue by various non-linear processes. A critical issue for all 1.3 micron-based approaches is gain competition; the performance of Q-switched Nd laser materials at 1.3 microns is constrained by competition from the higher gain $^5F_{3/2}$-$^4I_{11/2}$ transitions. The inversion (stored energy) which can be achieved in an isolated rod at 1.3μ is limited by the onset of parasitic oscillations at 1μ. The maximum gain at 1.3μ is the product of the inversion at the parasitic limit and the cross-section of the transition. Since the 1.3μ cross-section in Nd laser materials is much lower than at 1μ, a high parasitic threshold for the 1μ lines is essential.

The low index of refraction and uniaxial structure of Nd:YLF result in a remarkably high parasitic threshold for 1.047 oscillations in Nd:YLF[1]. Single pass gains in a .6 x 10 cm rod of up to 1800 have been demonstrated[1] corresponding to a small signal-gain coefficient of 0.84 cm^{-1}.

To determine the ratio of gain coefficients for the various lines in this material, a Q-switched oscillator operating on selected lines was constructed. Output pulses were attenuated to avoid saturation and propagated through a single pass .6 x 10 cm amplifier rod. Care was exercised to avoid detection of amplified spontaneous emission (ASE) and oscillator amplifier feedback. The results, shown in Fig. 1, indicate the ability to achieve reasonable gains in lines with substantially lower gain cross-section than the high gain transition. The single pass gain, G_0, is related to the small signal gain g_0 by $G_0=\exp g_0\ell$ where ℓ is the pump length of the amplifier rod. Using the values for G_0, as shown in Fig. 1, and with ℓ=8.9 cm, the small signal gain ratio of 1.3665 μm to 1.047 μm is 1:28.

Interest in the 1.3665 transition results from the ability to generate Cs resonant blue light (4555Å) with a frequency tripler. Although the gain of

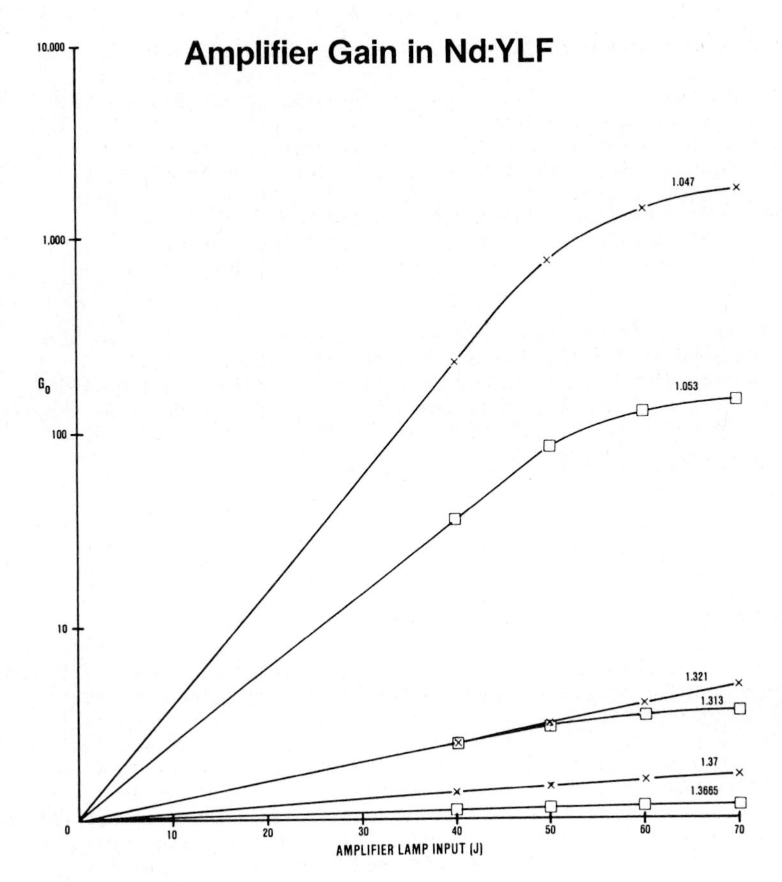

Fig.1. Amplifier gain in Nd:YLF

this transition is low and the gain competition formidable, substantial Q-switched output was obtained.

A two-head oscillator (Figure 2) was constructed to compensate for the low gain of the transition. With two 5 x 89mm rods in the oscillator, the effective single pass gain, G_0, at 1.047μ is in excess of 10^6 (1.3665μ G_0 = 1.7). The large gain ratio requires components exhibiting high 1μ

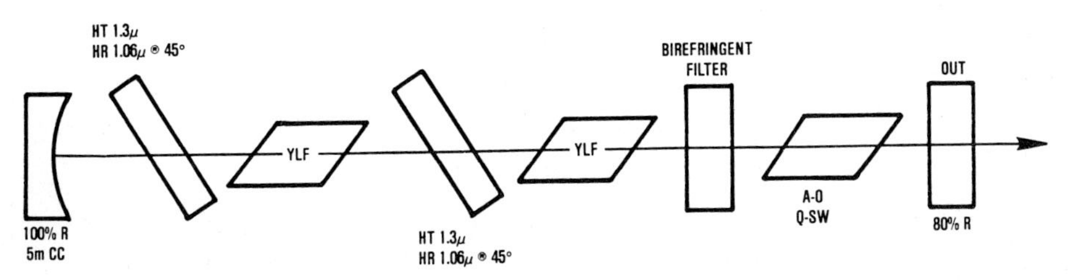

Fig.2. Two head oscillator

suppression and, due to the low gain, very low loss at 1.3665μ. Both resonator mirrors were coated for 1.3μ operation. In addition, two dielectric discrimination plates (HR at 1.047 microns, HT at 1.3 microns both 45°) were necessary for complete suppression. Gain competition from the 1.37μ (π) and 1.321μ (π) transitions was successfully managed by a carefully designed birefringent filter. The birefringent filter shown in Figure 2 consisted of a 3mm quartz waveplate to prohibit lasing of the higher gain 1.321μ line with two 1 cm-thick waveplates required for precise tuning within the 1.36-1.37μ manifold. The two thick waveplates were necessary to provide sufficient finesse and to maintain frequency stability during Q-switched operation.

In spite of the large number of surfaces and enormous gain ratio, we were able to Q-switch 90mJ at 1.3665μ with a .2% slope efficiency. 1.3665μ emission can then be directly tripled to generate Cs resonant light at 4555Å. However, the saturation fluence of this transition, determined from the cross-section ratio inferred from the gain measurements and the measured value at 1.047[1], is approximately 11 J/cm^2.

With more complex non-linear conversion schemes the Cs lines are accessible with the higher gain 1.3 micron lines. The 1.313μ line, for example, exhibits about 1/6th the gain of the 1.047 line and has an estimated saturation fluence of 3 J/cm^2. These parameters are far more desirable for high-power oscillator amplifiers than the 1.3665μ transition.

In order to investigate energy extraction and gain at 1.313μ, a Q-switched (40 nsec) multi-mode oscillator dual amplifier emitting .5J was constructed. The gain ratio at 1.313μ, 6 to 1, makes operation at this wavelength simpler than 1.3665μ. The resonator consisted of a 100% R 10m rear mirror, A-O Q-switch, 5 x 89mm Nd:YLF rod, 1 dielectric discrimination plate and a 70% R output coupler. Silicon at Brewster's angle, which exhibits low loss at 1.3μ and high loss at 1μ, was found to be an effective passive isolator between the amplifier stages. With the oscillator pre-amplifier providing 250 mJ, 310 mJ were extracted from the main amplifier when it was pumped to $g_o = .12\ cm^{-1}$.

Although Nd lasers are most efficient when run at the 1μ lines, operation at 1.3μ appears feasible with proper suppression. The 1.3μ transitions exhibit high-energy storage ($E_{sat}\ g_o = .36\ J/cm^3$) with efficient energy extraction possible at reasonable fluences (3J/cm^2).

References

1. M.G. Knights, H. Zenzie, G. Rines, M. Thomas, C. Carey, E.P. Chicklis and H. P. Jenssen, Cesium Filter Resonant Operation of Nd:YLF, presented at O-E Lase '86, Los Angeles, CA.

Index of Contributors

TA 1705 .O83 1986

OSA Topical Meeting (1986 :

Tunable solid-state lasers
II

PHYSICS

NON-CIRCULATING
UNTIL

MAR 30 1987